Green Chemistry and Sustainable Technology

Series editors

Prof. Liang-Nian He
State Key Laboratory of Elemento-Organic Chemistry, Nankai University, Tianjin, China

Prof. Robin D. Rogers
Department of Chemistry, McGill University, Montreal, Canada

Prof. Dangsheng Su
Shenyang National Laboratory for Materials Science, Institute of Metal Research, Chinese Academy of Sciences, Shenyang, China
and
Department of Inorganic Chemistry, Fritz Haber Institute of the Max Planck Society, Berlin, Germany

Prof. Pietro Tundo
Department of Environmental Sciences, Informatics and Statistics, Ca' Foscari University of Venice, Venice, Italy

Prof. Z. Conrad Zhang
Dalian Institute of Chemical Physics, Chinese Academy of Sciences, Dalian, China

Aims and Scope

The series *Green Chemistry and Sustainable Technology* aims to present cutting-edge research and important advances in green chemistry, green chemical engineering and sustainable industrial technology. The scope of coverage includes (but is not limited to):

- Environmentally benign chemical synthesis and processes (green catalysis, green solvents and reagents, atom-economy synthetic methods etc.)
- Green chemicals and energy produced from renewable resources (biomass, carbon dioxide etc.)
- Novel materials and technologies for energy production and storage (bio-fuels and bioenergies, hydrogen, fuel cells, solar cells, lithium-ion batteries etc.)
- Green chemical engineering processes (process integration, materials diversity, energy saving, waste minimization, efficient separation processes etc.)
- Green technologies for environmental sustainability (carbon dioxide capture, waste and harmful chemicals treatment, pollution prevention, environmental redemption etc.)

The series *Green Chemistry and Sustainable Technology* is intended to provide an accessible reference resource for postgraduate students, academic researchers and industrial professionals who are interested in green chemistry and technologies for sustainable development.

More information about this series at http://www.springer.com/series/11661

Mara G. Freire
Editor

Ionic-Liquid-Based Aqueous Biphasic Systems

Fundamentals and Applications

Editor
Mara G. Freire
CICECO – Aveiro Institute of Materials,
   Chemistry Department
University of Aveiro
Aveiro, Portugal

ISSN 2196-6982   ISSN 2196-6990 (electronic)
Green Chemistry and Sustainable Technology
ISBN 978-3-662-52873-0   ISBN 978-3-662-52875-4 (eBook)
DOI 10.1007/978-3-662-52875-4

Library of Congress Control Number: 2016953627

© Springer-Verlag Berlin Heidelberg 2016
This work is subject to copyright. All rights are reserved by the Publisher, whether the whole or part of the material is concerned, specifically the rights of translation, reprinting, reuse of illustrations, recitation, broadcasting, reproduction on microfilms or in any other physical way, and transmission or information storage and retrieval, electronic adaptation, computer software, or by similar or dissimilar methodology now known or hereafter developed.
The use of general descriptive names, registered names, trademarks, service marks, etc. in this publication does not imply, even in the absence of a specific statement, that such names are exempt from the relevant protective laws and regulations and therefore free for general use.
The publisher, the authors and the editors are safe to assume that the advice and information in this book are believed to be true and accurate at the date of publication. Neither the publisher nor the authors or the editors give a warranty, express or implied, with respect to the material contained herein or for any errors or omissions that may have been made.

Printed on acid-free paper

This Springer imprint is published by Springer Nature
The registered company is Springer-Verlag GmbH Berlin Heidelberg

Foreword

Is it aqueous biphasic systems or aqueous two-phase systems? What is in a name, an understanding or a narrow perspective? How you view a topic often reflects how you study a topic. In the 1950s, Albertson studied aqueous solutions of dextrans and polyethylene glycols (PEGs) which under certain concentrations and conditions formed two immiscible aqueous phases. The study of these "aqueous two-phase systems" (ATPS) flourished with both fundamental and applied studies of their use in, primarily, biotechnology where the gentle nature of the phases was seen as good for separations of proteins.

The use of aqueous solutions of salts to salt out aqueous polymer solutions followed, and terms such as "salting in" or "salting out" were used, along with the concept of the Hofmeister series to explain the behavior. Those who utilized these systems for separations found some measure of predictability, and those studying the fundamental science behind them found controversy, semantics, and issues of misunderstandings. This observation is not to cast a harsh light on the field; rather, it is to highlight how the words we use to describe a phenomenon may not translate across disciplines without a common language.

This point is further evident when one considers the relative recent phenomenon of "ionic liquids" (ILs) which everyone agrees (or almost agrees) are salts but can't seem to agree are "normal salts" or not. From the mid-1990s onward, a very strict following arose from a definition of ILs which one almost had to swear to uphold to be considered part of the field. The reality though of a class of materials defined by an arbitrary melting point, which include protic acids and bases which are obviously in equilibrium with both charged and neutral species, has led to countless academic arguments and indeed at the same time has had the positive effect of spurring a much deeper insight into the nature of these materials.

The huge phenomenon of ILs as solvents, however, completely overshadowed the study of ILs as salts. A mythology arose that ILs were not "salts," and a counter mythology arose saying ILs are "just" salts. But, if ILs are truly salts, why can't they be used in forming ABS? That is, even if the exact nature of these materials is not known, can't they be useful?

Indeed, they can and early studies of ABS with ILs led to some rather (at the time) surprising results; "traditional salts" within the Hofmeister series could salt out aqueous solutions of the new salts (aka ILs). Later, it was determined that the new salts (ILs) could also salt out PEG. Uh-oh! With the hundreds of new salts being studied in the IL field, new science was starting to show that misconceptions may have arisen by what we "know" defines a "salt"! The misunderstandings would start anew as the older, accepted terminology of salting in and salting out was still being used, although the specific mechanisms for how these worked were not necessarily the same.

From this back and forth, a small but strong field has been doing the work to actually try and make the distinctions of how ILs work in forming ABS (or ATPS). From their work, an understanding is arising that if the systems studied are limited in scope, our understanding will be limited in scope. If one always studies similar systems, it is possible to draw the wrong conclusions.

Water is the common (and often forgotten) component in "aqueous" biphasic systems. Could it not be the water which is key? Perhaps, but it does not have to always be the most important component. First, we know dextran and PEG form ABS, and we know high-melting salts form ABS with PEG; is the conclusion then that dextran and high-melting salts are the same or behave the same in water? Let's now add that we know high-melting salts form ABS with both PEG and some low-melting salts. Does that mean ILs and PEG are the same and have the same behavior? Of course not.

Complexity here is the key. These are complex, multicomponent systems and they require systems thinking. Life, as we know it, is a delicate balance of competing reactions and separations. It is the balance which tips in favor of health or illness. Even if we are sick, most of the healthy reactions and separations are still ongoing. To understand the entire body, one would need to understand all of the reactions, interactions, and separations and deduce which are the most important.

So let's start considering some of the competing interactions and separations in our study of ABS. Instead of only studying the macroscopic properties of phase separation and how we can use it, let's try and deduce the individual attractive and repulsive forces which might lead to the behavior we see. For example, mutual solubilities of the phase formers and their individual solubilities in water would provide a wealth of data from which we could draw more in-depth conclusions. Consider all of the possible interactions (and I mean all, not just the polar ones), and you can include the coulombic, polar, nonpolar, and popular examples of some of these such as hydrogen bonding (not a complete list).

Some of the major interactions which PEG can have with water include the hydrogen bond accepting from etheric oxygens, hydrogen bond donation and acceptance from alcoholic end groups, and indeed conformational flexibility of the PEG molecule as it curls or coils. A classic salt such as K_3PO_4 has ions which can accept coordination or hydrogen bonds from water, can react with water to hydrolyze, etc., and these are just the interactions with water, not those with the PEG where coordination, hydrogen bonding, etc., can occur, and not the interion attractions and repulsions!

Now we come to ILs. First, what is an IL? OK, I will leave the definition to others, but I will say that the same types of intermolecular interactions between the ions leading to low-melting behavior lead to typically weaker interactions between the ions and often weaker interactions with water. Importantly, however, the ions of an IL can be tuned to have very specific and strong interactions with water, or they can be tuned to have virtually no interactions with water and likely everything in between. This, I believe, is the key to their utility and why the groups represented in this book study their fascinating behavior in forming ABS.

In this book, ABS are demonstrated with ILs and traditional high-melting salts, with ILs and carbohydrates, and with ILs and polymers. The book also has some of the very interesting novel separations for which these unique ABS are being designed.

My call is for those looking at these separations to keep in mind the questions we have to close the holes in our understanding. What are the interactions which under any given set of conditions dominate and result in phase separation? How can we rank them in order of strength? How can we use this understanding to model and predict the formation of ABS in any complex system?

The future utility of the field of IL/ABS is bright. You can see it in the work being done today. There is a zeal to understand the systems at the most fundamental molecular level and to be able to translate this knowledge across disciplines by using the common language of chemistry at the molecular level. This newfound knowledge will lead to its use in novel separation technologies which will be found from that same understanding.

Enjoy the book; it is just the beginning.

Department of Chemistry   Robin D. Rogers
McGill University
Montreal, QC, Canada

Contents

1. **Introduction to Ionic-Liquid-Based Aqueous Biphasic Systems (ABS)** 1
   Mara G. Freire

2. **ABS Composed of Ionic Liquids and Inorganic Salts** 27
   José N. Canongia Lopes

3. **ABS Constituted by Ionic Liquids and Carbohydrates** 37
   André M. da Costa Lopes and Rafał Bogel-Łukasik

4. **ABS Composed of Ionic Liquids and Polymers** 61
   Rahmat Sadeghi

5. **Extraction of Amino Acids with ABS** 89
   Mohammed Taghi Zafarani-Moattar and Sholeh Hamzehzadeh

6. **Extraction of Proteins with ABS** 123
   Rupali K. Desai, Mathieu Streefland, Rene H. Wijffels, and Michel H.M. Eppink

7. **Extraction of Alcohols, Phenols, and Aromatic Compounds with ABS** 135
   María J. Trujillo-Rodríguez, Verónica Pino, and Juan H. Ayala

8. **Extraction of Natural Phenolic Compounds with ABS** 161
   Milen G. Bogdanov and Ivan Svinyarov

9. **Extraction of Metals with ABS** 183
   Isabelle Billard

10 **Surfactant Self-Assembly Within Ionic-Liquid-Based Aqueous Systems**.................................... 221
   Kamalakanta Behera, Rewa Rai, Shruti Trivedi, and Siddharth Pandey

11 **On the Hunt for More Benign and Biocompatible ABS**.................................... 247
   Jorge F.B. Pereira, Rudolf Deutschmann, and Robin D. Rogers

12 **Toward the Recovery and Reuse of the ABS Phase-Forming Components**.................................... 285
   Sónia P.M. Ventura and João A.P. Coutinho

Contributors

Juan H. Ayala Departamento de Química, Área de Química Analítica, Universidad de La Laguna (ULL), La Laguna (Tenerife), Spain

Kamalakanta Behera Department of Chemistry, Indian Institute of Technology Delhi, HauzKhas, New Delhi, India

Isabelle Billard University of Grenoble Alpes, LEPMI, Grenoble, France

CNRS, LEPMI, Grenoble, France

LEPMI, Saint Martin d'Hères, France

Milen G. Bogdanov Faculty of Chemistry and Pharmacy, University of Sofia "St. Kl. Ohridski", Sofia, Bulgaria

Rafał Bogel-Łukasik Unit of Bioenergy, National Laboratory of Energy and Geology, Lisbon, Portugal

José N. Canongia Lopes Centro de Química Estrutural, Instituto Superior Técnico, Universidade de Lisboa, Lisbon, Portugal

Instituto de Tecnologia Química e Biológica, Universidade Nova de Lisboa, Oeiras, Portugal

João A.P. Coutinho CICECO, Departamento de Química, Universidade de Aveiro, Aveiro, Portugal

André M. da Costa Lopes Unit of Bioenergy, National Laboratory of Energy and Geology, Lisbon, Portugal

LAQV-REQUIMTE, Departamento de Química, Faculdade de Ciências e Tecnologia, Universidade NOVA de Lisboa, Caparica, Portugal

Rupali K. Desai Bioprocess Engineering Department, Wageningen University, Wageningen, The Netherlands

Rudolf Deutschmann Department of Chemistry, McGill University, Montreal, QC, Canada

Michel H.M. Eppink Bioprocess Engineering Department, Wageningen University, Wageningen, The Netherlands

Mara G. Freire CICECO – Aveiro Institute of Materials, Chemistry Department, University of Aveiro, Aveiro, Portugal

Sholeh Hamzehzadeh Physical Chemistry Department, Chemistry and Chemical Engineering Research Center of Iran (CCERCI), Tehran, Iran

Siddharth Pandey Department of Chemistry, Indian Institute of Technology Delhi, HauzKhas, New Delhi, India

Jorge F.B. Pereira Department of Bioprocess and Biotechnology, School of Pharmaceutical Sciences, UNESP – Univ Estadual Paulista, Araraquara, SP, Brazil

Verónica Pino Departamento de Química, Área de Química Analítica, Universidad de La Laguna (ULL), La Laguna (Tenerife), Spain

Rewa Rai Department of Chemistry, Indian Institute of Technology Delhi, HauzKhas, New Delhi, India

Robin D. Rogers Department of Chemistry, McGill University, Montreal, QC, Canada

Rahmat Sadeghi Department of Chemistry, University of Kurdistan, Sanandaj, Iran

Mathieu Streefland Bioprocess Engineering Department, Wageningen University, Wageningen, The Netherlands

Ivan Svinyarov Faculty of Chemistry and Pharmacy, University of Sofia "St. Kl. Ohridski", Sofia, Bulgaria

Shruti Trivedi Department of Chemistry, Indian Institute of Technology Delhi, HauzKhas, New Delhi, India

María J. Trujillo-Rodríguez Departamento de Química, Área de Química Analítica, Universidad de La Laguna (ULL), La Laguna (Tenerife), Spain

Sónia P.M. Ventura CICECO, Departamento de Química, Universidade de Aveiro, Aveiro, Portugal

Rene H. Wijffels Bioprocess Engineering Department, Wageningen University, Wageningen, The Netherlands

Mohammed Taghi Zafarani-Moattar Physical Chemistry Department, Faculty of Chemistry (Excellence Center of New Materials and Clean Chemistry), University of Tabriz, Tabriz, Iran

Chapter 1
Introduction to Ionic-Liquid-Based Aqueous Biphasic Systems (ABS)

Mara G. Freire

Abstract During the past 13 years, ionic-liquid-based aqueous biphasic systems (IL-based ABS) have been the focus of remarkable interest and research. They have shown to be promising separation strategies for the most diverse compounds, resulting mainly from their tailoring ability offered by the current large number of IL chemical structures. A significant number of scientific manuscripts on IL-based ABS have been reported up to date, either on their characterization by attempting the determination of their phase diagrams or by exploring their viability on the separation of target compounds. The molecular-based scenario which rules the phase demixing in these systems and the comprehension of the best conditions and systems for improved separation performance have been ascertained. Both IL and other phase-forming components chemical structures, as well as pH and temperature effects, have been deeply evaluated in order to infer on their liquid–liquid demixing aptitude. On the other hand, possible and promising applications of IL-based ABS have been disclosed by investigating their role on the extraction of a wide plethora of biomolecules and compounds, e.g. amino acids, proteins, alkaloids, phenolic acids and antibiotics, amongst others. In fact, IL-based ABS proved to be outstanding separation platforms compared to more traditional polymer-based systems due to their wider hydrophilic–hydrophilic range which allows enhanced and selective extractions. Concentration factors up to 1000-fold and purification factors up to 245 have been reported with IL-based ABS. In this chapter we review and summarize the definition of IL-based ABS; describe the main phase-forming components used for their creation; define the fundamentals behind the formation of two aqueous-rich phases; demonstrate and show examples of their applications at the extraction, purification and concentration levels; evaluate the most promising applications of IL-based ABS; and discuss their future applicability.

Keywords Aqueous biphasic system • Ionic liquid • Phase diagram • Liquid–liquid extraction • Separation • Purification • Concentration

---

M.G. Freire (✉)
CICECO – Aveiro Institute of Materials, Chemistry Department, University of Aveiro, 3810-193 Aveiro, Portugal
e-mail: maragfreire@ua.pt

1.1 Introduction

In 1958, Albertsson [1, 2] proposed aqueous biphasic systems (ABS) as effective separation and purification platforms for biologically active products, while demonstrating their potential in the separation of proteins, cells, cell organelles, viruses, etc. ABS are formed by mixing different pairs of solutes (polymer–polymer, polymer–salt or salt–salt) in water, in that although water miscible, above given concentrations separate into two aqueous-rich phases, each one enriched in one of the solutes. Because of the non-volatile nature of the phase-forming components, in addition to being water-rich, ABS represent a more benign and biocompatible alternative to traditional liquid–liquid extractions which use volatile and hazardous organic solvents. Based on this main advantage, traditional polymer-based ABS have been widely investigated [2]. However, the wide performance of these systems, which has been restricted to narrow extraction yields and purification factors, has been blocked by the limited relative difference in polarities between the two phases. ABS constituted by two polymers display coexisting phases of similar polarities, whereas polymer–salt ABS consist of two phases with highly distinct characteristics. This limited difference in polarities, aiming at improving the extraction and purification potential of polymer-based ABS, has led to studies on polymer derivatization or on the use of additives [3–6]. Nevertheless, and particularly, the polymer functionalization makes these separation processes more costly and complex.

ABS composed of ionic liquids (ILs) were proposed by Rogers and co-workers [7] in the past decade, and as shown in this book, they represent an outstanding alternative to the more traditional polymer-based systems. After this pioneering work, it was observed a boom on the exploitation of IL-based ABS for the most diverse applications [8]. As described in the original manuscript [7], IL-based ABS can be used: (*i*) for the recovery and recycling of ILs from aqueous solutions, (*ii*) for carrying out metathesis reactions in the formation of new ILs and (*iii*) as new separation techniques.

Ionic liquids are comprised entirely of ions, thus salts, which are liquid at temperatures below 100 °C [9], a well-accepted temperature limit amongst the IL scientific community although with no specific/scientific meaning connected to ILs – it is the boiling temperature of water. Due to their ionic nature, ILs display exceptional properties, such as a negligible vapour pressure, a wide liquid temperature range and high thermal and chemical stabilities [10]. ILs are also improved extraction/solubilization solvents, both in neat and in aqueous solutions [11], for a wide variety of compounds or materials, and are able to increase the stability of added-value biomolecules, namely, proteins, enzymes and DNA, amongst others [12, 13]. Room temperature ILs have been investigated over the past two decades as green solvents for industrial applications [9], including their use in the petrochemical industry, via the production/recovery of heavy chemicals, fine chemicals, agrochemicals, etc., and in the chemical, nuclear and pharmaceutical industries. ILs have been claimed as representatives of an innovative approach to green chemistry. Although a large controversy still exists since these fluids can also display high toxicity and low biodegradability and require complex synthesis processes [14, 15], their non-volatile nature should always be seen as one of the major advantages compared to the use of volatile and often toxic solvents currently used by industries. To evaluate their performance in a given process compared to usual solvents, certainly, it is always required to a priori define their "green and biocompatible" nature, by evaluating their toxicity, biodegradability and through their life cycle analysis, and by attempting their recycling and reuse towards the development of more sustainable processes.

It is estimated to be possible to synthesize at least 1 million of different ILs [9] by the combination of different chemical structures of cations and anions. Due to this large array of ions combinations, ILs have been claimed as "designer solvents" – one of their most striking features. This designer ability is directly transferrable to IL-based ABS, and ABS with tailored properties and characteristics have been created, overcoming thus the main limitations encountered in typical polymer-based ABS [8]. ABS formed by ILs cover a wider hydrophilic–hydrophobic range and remarkable extractions and selectivities have been attained [16]. As a result of their high and tailored performance, IL-based ABS have been studied in a variety of applications [8], including their use in the recovery and recycling of ILs from aqueous solutions and as new separation/purification/concentration strategies, as highlighted in the pioneering work of Rogers and co-workers [7]. Nevertheless, and to the best of our knowledge, no applications of IL-based ABS for carrying out metathesis reactions in the formation of new ILs have been found in the open literature, leaving space to explore new applications of these promising liquid–liquid systems. IL-based ABS also display a low viscosity, when compared with polymer-based systems [17, 18], which favours the solutes mass transfer and reduces energetic inputs.

Up to date, circa 183 scientific reports regarding IL-based ABS were published. Figure 1.1 depicts the number of published scientific reports and number of citations per year under the subject of IL-based ABS. It is clear that studies on IL-based ABS went through an exponential growth in the past 13 years, also confirmed by the same type of increase in the number of citations per year. After the initial work of Rogers and co-workers in 2003 [7], only in 2007 did a larger number of manuscripts appear in the literature. Since 2009 the number of manuscripts has largely increased and a higher number of both ILs and second phase-forming components have been evaluated. Amongst the ILs investigated, 1-alkyl-3-methylimidazolium-based ILs have been the preferred choice combined with halogens, sulphates, sulphonates, alkanoates, tetrafluoroborate and triflate anions. On the other hand, and as the second phase-forming component, inorganic salts, such as K_3PO_4, K_2CO_3 and K_2HPO_4, have been the most studied, while a lower number of studies appear regarding the substitution of high-charge-density salts by more benign species, such as organic salts, amino acids, carbohydrates and polymers [8]. Most of these manuscripts deal with the characterization of their phase diagrams and their ability to form ABS. Investigations on their potential use as separation systems have been mainly demonstrated with model molecules, such

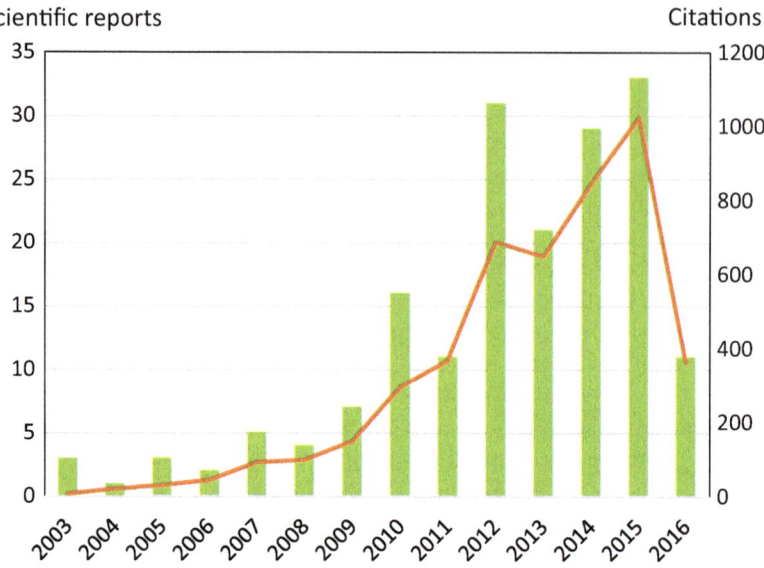

Fig. 1.1 Number of scientific reports (*bars, left scale*) and number of citations (*lines, right scale*) per year on IL-based ABS, from 2003 to 2016. Values based on a search on ISI Web of Knowledge in 27 May 2016, with the following topics: "aqueous biphasic system*" and "ionic liquid*" or "aqueous two-phase system*" and "ionic liquid*"

as alkaloids, phenolic acids, steroids, antibiotics, amino acids and proteins [8], by the determination and evaluation of the respective partition coefficients or extraction efficiencies, i.e. the partitioning extent of each molecule between the two phases. Some investigations on the use of IL-based ABS to concentrate and recover ILs from aqueous phases were also carried out [8]. In the past few years, more promising applications have appeared, particularly by starting evaluating their use to separate and purify value-added compounds from real/raw matrices, as well as concentration platforms to increase the levels of target compounds (drugs, endocrine disruptors, etc.) from human fluids and wastewaters, allowing thus a more vast application of these systems in the chemistry, biochemistry, biotechnology and analytical fields.

IL-based ABS are promising liquid–liquid extraction alternatives since they allow the extraction, purification and concentration, most of the times in a single step, of a wide variety of solutes, with tailored extraction efficiencies and selectivities. Their high performance opens thus new horizons to implement such processes in several areas of research to real applications at an industrial level. ABS meet the requirements for an easy design and scale-up of a wide range of separation processes and could, undoubtedly, become a target of high interest in industrial separation steps. ABS can be scaled-up by their implementation in counter current or centrifugal partition chromatography large-scale devices.

1.2 Fundamentals

Rogers and co-workers [8] were the first to show that a mixture of an aqueous solution of a hydrophilic IL and a concentrated aqueous solution of K_3PO_4 leads to the creation of an aqueous biphasic system formed by an upper IL-rich phase and a lower inorganic-salt-rich phase. An example of the macroscopic appearance of IL-based ABS is shown in Fig. 1.2.

ABS are commonly defined as water-rich systems formed by the addition of two water-soluble compounds above given concentrations at which the phase separation occurs, i.e. they are ternary systems composed of water and two solutes. Since ABS refer to two aqueous-rich phases, according to the authors' opinion, only ILs miscible with water can be considered in the formation of IL-based ABS at given pressure and temperature conditions. When dealing with highly hydrophobic or non-water miscible ILs, two phases already exist before the addition of any salting-out agent, and one of the phases is far from being aqueous rich due to these ILs' low solubility in water and vice versa [19]. Furthermore, hydrophobic ILs are limited to few cation/anion combinations, and most of them contain fluorinated and long alkyl side chain ions, which are more expensive, less environmentally benign and sometimes less stable in water [20]. On the other hand, the number of water-soluble ILs is much higher meaning that their environmental impact and biocompatible feature can be controlled in a more versatile way. It is already well accepted that the toxicity of ILs mainly depends on their hydrophobicity [21], and thus, most of water-soluble ILs used to prepare ABS are more environmentally-friendly than hydrophobic-based counterparts. Still, the widespread application of IL-based ABS will inevitably result in IL losses, requiring thus the search on more eco-friendly and biocompatible fluids as well as in the development of processes for their recovery and reuse.

As stated before, ABS are ternary systems composed of water and two solutes. In order to identify the mixture points at which they can be used as liquid–liquid

Fig. 1.2 Macroscopic appearance of ABS composed of ILs. Dyes were added to colour preferentially one of the phases

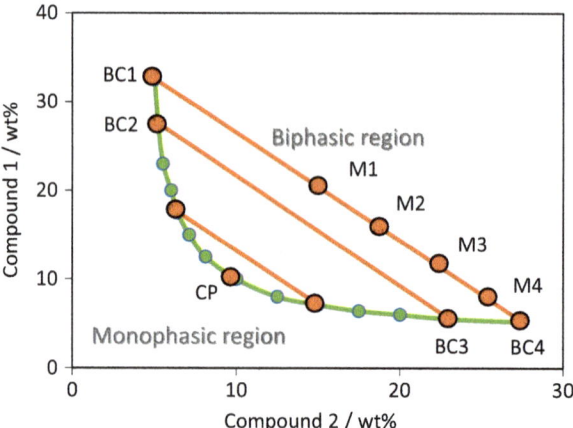

Fig. 1.3 Orthogonal ternary phase diagram of a hypothetical system composed of compound 1 + compound 2 + water (in weight fraction percentage)

separation routes, it is first required to determine their phase diagrams. In fact, most of the IL-based ABS research community has been dedicated to the determination of their phase diagrams. In literature, most of ABS ternary phase diagrams are depicted in an orthogonal representation, although being ternary systems, in which the water concentration is omitted (pure water becomes the origin of the orthogonal axes and corresponds to the difference required to reach 100 wt% in a given mixture composition). The major reason for such a decision relays on the easier interpretation of the phase diagrams and thus on more direct comparisons on the performance between different ILs and other phase-forming components in what concerns their ability to form ABS. Figure 1.3 illustrates an example of a phase diagram in an orthogonal representation where the water concentration is omitted.

In Fig. 1.3, the binodal curve (BC1-BC2-BC3-BC4) confines the monophasic and biphasic regions. For mixture compositions above the binodal curve (e.g. M1 to M4), there is the formation of a two-phase system, while mixture compositions below this curve fit into the monophasic region and will not phase separate. For total mixture compositions M1 to M4, the composition of each phase in equilibrium is given by the points BC1 and BC4, which are the end points (nodes) of a specific tie line (TL). Mixtures M1–M4 are along the same tie line and thus have the same composition at each phase. However, by changing from mixture M1 to M4, there is a decrease on the volume/weight of the phase rich in compound 1. This particularity allows to use ABS as concentration platforms since it is possible to maintain the partition behaviour of a given solute between the two phases while just changing their volume/weight ratio (if not considering saturation effects). The tie line length (TLL) is a numerical indicator of the composition difference between the two phases and is commonly used to correlate trends in the solutes partitioning amongst the coexisting phases. The critical point of the ternary system, where the compositions of the two phases would become equal and the biphasic system ceases to exist, is identified as CP in Fig 1.3. Although from an application perspective it is more useful to represent phase diagrams in weight fraction, in most studies, phase diagrams are shown in molality units to avoid inconsistencies that may result from different molecular weights of ILs and of second phase-forming components. This representation allows for a more chemistry-oriented analysis of the effect of individual phase-forming components in promoting liquid–liquid demixing [8].

In the literature, the determination of the binodal/solubility curves is mainly carried out by the cloud-point titration method [8]. Briefly, the experimental procedure consists in two major steps: (*i*) dropwise addition of an aqueous solution of the salting-out agent to the other aqueous solution rich in the second phase-forming component until the detection of a cloudy and biphasic solution, or vice versa, and (*ii*) dropwise addition of ultrapure water until the formation of a clear and limpid solution corresponding to the monophasic regime. In these, the systems compositions are most of the times gravimetrically determined [8].

In the majority of IL-based ABS studies [8], a three-parameter equation is usually applied to fit the experimental points, according to

$$Y = A \exp\left[\left(B \times X^{0.5}\right) - \left(C \times X^3\right)\right] \quad (1.1)$$

where Y and X are the weight fraction percentages of the IL and the second phase-forming component and A, B and C are fitting parameters obtained by least squares regression.

Rogers and co-workers [7] were the first to apply this type of mathematical description (Eq. 1.1) to correlate the experimental coexisting curve which was initially proposed by Merchuk et al. [22] to describe polymer-based systems. As initially proposed [22], the fitting of the binodal data by Eq. (1.1) allows to determine TLs by a weight balance relationship. The compositions of the top and bottom phases and the overall system composition are determined by the lever-arm rule. TLs are determined according to the solution of the following system of four unknown values:

$$Y_T = A \exp\left[\left(B \times X_T^{0.5}\right) - \left(C \times X_T^3\right)\right] \quad (1.2)$$

$$Y_B = A \exp\left[\left(B \times X_B^{0.5}\right) - \left(C \times X_B^3\right)\right] \quad (1.3)$$

$$Y_T = \frac{Y_M}{\alpha} - \frac{1-\alpha}{\alpha} \times Y_B \quad (1.4)$$

$$X_T = \frac{X_M}{\alpha} - \frac{1-\alpha}{\alpha} \times X_B \quad (1.5)$$

where T, B and M designate the top phase, the bottom phase and the initial mixture, respectively; X and Y represent the weight fraction percentage of the second phase-forming component and of the IL, respectively; α is the ratio between the weight of the top phase and the total weight of the mixture; and A, B and C are the fitted constants obtained from Eq. (1.1).

Most authors used this mathematical approach (Eqs. 1.1, 1.2, 1.3, 1.4, and 1.5) to describe the binodal/solubility curve and to determine the corresponding TLs [8].

However, it should be remarked that other equations to fit the experimental binodal curves also have been proposed, that additionally allow to describe the effect of pH and temperature [23–25]. Moreover, other attempts to determine the TLs by analytical techniques have been reported. For instance, Zafarani-Moattar and Hamzehzadeh [23–25] determined the concentration of salt ions by flame photometry, the IL concentration by refractive index measurements or by nitrogen elemental analysis, and the water content by Karl Fischer titration. Chen et al. [26, 27] used high-performance liquid chromatography to quantify the IL and Karl Fischer titration to quantify water. Han et al. [28, 29] quantified conventional salts in both phases by refractive index or density while the amount of ILs was directly taken from the fitting using Eq. (1.1). Pei et al. [30] quantified ILs by UV–vis spectroscopy and water by a drying procedure, while Yu et al. [31] determined the IL content by UV–vis spectroscopy and the amount of salts by acid–base titration.

The reliability of TLs also has been ascertained by several authors [8], using the Othmer-Tobias and Bancroft equations [32] described below:

$$\frac{1-Y_T}{Y_T} = k_1 \left(\frac{1-X_B}{X_B}\right)^n \quad (1.6)$$

$$\frac{Z_B}{X_{BT}} = k_2 \left(\frac{Z_T}{Y_T}\right)^r \quad (1.7)$$

where X, Y and Z represent the weight fraction percentage of the second phase-forming component, the IL and water, respectively, in the top (T) and bottom (B) phases and k_1, k_2, n and r are adjustable parameters from the fitting.

The TLL and TL slope (STL) have also been addressed for IL-based ABS [8], being described by the following equations:

$$\text{TLL} = \sqrt{(X_T - X_B)^2 + (Y_T - Y_B)^2} \quad (1.8)$$

$$S_{TL} = \frac{Y_T - Y_B}{X_T - X_B} \quad (1.9)$$

where T, B and M correspond to the top phase, the bottom phase and the initial mixture, respectively, and X and Y represent the compositions in equilibrium.

Although less explored, the location of the critical point also has been addressed [8], and it is estimated from the TL composition extrapolation applying the following equation:

$$Y = f + gX \quad (1.10)$$

where Y and X are the IL and second phase-forming component compositions and f and g are fitting parameters.

Up to date, IL-based ABS have been created with inorganic/organic salts, amino acids, carbohydrates, polymers and organic buffers [8, 33]. Amongst these, the most studied class is the IL–salt one due to the high ability of salt ions to induce the salting-out of the IL and therefore to create two aqueous phases. Figure 1.4 depicts the combination of phase-forming components used to form IL-based ABS.

The basis behind the formation of ABS comprising ILs and salts is well accepted as a salting-out mechanism resulting from the creation of water-ion complexes that cause the dehydration of the solute and the increase of the surface tension of the cavity in aqueous media [34]. The addition of high-charge-density salts to aqueous solutions of ILs leads to liquid–liquid demixing due to a preferential hydration of the high-charge-density salt ions over IL ions, leading thus to the salting-out (exclusion) of the IL to the IL-rich phase. In presence of water, ILs will be weakly hydrated species as compared to common inorganic/salting-out salts due to their low-symmetry charge-delocalized ions. However, in IL–salt ABS, the two phase-forming components are ionic, and some ion exchange could be expected. Neves et al. [35] demonstrated that, at least in ABS composed of ILs and high-charge-density ions, i.e. strong salting-out species, the ion exchange between the coexisting phases is negligible.

The closer to the axis origin a binodal curve is, *cf.* Fig. 1.3, the higher the IL ability to phase split and to create an ABS. Regarding IL–salt ABS, it has been found that quaternary ammonium- and phosphonium-based ILs display a higher tendency to phase separate from aqueous media when compared, for instance, with aromatic ILs, namely, imidazolium- and pyridinium-based fluids [36, 37]. Furthermore, ILs with cations comprising longer alkyl side chains require lower quantities of IL and salt to form an ABS (at least up to alkyl chains up to hexyl since for longer ones there is the formation of IL aggregates in aqueous media and the ABS creation phenomenon becomes more complex) [38]. On the other hand, it has been demonstrated that the ability of an IL anion to create ABS closely follows the decrease in

Fig. 1.4 Description of phase-forming components reported up to date for the creation of IL-based ABS

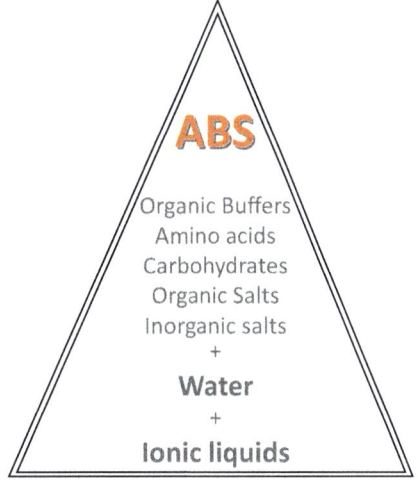

their hydrogen bond accepting strength or electron pair donation ability [39]. This is a direct result of the competition between the salting-out ions and IL ions for the formation of hydration complexes. Shahriari et al. [40] published the most complete study on the effect of salts through the liquid–liquid demixing of 1-butyl-3-methylimidazolium-triflate–water mixtures. Authors [40] evaluated both the cation and anion influence of the salt on ABS formation. In summary, the salting-out strength of sodium-based salts to induce the formation of a second aqueous phase follows the rank $PO_4^{3-} > C_6H_5O_7^{3-} > HPO_4^{2-} \approx CO_3^{2-} > SO_4^{2-} \approx SO_3^{2-} > C_4H_4O_6^{2-} >> H_2PO_4^- > OH^- > CH_3CO_2^- \approx HSO_4^- \approx HCO_3^- > Cl^-$, while the effect of the salt cation with chloride-based salts follows the order $Mg^{2+} \approx Ni^{2+} \approx Sr^{2+} > Ca^{2+} >> Na^+ > K^+ > Cs^+$. These trends follow the well-known Hofmeister series. Shahriari et al. [40] used this large compilation of data on the effect of salts on the creation of ABS and demonstrated that there is a close correlation between the ILs molality required for the formation of a given ABS and the molar entropy of hydration of the ions. These results support the idea that the salting-out in IL-based ABS is an entropically driven phenomenon, as previously proposed [34] for hydrophobic ILs and a wide range of salts.

The effects of temperature and pH on IL–salt ABS have also been investigated [23, 25, 41]. In general, the locus for the experimental binodal curves obtained by different authors [23, 41] demonstrates that the two-phase region decreases with the increase in temperature. The higher the temperature, the higher the salt and IL concentrations required for phase separation, and thus lower temperatures are favourable for the creation of IL–salt-based ABS. Distinct salts generate different pH values at the coexisting phases of IL-based ABS [25]. Most works in the literature have dealt with alkaline aqueous solutions since, in general, the formation of biphasic systems is more easily achieved under such conditions. If salt ions that can suffer speciation at different pH values are used, they will become more protonated at lower pH values and thus become weaker salting-out species. In summary, the number of ILs that are able to suffer phase splitting is largely reduced at lower pH values. Chapter 2 provides an overview on the fundamentals behind IL–salt ABS.

In addition to the well-studied IL–salt ABS, some researchers [26, 27, 42–45] addressed the finding of more benign ABS promoters and shifted to ternary phase systems formed by ILs + carbohydrates + water. In this case, the ionic exchange amongst phases can be fully avoided. Since carbohydrates are non-toxic and renewable resources, greener processes involving IL-based ABS can be directly envisaged. However, carbohydrates are weaker salting-out agents compared to salts and thus the number of ILs able to form ABS is more limited. ILs with fluorinated anions (lower hydrogen-bond basicity) were combined with sucrose, glucose, fructose, xylose, maltose, galactose, mannose, arabinose, maltitol, sorbitol and xylitol aqueous solutions and demonstrated to be able to form ABS [26, 27, 42–45]. From all results, the formation of IL–carbohydrate ABS was demonstrated to be a direct result of the competition between the two solutes to be hydrated. Since the ability of a specific IL ion to salt-out from aqueous media depends on its affinity for water, an increase in the size of the aliphatic moiety at the cation and a decrease on the hydrogen-bond basicity of the anion increase the IL hydrophobicity and, therefore, the ability of the IL to be separated from aqueous media. This trend closely correlates with that displayed by systems constituted by ILs and salts discussed above. On the other hand, Freire et al. [42] demonstrated that the preferential formation of carbohydrate hydrated complexes plays the major role in the formation of ABS, since it defines the carbohydrate salting-out ability, although the sugars stereochemistry cannot be discarded. The effect of temperature on IL–carbohydrate-based ABS has also been investigated, where it was found that a decrease in temperature leads to an expansion of the two-phase region of the phase diagrams [8, 44] – analogous trend to that observed in ABS formed by ILs and salts. Chapter 3 provides an overview on the fundamentals and application of ABS formed by ILs and carbohydrates.

IL-based ABS formed with amino acids and organic buffers have also been reported [33, 46, 47]. As happens with carbohydrates, few ILs are able to form ABS with amino acids or organic buffers due to their low salting-out capacity. It was found that the formation of IL-amino acid and IL-buffer ABS is a direct result of the competition between the two solutes to be hydrated, where an increase in the size of the aliphatic chain at the cation and a decrease on the hydrogen-bond basicity of the anion increase the IL hydrophobicity and, therefore, the ability of the IL to undergo phase separation. Moreover, it was found that a decrease in temperature leads to a higher ability for the liquid–liquid demixing of IL-amino acid systems [46], as observed in ABS formed by ILs and salts or ILs and carbohydrates.

Some works [48–52] have been published regarding ABS formed by polymers and ILs. Polymer–polymer ABS usually display restricted differences in polarities that mainly depend on the content of water at each phase. On the other hand, polymer–salt ABS present a more hydrophobic phase majorly composed of the polymer and a highly-charged and more hydrophilic aqueous phase enriched in the salt. Therefore, the substitution of high-charge-density salts by more amenable ILs will allow a more close-fitting control of the phases' polarities. Furthermore, salt crystallization problems can be avoided with ILs which melt below room temperature, because their saturation levels in aqueous solutions are usually higher than those observed with common salts. Rebelo and co-workers [53] reported the pioneer work on the salting-in/salting-out effects of ILs over polymers dissolved in aqueous media. This pioneering work provided evidences that ABS formed by polymers and ILs can be created. ABS composed of ILs and poly(propylene glycol) (PPG) [48, 49] and poly(ethylene glycol) (PEG) [50] have been published, and their applications also have been explored [51, 54]. In general, ILs comprising cations with shorter alkyl side chains and anions of high hydrogen-bond basicity or ability to accept protons enhance the ABS creation ability [48–50]. This is the opposite behaviour to that found in IL–salt ABS, suggesting that, in these examples, ILs seem to act as salting-out species. However, when evaluating the IL cation core effect [50], it was found that more hydrophobic ILs display cloud points at lower concentrations of polymer. This pattern is in good agreement with that observed in ABS formed by ILs and salts, meaning that this cation trend is in stark contradiction to the initially proposed salting-out phenomenon exerted by ILs [48, 49]. Based on these striking results, Freire and co-workers [50] concluded that the formation of polymer–IL-based ABS is not only a direct consequence of the IL ions' ability to form hydration complexes. Instead, the interactions that occur in IL–polymer pairs of solutes are the key driving forces in the formation of the respective ABS. The larger the immiscibility between the IL and the polymer the greater is the ability of the IL and polymer to phase separate in aqueous media [50]. In particular, it has been shown that some ternary IL–PEG–water systems are of type 0, i.e. the mutual miscibility amongst all the binary pairs is observed, while an immiscibility regime appears only in the ternary mixture, and that the formation of PEG–IL-based ABS is a result of a "washing-out" phenomenon [55].

PPG and PEG are able to lead to liquid–liquid demixing in IL aqueous solutions, where an increase in the molecular weight of the polymer facilitates the formation of ABS [50]. When comparing the data between PPG- and PEG-based systems [8], it is clear that PPG is much more likely to undergo phase separation in aqueous media than PEG due to its more hydrophobic character. The effect of temperature on the phase diagrams of water–polymer–IL systems was also addressed [50]. An increase in temperature decreases the immiscibility region of aqueous systems containing PEG and ILs [50], while those constituted by PPG and ILs display an opposite behaviour [49]. These results seem thus to indicate that a more complex phenomenon governs the phase behaviour of ABS containing polymers and ILs, although new data concerning these particular systems are obviously needed to gather a broader picture of the mechanisms involved in the formation of such type of ABS.

Although all works discussed above correspond to ternary systems, quaternary ABS composed of ILs have also been reported [56, 57]. In these, ILs were mainly investigated as adjuvants (in low contents) in ABS formed by polymers and salts. These systems have demonstrated to be quite promising in controlling the coexisting phases' polarities and have particular interest in biotechnological separation processes aiming at replacing the approach of PEG functionalization. Different ILs were investigated demonstrating that an increase in the cation side alkyl chain length reduces the phases' miscibility when compared with the system with no IL, whereas IL anions with higher hydrogen-bond basicity increase the ability for phase separations. Authors [56, 57] also demonstrated that almost all ILs investigated preferentially partition to or are enriched at the polymer-rich phase. Chapter 4 summarizes and discusses the main mechanisms governing the formation of IL–polymer-based ABS.

1.3 Applications

Albeit most of the works in the literature refer to the characterization and/or determination of the IL-based ABS phase diagrams, these systems have been further investigated in the extraction/separation of a wide plethora of compounds. Figure 1.5 depicts the major classes of compounds separated with IL-based ABS

Fig. 1.5 Summary of the major classes of compounds extracted, purified or concentrated with IL-based ABS

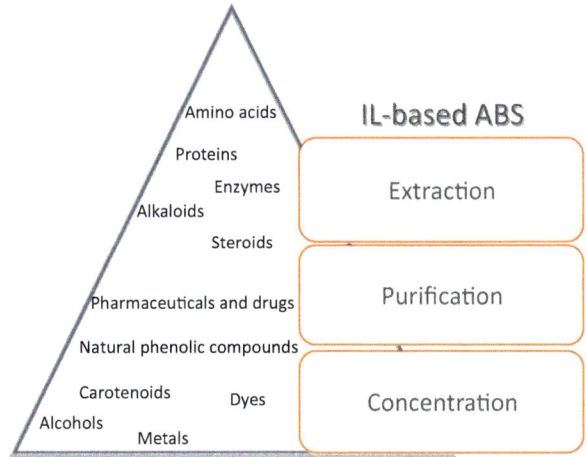

connected to their three major applications: extraction, purification and concentration.

The majority of initial works reporting to the application of IL-based ABS as extraction and purification platforms were carried out with amino acids [8, 58, 59]. Amino acids are important compounds in many biotechnological processes, and the development of methods for their separation and purification is still the subject of intensive research. Some amino acids, such as L-tryptophan, L-phenylalanine and L-tyrosine may be produced by bacterial fermentation [60]. In this context, the application of ABS composed of ILs could be seen as a promising alternative for the extraction of amino acids within an integrated continuous fermentation process. Ventura et al. [58] and Neves et al. [59] studied the ability of ABS formed by a wide variety of ILs and K_3PO_4 to extract L-tryptophan. Authors [58, 59] demonstrated that, depending on the nature of the IL, partition coefficients between 10 and 120 for the IL-rich phase can be obtained and that these values are significantly higher than those obtained with conventional ABS formed by polymers (in the order of 1–7) [61], as well as compared with systems formed by hydrophobic ILs [62]. ABS formed by ILs and organic salts have also been investigated for amino acids extraction [37]. In these systems, amino acids preferentially partition to the IL-rich phase with extraction efficiencies ranging between 72 % and 99 % in a single step [37]. Pereira et al. [56] investigated polymer–salt ABS using ILs as adjuvants in the extraction of L-tryptophan. The results obtained demonstrate that the addition of small amounts of ILs rules the partitioning trend and allows to control the extraction efficiency of various systems by a proper selection of the chemical structure of the ILs. With the aim of replacing inorganic salts in the formation of ABS, systems comprising ILs and carbohydrates or organic buffers [33, 42] were investigated for the extraction of amino acids. The first set of systems studied allowed ca. 50 % of extraction of amino acids for the IL-rich phase in a single step, while in the second type of ABS, amino acids preferentially migrate to the organic-buffer-rich phase with extraction efficiencies ranging between 22 %

and 100 %. Based on these results, authors [33] proposed the use of these systems for the fractionation of complex mixtures of amino acids by optimizing the composition and pH of the mixture. In summary, the wide range of values on the extraction efficiency offered by different IL-based ABS [33, 37, 42, 56, 58, 59] suggest that the extraction of amino acids can be tailored by properly selecting the IL and the second phase-forming component chemical structures. Chapter 5 summarizes the main advances found in the literature on the application of IL-based ABS for the extraction of amino acids.

In recent years, a large interest has been noticed on the extraction and purification of biomolecules with therapeutic and diagnostic applications, in particular for proteins [63, 64]. Traditional methods for the purification of proteins are highly complex, time-consuming and expensive since proteins easily undergo denaturation and lose their biological activity when in contact with organic solvents and because their separation is mainly achieved by chromatographic approaches [63, 64]. As a potential alternative, typical ABS have been investigated for the purification of proteins [2, 63, 64] and, more recently, IL-based ABS [8]. Proteins and enzymes mainly studied are albumin, cytochrome c, lysozyme, myoglobin, trypsin and lipases [8]. Ruiz-Angel et al. [65] evaluated the extraction efficiency of ABS formed by ILs for four proteins (cytochrome c, myoglobin, albumin and haemoglobin). Authors [65] concluded that the partition coefficients obtained are usually 2 to 3 orders of magnitude higher than those achieved with traditional polymer–polymer or polymer–salt ABS. Several works [54, 66] additionally demonstrated that it is possible to completely extract proteins to the IL-rich phase in a single step by an appropriate manipulation of the IL chemical structure and composition of the system. Some investigations [67–69] reported on the use of ILs with buffer characteristics to form ABS envisaging their use as extraction/purification platforms at controlled pH – an essential requirement to maintain the stability of proteins in solution. In addition to the complete extraction achieved in a single step, it was demonstrated that proteins retain their secondary structure in the IL-rich phase after the extraction process [69]. The application of ABS formed by ILs can also be seen as a good option for extracting enzymes from their fermentation media. Nevertheless, when considering enzymes, it should be taken into account that it is essential to maintain their catalytic activity after the extraction/purification step. IL-based ABS formed by quaternary ammonium ILs allow extraction efficiencies of 90 % and an increase of 400 % in the specific activity of two dehydrogenases [70]. Further to extraction/partition studies of model proteins, IL-based ABS have been studied in the extraction/purification of proteins and enzymes from real matrices, such as bovine serum [54] and extracellular media [71]. The complete removal of albumin from bovine serum, with no losses of protein [54], and lipase purification factors between 26 and 51, with recovery factors between 91 and 96 %, from *Bacillus* sp. PTI-001 media [71], were obtained with IL-based ABS. Later, the same research group [72] demonstrated that ABS containing ILs as adjuvants lead to higher purification factors of lipase as a result of a preferential partitioning of contaminating proteins to the top phase (rich in polymer), while the target enzyme preferentially migrates to the bottom phase (rich in salt). More recently, IL-based ABS have demonstrated to be outstanding purification platforms for the purification of high-value proteins, namely, antibodies/biopharmaceuticals [68]. Authors [68] synthesized a series of new ILs derived from natural sources and with buffering characteristics and used them in the formation of ABS with biocompatible polymers. With these systems, extraction efficiencies ranging between 79 and 94 % were achieved in a single step. The high selectivity of these systems for antibodies was demonstrated supporting their further use in the purification of antibodies from real matrices [68]. Chapter 6 describes the main achievements gathered with IL-based ABS on the extraction and purification of proteins.

IL-based ABS have also been investigated for the extraction/separation of alcohols, phenols and other synthetic aromatic compounds. These applications are summarized and discussed in Chap. 7. Ruiz-Angel and co-workers [65] investigated IL-based ABS combined with countercurrent chromatography on the separation of a homologous series of linear alcohols (methanol up to hexanol). Authors [65] demonstrated the significant polarity difference between PEG- and IL-based systems and concluded that all compounds with an octanol–water partition coefficient higher than 0.22, which include the vast majority of compounds investigated, will partition preferably into the IL-rich phase.

Aromatic compounds are common contaminants in wastewater streams. For instance, phenol derivatives are listed as top pollutants by the US Environmental Protection Agency (EPA) [73]. These compounds are harmful to organisms, even at low concentrations, and many of them have been classified as hazardous pollutants given their potential harm to human health. Strict EPA regulations request for lowering phenol contents in wastewater streams, to less than 1 mg/L [73]. Therefore, a large interest has been devoted to IL-based ABS as a simple and effective method for extracting aromatic compounds, such as phenol [27], nitrobenzene [74], 4-nitrophenol [74], aniline [74] and textile aromatic dyes [75], before their disposal into water cycles. Zhang and co-workers [74] reported the enrichment of nitrobenzene, 4-nitrophenol, phenol and aniline from water, using both hydrophobic and hydrophilic ILs. In these, the implementation of extraction processes based on hydrophilic ILs in a water-rich media appears as a more promising alternative. Aromatic dyes display high toxicity and suffer from non-biodegradability issues, and IL-based ABS demonstrated to be efficient alternatives for the removal of aromatic dyes from environmental waters [75–77]. Several works studied IL–salt combinations [75, 76], whereas others focused on IL–polymer pairs [11, 17]. Sheikhian et al. [76] studied ABS composed of ILs for the extraction of three anionic dyes, namely, reactive red 120, 4-(2-pyridylazo)-resorcinol and methyl orange. An enrichment of the dyes in the IL-rich phase (top phase) was observed in all systems and for all dyes. In this work [76], it was also demonstrated the possibility of recycling the hydrophilic IL by the addition of a hydrophobic (non-water miscible) one, further allowing the recovery of dyes in an aqueous "IL-free" solution. Ferreira et al. [75] reported IL–salt ABS, based on phosphonium- and imidazolium-based ILs combined with inorganic and organic salts, for the extraction of chloranilic acid, indigo blue and Sudan III. It was demonstrated that phosphonium-based ILs are more effective and allow the complete extraction of dyes in a single step [75]. Finally, with the aim of recovering dyes and of recycling the IL phase, authors [75] took advantage of the hydrotropic effect displayed by ILs [11] and added water and reduced the temperature, which led to a remarkable reduction of the dyes' solubility while inducing their precipitation. Pereira et al. [16] applied ABS composed of the PEG and different chloride-based ILs for the extraction of chloranilic acid, indigo carmine and indigo blue. The gathered results demonstrated an inversion on the dyes' partition trend by changing the IL cation and allowed authors [16] to conclude that it is possible to tune the ABS coexisting phases, in terms of hydrophobicity/hydrophilicity, by simply modifying the IL chemical structure.

Still within aromatic compounds, and in order to overcome the detection limits of analytical equipment commonly used for the quantification of endocrine disruptors in human fluids, while envisaging their more accurate monitoring, ABS formed by ILs were recently proposed as a combined extraction and concentration technique [78]. To this end, authors [78] tested ABS formed by a wide range of ILs and K_3PO_4 in order to find systems that could allow the complete extraction and concentration of bisphenol A from human fluids in a single step. Passos et al. [78] demonstrated that adequate IL-based ABS allow the complete extraction of bisphenol A to the IL-rich phase and its concentration up to 100-fold. In fact, it is possible to concentrate several products with ABS using mixtures with different initial compositions along the same tie line (*cf.* Fig. 1.3). After this demonstration, it has been shown that these systems can lead to concentrations of up to 1000 times of estrogens [79], useful for a more accurate monitoring of persistent pollutants in wastewaters.

In addition to the application of IL-based ABS for the extraction/purification of products from biotechnological processes and reaction media, in recent years, there has been a great interest in applying such systems for the purification of natural products extracted from biomass. These include, for instance, alkaloids and phenolic compounds with relevant properties (antifungal, anti-inflammatory and antioxidant) with interest for the pharmaceutical, food and cosmetic industries. ABS consisting of ILs have been studied in the extraction and separation of the following alkaloids: codeine, papaverine, caffeine, nicotine, xanthine, theophylline and theobromine [80–82]. In particular, the ability of ABS composed of ILs and inorganic salts, carbohydrates and amino acids can be compared in terms of extraction performance for caffeine [42, 47, 81]. In general, higher extraction efficiencies are obtained with ABS formed by ILs and salts due to salting-out effect exerted by the salt which promotes the migration of alkaloids for the opposite phase–IL-rich phase. Apart from ABS consisting of ILs and salts, Pereira et al. [83] studied ABS formed by ILs and polymers for extracting three alkaloids. In almost all systems, a preferential migration of caffeine to the polymer-rich phase was observed, while nicotine and xanthine partition to the opposite phase (IL-rich) [83]. Therefore, despite the structural similarity of the three alkaloids, these results confirm the high selectivity of ABS formed by ILs.

In addition to studies carried out with model systems, more recently, we have been facing a turnover on the application of IL-based ABS, where their direct application to the extraction/purification of target compounds from real matrices started to appear. The pioneering work on the application of ABS to real matrices was reported by Li et al. [80]. Authors [80] used ABS comprising ILs as a pretreatment extraction strategy for the analysis of opium alkaloids in human fluids (codeine and papaverine). After the optimization of operating conditions, extraction efficiencies for papaverine and codeine of 93 and 65 %, respectively, were obtained. In order to develop pre-concentration methods for more efficient doping control, Freire et al. [81] demonstrated that ABS formed by ILs and strong salting-out salts allow the complete extraction of alkaloids (caffeine and nicotine) from urine samples to the IL-rich phase in a single step. In fact, the wide variety of ILs available today allows to manipulate their chemical structures in order to optimize extraction and separation approaches.

Within the compounds obtained from biomass sources, IL-based ABS have also been investigated for the extraction of a wide range of phenolic compounds with antioxidant and anti-inflammatory activities, namely, vanillin [17]; gallic, vanillic and syringic acids [17, 58, 84], eugenol [85]; and propyl gallate [85]. Cláudio et al. [17, 84] studied IL-based ABS in order to avoid the use of organic solvents in the purification of extracts containing antioxidants, in particular vanillin and gallic acid. Authors [17] evaluated the effect of the IL chemical structure, the equilibrium temperature and the initial concentration of solute and demonstrated that antioxidants preferentially migrate to the IL-rich phase [17], although this trend can be tailored by changing the pH of the aqueous medium since both solutes present low acidic dissociation constants [84]. Based on the ILs' ability to manipulate the extraction efficiencies, more recently, the use of ILs as additives (5 to 10 wt %) in polymer–salt ABS for the extraction of gallic, vanillic and syringic acids was also proposed [58]. All antioxidants preferentially migrate to the polymer-rich phase, which in turn corresponds to the phase where the IL concentrates. The authors [58] further showed that the addition of small amounts of ILs leads to increased extraction efficiencies of all studied phenolic acids, ranging between 80 % and 99 %, thus supporting the ILs ability to adjust the polarity/affinity of the ABS coexisting phases. In the same line of investigation, the possibility of using ILs as adjuvants in polymer–salt ABS to extract other antioxidants, namely, eugenol and propyl gallate, has also been demonstrated, allowing complete extractions in a single step [85].

In addition to the studies discussed before carried out with model solutes, the extraction of saponins and polyphenols from mate tea leaves and their subsequent purification with ABS involving ILs has been suggested [86]. The authors [86] performed the direct extraction of the desired compounds from dried tea leaves with aqueous solutions of ILs. Under the conditions tested, most of aqueous solutions of ILs lead to higher extraction efficiencies than the method traditionally used (30 % ethanol aqueous solution). After the extraction step, authors [86] used aqueous solutions containing the extracts, concentrated in saponins and phenolic compounds, and created ABS for their separation. Finally, by taking advantage of the miscibility between most ILs, the saponins and phenolic compounds at the IL phase were recovered by the addition of a second IL immiscible in water, leading to the recovery of target compounds in an aqueous "IL-free" medium [16]. Chapter 8 summarizes the main advances carried out on the application of IL-based ABS for the extraction and separation of natural phenolic compounds.

IL-based ABS could also represent great potential on the extraction of metal ions – see Chap. 9. For instance, the maximum concentration of metal ions is usually regulated by World Health Organization guidelines for drinking water. However, few articles have been published addressing this subject. Akama et al. [87, 88] used ABS composed of tetrabutylammonium bromide and ammonium sulphate to selectively extract Cr(IV) from Cr(III), and Cd^{2+} from Co^{2+}, Cu^{2+}, Fe^{3+} and Zn^{2+}. The formation of ion pairs between the metal cation and the IL anion was proposed as the main extraction mechanism in both examples [87, 88]. Authors [88] showed that ABS can be combined with atomic absorption to determine traces of Cr(IV) in wastewater samples. The determination limit found was of 60 μgL^{-1}, while recovery efficiencies of 90 % have been reported [88]. On the other hand, the extraction of metals from aqueous phases is a relevant issue envisaging the recycling of precious metals. For this approach, hydrophobic (non-water miscible and fluorinated ones) have been the preferred choice. Outstanding extraction efficiencies have been reported; however, no studies on IL-based ABS have been found in the open literature, leaving an open door to further boost the potential applications of IL-based ABS.

A single work [89] on ABS composed of ILs and surfactants is available in the open literature. Surfactant-based systems can be of high interest since they allow the dispersion of highly hydrophobic substances in aqueous media. Wei et al. [89] have shown that aqueous mixtures of sodium dodecyl benzene sulphonate and ILs can spontaneously separate and form an ABS. Authors [89] used transmission electron microscopy to identify the presence of aggregates in both phases: large micelles are present in the upper phase (surfactant-rich phase) while relatively smaller micelles appear in the bottom phase (IL-rich phase). Chapter 10 presents a summary on the surfactant self-assembly behaviour within IL-based aqueous systems aiming at leaving clues for the further development of surface-active-based ABS.

Although most ILs were originally considered as green solvents due to their negligible volatility, the inevitably release of these fluids into aqueous streams and environment has raised several concerns. Furthermore, when envisaging the use of ILs for the separation and purification of compounds for human's consumption, strong apprehensions appear regarding their cytotoxicity and biocompatible nature. Thus, more recent ABS investigations started to focus on more benign and biocompatible ILs and second phase-forming components foreseeing the development of more green and biocompatible IL-based ABS. Regarding concerns related to the second phase-forming component, more eco- and bio-friendly ABS were suggested, namely, those composed of amino acids, carbohydrates and polymers

[8, 26, 27, 42–47, 49–53]. On the other hand, while considering the IL nature, ABS constituted by cholinium-based ILs combined with several polymers or salts were the most thoroughly studied [8, 52, 53, 55, 90]. Chapter 11 reviews and summarizes the available information on more benign ABS published so far, highlights the most important differences between these and more conventional IL-based ABS, provides examples on nontoxic and biodegradable phase-forming components and evaluates future industrial perspectives of IL-based ABS. However, and even if more benign ABS phase-forming components are used, it is critical to develop effective approaches for their recovery and reuse, both to decrease the overall environmental footprint and to decrease the associated costs of the foreseen large-scale and sustainable ABS-related technologies. In the past few years, authors started looking into the recovery of the target compounds, followed by the recovery and recycling of ILs [66, 75, 84, 91–93]. This trend is not surprising given current sustainability concerns and moderately high cost of ILs. In these works [66, 75, 84, 91–93] it was demonstrated that IL-based ABS maintain their extraction and separation performance after their recovery and further reuse. Still, there is a large number of approaches that can be developed, as well as a wide number of conditions to optimize, in order to develop cost-effective and sustainable IL-based ABS with potential to be applied at an industrial scale. Chapter 12 summarizes the approaches and strategies reported up to date regarding the recovery, recycling and reuse of ILs and second phase-forming components of IL-based ABS.

1.4 Conclusions and Future Perspectives

IL-based ABS were first proposed in 2003, and since then we faced a boom on the research regarding their characterization and applications. ILs and ABS share relevant advantages, namely, enriched aqueous media and no use of volatile organic solvents. Nevertheless, the most remarkable characteristic of IL-based ABS comes from their designer solvents ability. ABS composed of ILs exhibit a much wider hydrophilic–hydrophobic range than conventional polymer-based systems since their phase polarities can be easily tuned by the choice of appropriate IL chemical structures. In summary, IL-based ABS are more amenable to be tuned, as demonstrated by their outstanding extraction and purification performance for a wide variety of molecules and products.

IL-based ABS can be formed with a large range of inorganic and organic salts, amino acids, sugars/polyols and polymers. Before their application, it is crucial to determine the liquid–liquid phase diagrams of IL-based ABS aiming at identifying mixture compositions at which they phase separate as well as to infer on the compositions of the coexisting phases. Most scientific works found in the literature are devoted to the characterization of novel IL-based ABS by the determination of the corresponding phase diagrams. The effects of temperature and pH on the phase diagrams behaviour have also been ascertained. In general, the understanding of the impact of the various phase-forming components in the creation of ABS further permits a more conveniently tailoring of the corresponding phase diagrams. On the other hand, a large interest has been found on the understanding of the main mechanisms which rule the phase demixing. Although ABS composed of salts, amino acids or carbohydrates are ruled by a competition between both classes of solutes to be hydrated and thus of the latter to induce the ILs salting-out from aqueous media, ABS formed by ILs and polymers seem to be more complex than initially admitted, where the mutual miscibility between both solutes seems to play the primary role. In addition to this more fundamental perspective, a large amount of works focused on evaluating the IL-based ABS extraction/separation performance by carrying out experiments on the partitioning of biomolecules, i.e. their partitioning extent between the coexisting phases, including amino acids, proteins, phenolic compounds, alkaloids, steroids, etc. As stated before, the major advantage of IL-based ABS stands on the possibility of tailoring the phases' polarities and their selectivity for a target biomolecule. In fact, it was demonstrated that a proper manipulation of the system constituents and respective composition allows the pre-concentration, complete extraction or purification of the most diverse compounds. Concentration factors up to 1000-fold in a single step and purification factors up to 245 have been reported using IL-based ABS.

Although ILs can be viewed as green solvents due to their negligible volatility, some concerns still exist regarding their discharge into water ecosystems. Therefore, the search on more benign and biocompatible ILs in ABS formulations is a critical requirement, as well as the use of biodegradable organic salts, amino acids, carbohydrates and polymers. Furthermore, while foreseeing the large-scale application of these promising systems, it is critical to develop novel approaches for their recovery and reuse in order to develop more cost-effective and sustainable separation technologies.

IL-based ABS are remarkable separation systems; yet, at this stage, it is required to invest on more biocompatible phase-forming components (including ILs), to infer on their switchable behaviour, to develop novel approaches for their recovery and reuse and to prove their scale-up viability. Finally, if IL-based ABS are so promising systems, it makes sense to move further and to leave the comfort zone on the partitioning of model biomolecules and to proceed with their applications to real matrices and particularly to explore their use on the purification of currently added-value compounds. The best is yet to come!!!

Acknowledgements This work was developed within the scope of the project CICECO - Aveiro Institute of Materials, POCI-01-0145-FEDER-007679 (FCT Ref. UID/CTM/50011/2013), financed by national funds through the FCT/MEC and when appropriate co-financed by FEDER under the PT2020 Partnership Agreement. The research leading to reported results has received funding from the European Research Council under the European Union's Seventh Framework Programme (FP7/2007-2013)/ERC grant agreement no. 337753.

References

1. Albertsson PA (1958) Partition of proteins in liquid polymer-polymer two-phase systems. Nature 182:709–711
2. Albertsson PA (1986) Partitioning of cell particles and macromolecules, 3rd edn. Wiley-Interscience, New York
3. Zalipsky S (1995) Functionalized poly(ethylene glycols) for preparation of biologically relevant conjugates. Bioconjug Chem 6:150–165
4. Li J, Kao WJ (2003) Synthesis of polyethylene glycol (PEG) derivatives and PEGylatedpeptide biopolymer conjugates. Biomacromolecules 4:1055–1067
5. Rosa PAJ, Azevedo AM, Ferreira IF, de Vries J, Korporaal R, Verhoef HJ, Visser TJ, Aires-Barros MR (2007) Affinity partitioning of human antibodies in aqueous two-phase systems. J Chromatogr A 1162:103–113
6. da Silva NR, Ferreira LA, Madeira PP, Teixeira JA, Uversky VN, Zaslavsky BY (2015) Effect of sodium chloride on solute–solvent interactions in aqueous polyethylene glycol–sodium sulfate two-phase systems. J Chromatogr A 1425:51–61
7. Gutowski KE, Broker GA, Willauer HD, Huddleston JG, Swatloski RP, Holbrey JD, Rogers RD (2003) Controlling the aqueous miscibility of ionic liquids: aqueous biphasic systems of water-miscible ionic liquids and water-structuring salts for recycle, metathesis, and separations. J Am Chem Soc 125:6632–6633
8. Freire MG, Cláudio AFM, Araújo JMM, Coutinho JAP, Marrucho IM, Canongia Lopes JN, Rebelo LPN (2012) Aqueous biphasic systems: a boost brought about by using ionic liquids. Chem Soc Rev 41:4966–4995
9. Plechkova NV, Seddon KR (2008) Applications of ionic liquids in the chemical industry. Chem Soc Rev 37:123–150
10. Wasserscheid P, Welton T (2002) Ionic liquids in synthesis. Wiley-VCH Verlag GmbH & Co. KGaA, Weinheim
11. Cláudio AFM, Neves MC, Shimizu K, Canongia Lopes JN, Freire MG, Coutinho JAP (2015) The magic of aqueous solutions of ionic liquids: ionic liquids as a powerful class of catanionic hydrotropes. Green Chem 17:3948–3963
12. Debeljuh N, Barrow CJ, Henderson L, Byrne N (2011) Structure inducing ionic liquids-enhancement of alpha helicity in the Abeta(1–40) peptide from Alzheimer's disease. Chem Commun 47:6371–6373
13. Khimji I, Doan K, Bruggeman K, Huang PJJ, Vajha P, Liu J (2013) Extraction of DNA staining dyes from DNA using hydrophobic ionic liquids. Chem Commun 49:4537–4539
14. Phama TPT, Choa C-W, Yuna Y-S (2010) Environmental fate and toxicity of ionic liquids: a review. Water Res 44:352–372
15. Petkovic M, Seddon KR, Rebelo LPN, Pereira CS (2011) Ionic liquids: a pathway to environmental acceptability. Chem Soc Rev 40:1383–1403
16. Pereira JFB, Rebelo LPN, Rogers RD, Coutinho JAP, Freire MG (2013) Combining ionic liquids and polyethylene glycols to boost the hydrophobic-hydrophilic range of aqueous biphasic systems. Phys Chem Chem Phys 15:19580–19583
17. Cláudio AFM, Freire MG, Freire CSR, Silvestre AJD, Coutinho JAP (2010) Extraction of vanillin using ionic-liquid-based aqueous two-phase systems. Sep Purif Technol 75:39–47
18. Quental MV, Passos H, Kurnia KA, Coutinho JAP, Freire MG (2015) Aqueous biphasic systems composed of ionic liquids and acetate-based salts: phase diagrams, densities, and viscosities. J Chem Eng Data 60:1674–1682
19. Freire MG, Neves CMSS, Carvalho PJ, Gardas RL, Fernandes AM, Marrucho IM, Santos LMNBF, Coutinho JAP (2007) Mutual solubilities of water and hydrophobic ionic liquids. J Phys Chem B 111:13082–13089
20. Freire MG, Neves CMSS, Marrucho IM, Coutinho JAP, Fernandes AM (2010) Hydrolysis of tetrafluoroborate and hexafluorophosphate counter ions in imidazolium-based ionic liquids. J Phys Chem A 114:3744–3749

21. Ranke J, Stolte S, Störmann R, Arning J, Jastorff B (2007) Design of sustainable chemical products: the example of ionic liquids. Chem Rev 107:2183–2206
22. Merchuk JC, Andrews BA, Asenjo JA (1998) Aqueous two-phase systems for protein separation. Studies on phase inversion. J Chromatogr B 711:285–293
23. Zafarani-Moattar MT, Hamzehzadeh S (2009) Phase diagrams for the aqueous two-phase ternary system containing the ionic liquid 1-butyl-3-methylimidazolium bromide and tri-potassium citrate at T = (278.15, 298.15, and 318.15) K. J Chem Eng Data 54:833–841
24. Zafarani-Moattar MT, Hamzehzadeh S (2010) Salting-out effect, preferential exclusion, and phase separation in aqueous solutions of chaotropic water-miscible ionic liquids and kosmotropic salts: effects of temperature, anions, and cations. J Chem Eng Data 55:1598–1610
25. Zafarani-Moattar MT, Hamzehzadeh S (2011) Effect of pH on the phase separation in the ternary aqueous system containing the hydrophilic ionic liquid 1-butyl-3-methylimidazolium bromide and the kosmotropic salt potassium citrate at T = 298.15 K. Fluid Phase Equilib 304:110–120
26. Chen YH, Wang YG, Cheng QY, Liu XL, Zhang SJ (2009) Carbohydrates-tailored phase tunable systems composed of ionic liquids and water. J Chem Thermodyn 41:1056–1059
27. Chen YH, Meng YS, Zhang SM, Zhang Y, Liu XW, Yang J (2010) Liquid – liquid equilibria of aqueous biphasic systems composed of 1-butyl-3-methyl imidazolium tetrafluoroborate + sucrose/maltose + water. J Chem Eng Data 55:3612–3616
28. Han J, Yu C, Wang Y, Xie X, Yan Y, Yin G, Guan W (2010) Liquid–liquid equilibria of ionic liquid 1-butyl-3-methylimidazolium tetrafluoroborate and sodium citrate/tartrate/acetate aqueous two-phase systems at 298.15 K: experiment and correlation. Fluid Phase Equilib 295:98–103
29. Han JA, Pan R, Xie XQ, Wang Y, Yan YS, Yin GW, Guan WX (2010) Liquid – liquid equilibria of ionic liquid 1-butyl-3-methylimidazolium tetrafluoroborate + sodium and ammonium citrate aqueous two-phase systems at (298.15, 308.15, and 323.15) K. J Chem Eng Data 55:3749–3754
30. Pei YC, Wang JJ, Liu L, Wu K, Zhao Y (2007) Liquid – liquid equilibria of aqueous biphasic systems containing selected imidazolium ionic liquids and salts. J Chem Eng Data 52:2026–2031
31. Yu C, Han J, Hu S, Yan Y, Li Y (2011) Phase diagrams for aqueous two-phase systems containing the 1-ethyl-3-methylimidazolium tetrafluoroborate/1-propyl-3-methylimidazolium tetrafluoroborate and trisodium phosphate/sodium sulfite/sodium dihydrogen phosphate at 298.15 K: experiment and correlation. J Chem Eng Data 56:3577–3584
32. Othmer DF, Tobias PE (1942) Liquid-liquid extraction data – toluene and acetaldehyde systems. Ind Eng Chem 34:690–692
33. Luís A, Dinis TBV, Passos H, Taha M, Freire MG (2015) Good's buffers as novel phase-forming components of ionic-liquid-based aqueous biphasic systems. Biochem Eng J 101:142–149
34. Freire MG, Carvalho PJ, Silva AMS, Santos LMNBF, Rebelo LPN, Marrucho IM, Coutinho JAP (2009) Ion specific effects on the mutual solubilities of water and hydrophobic ionic liquids. J Phys Chem 113:202–211
35. Neves CMSS, Freire MG, Coutinho JAP (2012) Improved recovery of ionic liquids from contaminated aqueous streams using aluminium-based salts. RSC Adv 2:10882–10890
36. Bridges NJ, Gutowski KE, Rogers RD (2007) Investigation of aqueous biphasic systems formed from solutions of chaotropic salts with kosmotropic salts (salt–salt ABS). Green Chem 9:177–183
37. Passos H, Ferreira AR, Cláudio AFM, Coutinho JAP, Freire MG (2012) Characterization of aqueous biphasic systems composed of ionic liquids and a citrate-based biodegradable salt. Biochem Eng J 67:68–76
38. Freire MG, Neves CMSS, Canongia Lopes JN, Marrucho IM, Coutinho JAP, Rebelo LPN (2012) Impact of self-aggregation on the formation of ionic-liquid-based aqueous biphasic systems. J Phys Chem B 116:7660–7668

39. Cláudio AFM, Ferreira AM, Shahriari S, Freire MG, Coutinho JAP (2011) Critical assessment of the formation of ionic-liquid-based aqueous two-phase systems in acidic media. J Phys Chem B 115:11145–11153
40. Shahriari S, Neves CMSS, Freire MG, Coutinho JAP (2012) Role of the Hofmeister series in the formation of ionic-liquid-based aqueous biphasic systems. J Phys Chem B 116:7252–7258
41. Sadeghi R, Golabiazar R, Shekaari H (2010) The salting-out effect and phase separation in aqueous solutions of tri-sodium citrate and 1-butyl-3-methylimidazolium bromide. J Chem Thermodyn 42:441–453
42. Freire MG, Louros CLS, Rebelo LPN, Coutinho JAP (2011) Aqueous biphasic systems composed of a water-stable ionic liquid + carbohydrates and their applications. Green Chem 13:1536–1545
43. Wu B, Zhang YM, Wang HP (2008) Phase behavior for ternary systems composed of ionic liquid + saccharides + water. J Phys Chem B 112:6426–6429
44. Wu B, Zhang YM, Wang HP (2008) Aqueous biphasic systems of hydrophilic ionic liquids + sucrose for separation. J Chem Eng Data 53:983–985
45. Okuniewski M, Paduszyński K, Domańska U (2016) Effect of cation structure in trifluoromethanesulfonate-based ionic liquids: density, viscosity, and aqueous biphasic systems involving carbohydrates as "salting-out" agents. J Chem Eng Data 61:1296–1304
46. Zhang M, Zhang YQ, Chen YH, Zhang SJ (2007) Mutual coexistence curve measurement of aqueous biphasic systems composed of [bmim][BF4] and glycine, L-serine, and L-proline, respectively. J Chem Eng Data 52:2488–2490
47. Domínguez-Pérez M, Tomé LIN, Freire MG, Marrucho IM, Cabeza O, Coutinho JAP (2010) (Extraction of biomolecules using) aqueous biphasic systems formed by ionic liquids and aminoacids. Sep Purif Technol 72:85
48. Wu CZ, Wang JJ, Pei YC, Wang HY, Li ZY (2010) Salting-out effect of ionic liquids on poly (propylene glycol) (PPG): formation of PPG + ionic liquid aqueous two-phase systems. J Chem Eng Data 55:5004–5008
49. Zafarani-Moattar MT, Hamzehzadeh S, Nasiri S (2012) A new aqueous biphasic system containing polypropylene glycol and a water-miscible ionic liquid. Biotechnol Prog 28:146–156
50. Freire MG, Pereira JFB, Francisco M, Rodríguez H, Rebelo LPN, Rogers RD, Coutinho JAP (2012) Insight into the interactions that control the phase behaviour of new aqueous biphasic systems composed of polyethylene glycol polymers and ionic liquids. Chem Eur J 18:1831–1839
51. Li Z, Liu X, Pei Y, Wang J, He M (2012) Design of environmentally friendly ionic liquid aqueous two-phase systems for the efficient and high activity extraction of proteins. Green Chem 14:2941–2950
52. Pereira JFB, Kurnia KA, Freire MG, Coutinho JAP, Rogers RD (2015) Controlling the formation of ionic-liquid-based aqueous biphasic systems by changing the hydrogen-bonding ability of polyethylene glycol end groups. ChemPhysChem 16:2219–2225
53. Visak ZP, Canongia Lopes JN, Rebelo LPN (2007) Ionic liquids in polyethylene glycol aqueous solutions: salting-in and salting-out effects. Monatsh Chem 138:1153–1157
54. Quental MV, Caban M, Pereira MM, Stepnowski P, Coutinho JAP, Freire MG (2015) Enhanced extraction of proteins using cholinium-based ionic liquids as phase-forming components of aqueous biphasic systems. Biotechnol J 10:1457–1466
55. Tomé L, Pereira JFB, Rogers RD, Freire MG, Gomes JRB, Coutinho JAP (2014) "Washing-out" ionic liquid from polyethylene glycol to form aqueous biphasic systems. Phys Chem Chem Phys 16:2271–2274
56. Pereira JFB, Lima ÁS, Freire MG, Coutinho JAP (2010) Ionic liquids as adjuvants for the tailored extraction of biomolecules in aqueous biphasic systems. Green Chem 12:1661–1669
57. Almeida MR, Passos H, Pereira M, Lima ÁS, Coutinho JAP, Freire MG (2014) Ionic liquids as additives to enhance the extraction of antioxidants in aqueous two-phase systems. Sep Purif Technol 128:1–10

58. Ventura SPM, Neves CMSS, Freire MG, Marrucho IM, Oliveira J, Coutinho JAP (2009) Evaluation of anion influence on the formation and extraction capacity of ionic-liquid-based aqueous biphasic systems. J Phys Chem B 113:9304–9310
59. Neves CMSS, Ventura SPM, Freire MG, Marrucho IM, Coutinho JAP (2009) Evaluation of cation influence on the formation and extraction capability of ionic-liquid-based aqueous biphasic systems. J Phys Chem B 113:5194–5199
60. Ikeda M (2006) Towards bacterial strains overproducing L-tryptophan and other aromatics by metabolic engineering. Appl Microbiol Biotechnol 69:615–626
61. Lu M, Tjerneld F (1997) Interaction between tryptophan residues and hydrophobically modified dextran: effect on partitioning of peptides and proteins in aqueous two-phase systems. J Chromatogr A 766:99–108
62. Tomé LIN, Catambas VR, Teles ARR, Freire MG, Marrucho IM, Coutinho JAP (2010) Tryptophan extraction using hydrophobic ionic liquids. Sep Purif Technol 72:167–173
63. Rosa PAJ, Azevedo AM, Sommerfeld S, Bäcker W, Aires-Barros MR (2011) Aqueous two-phase extraction as a platform in the biomanufacturing industry: economical and environmental sustainability. Biotechnol Adv 29:559–567
64. Rosa PAJ, Ferreira IF, Azevedo AM, Aires-Barros MR (2010) Aqueous two-phase systems: a viable platform in the manufacturing of biopharmaceuticals. J Chromatogr A 1217:2296–2305
65. Ruiz-Angel MJ, Pino V, Carda-Broch S, Berthod A (2007) Solvent systems for countercurrent chromatography: an aqueous two phase liquid system based on a room temperature ionic liquid. J Chromatogr A 1:65–73
66. Pereira MM, Pedro SN, Quental MV, Lima ÁS, Coutinho JAP, Freire MG (2015) Enhanced extraction of bovine serum albumin with aqueous biphasic systems of phosphonium- and ammonium-based ionic liquids. J Biotechnol 206:17–25
67. Taha M, e Silva FA, Quental MV, Ventura SPM, Freire MG, Coutinho JAP (2014) Good's buffers as a basis for developing self-buffering and biocompatible ionic liquids for biological research. Green Chem 16:3149–3159
68. Taha M, Almeida MR, e Silva FA, Domingues P, Ventura SPM, Coutinho JAP, Freire MG (2015) Novel biocompatible and self-buffering ionic liquids for biopharmaceutical applications. Chem Eur J 21:4781–4788
69. Taha M, Quental MV, Correia I, Freire MG, Coutinho JAP (2015) Extraction and stability of bovine serum albumin (BSA) using cholinium-based Good's buffers ionic liquids. Process Biochem 50:1158–1166
70. Dreyer S, Kragl U (2008) Ionic liquids for aqueous two-phase extraction and stabilization of enzymes. Biotechnol Bioeng 99:1416–1424
71. Ventura SPM, de Barros RLF, de Pinho Barbosa JM, Soares CMF, Lima ÁS, Coutinho JAP (2012) Production and purification of an extracellular lipolytic enzyme using ionic liquid-based aqueous two-phase systems. Green Chem 14:734–740
72. Souza RL, Ventura SPM, Soares CMF, Coutinho JAP, Lima ÁS (2015) Lipase purification using ionic liquids as adjuvants in aqueous two-phase systems. Green Chem 17:3026–3034
73. Dutta NN, Borthakur S, Patil GS (1992) Phase transfer catalyzed extraction of phenolic substances from aqueous alkaline stream. Sep Sci Technol 27:1435–1448
74. Zhang DL, Deng YF, Chen J (2010) Enrichment of aromatic compounds using ionic liquid and ionic liquid-based aqueous biphasic systems. Sep Sci Technol 45:663–669
75. Ferreira AM, Coutinho JAP, Fernandes AM, Freire MG (2014) Complete removal of textile dyes from aqueous media using ionic-liquid-based aqueous two-phase systems. Sep Purif Technol 128:58–66
76. Sheikhian L, Akhond M, Absalan G (2014) Partitioning of reactive red–120, 4– (2–pyridylazo) –resorcinol, and methyl orange in ionic liquid–based aqueous biphasic systems. J Environ Chem Eng 2:137–142
77. de Souza RL, Campos VC, Ventura SPM, Soares CMF, Coutinho JAP, Lima AS (2014) Effect of ionic liquids as adjuvants on PEG–based ABS formation and the extraction of two probe dyes. Fluid Phase Equilib 375:30–36

78. Passos H, Sousa ACA, Pastorinho MR, Nogueira AJA, Rebelo LPN, Coutinho JAP, Freire MG (2012) Ionic-liquid-based aqueous biphasic systems for improved detection of bisphenol A in human fluids. Anal Methods 4:2664–2667
79. Dinis TBV, Passos H, Lima DLD, Esteves VI, Coutinho JAP, Freire MG (2015) One-step extraction and concentration of estrogens for an adequate monitoring of wastewater using ionic-liquid-based aqueous biphasic systems. Green Chem 17:2570–2579
80. Li S, He C, Liu H, Li K, Liu F (2005) Ionic liquid-based aqueous two-phase system, a sample pretreatment procedure prior to high-performance liquid chromatography of opium alkaloids. J Chromatogr B 826:58–62
81. Freire MG, Neves CMSS, Marrucho IM, Canongia Lopes JN, Rebelo LPN, Coutinho JAP (2010) High-performance extraction of alkaloids using aqueous two-phase systems with ionic liquids. Green Chem 12:1715–1718
82. Passos H, Trindade MP, Vaz TSM, da Costa LP, Freire MG, Coutinho JAP (2013) The impact of self-aggregation on the extraction of biomolecules in ionic-liquid-based aqueous two-phase systems. Sep Purif Technol 108:174–180
83. Pereira JFB, Ventura SPM, e Silva FA, Shahriari S, Freire MG, Coutinho JAP (2013) Aqueous biphasic systems composed of ionic liquids and polymers: a platform for the purification of biomolecules. Sep Purif Technol 113:83–89
84. Cláudio AFM, Marques CFC, Boal-Palheiros I, Freire MG, Coutinho JAP (2014) Development of back-extraction and recyclability routes for ionic-liquid-based aqueous two-phase systems. Green Chem 16:259–268
85. Santos JH, e Silva FA, Ventura SPM, Coutinho JAP, de Souza RLS, Soares CM, Lima ÁS (2015) Ionic liquid-based aqueous biphasic systems as a versatile tool for the recovery of antioxidant compounds. Biotechnol Prog 31:70–77
86. Ribeiro BD, Coelho MAZ, Rebelo LPN, Marrucho IM (2013) Ionic liquids as additives for extraction of saponins and polyphenols from mate (Ilex paraguariensis) and tea (Camellia sinensis). Ind Eng Chem Res 52:12146–12153
87. Akama Y, Ito M, Tanaka S (2000) Selective separation of cadmium from cobalt, copper, iron (III) and zinc by water-based two-phase system of tetrabutylammonium bromide. Talanta 53:645–650
88. Akama Y, Sali A (2002) Extraction mechanism of Cr(VI) on the aqueous two-phase system of tetrabutylammonium bromide and $(NH_4)_2SO_4$ mixture. Talanta 57:681–686
89. Wei XL, Wei ZB, Wang XH, Wang ZN, Sun DZ, Liu J, Zhao HH (2011) Phase behavior of new aqueous two-phase systems: 1-butyl-3-methylimidazolium tetrafluoroborate + anionic surfactants + water. Soft Matter 7:5200–5207
90. Shahriari S, Tomé LC, Araújo JMM, Rebelo LPN, Coutinho JAP, Marrucho IM, Freire MG (2013) Aqueous biphasic systems: a benign route using cholinium-based ionic liquids. RSC Adv 3:1835–1843
91. Almeida HFD, Freire MG, Marrucho IM (2016) Improved extraction of fluoroquinolones with recyclable ionic-liquid-based aqueous biphasic systems. Green Chem 18:2717. doi:10.1039/C5GC02464A
92. Zawadzki M, e Silva FA, Domańska U, Coutinho JAP, Ventura SPM (2016) Recovery of an antidepressant from pharmaceutical wastes using ionic liquid-based aqueous biphasic systems. Green Chem 18:3527–3536. doi:10.1039/C5GC03052H
93. Passos H, Luís A, Coutinho JAP, Freire MG (2016) Thermoreversible (ionic-liquid-based) aqueous biphasic systems. Sci Rep 6:20276

Chapter 2
ABS Composed of Ionic Liquids and Inorganic Salts

José N. Canongia Lopes

Abstract Aqueous biphasic systems (ABS) composed of ionic liquids (ILs) and inorganic salts (ISs) have been by far the most investigated class of IL-based ABS. In the present chapter, the formation of this type of systems is discussed keeping in mind that the IS acts as a salting-out agent, while ILs are very special salted-out solutes. The task at hand is rather straightforward due to two important circumstances: (*i*) the start of this particular research front one decade ago has been defined by a groundbreaking paper in 2003; and (*ii*) the development of ABS brought about by the use of ILs has been thoroughly described in a recent review paper in 2012. Therefore, this chapter is not about an exhaustive listing of (IL + IS) ABS but rather a general synopsis of the evolution of the field, highlighting different types of experiments that allowed the rationalization of salting-out/salting-in effects in the context of ABS formation mediated by ILs and ISs.

Keywords Ionic liquids • Inorganic salts • Salting-out effects • Ternary diagrams • Aqueous biphasic systems

2.1 Introduction

In 2003 Rogers and co-workers [1] demonstrated that it is possible to form aqueous biphasic systems (ABS), i.e., to cause the coexistence of two immiscible aqueous phases, using a water-soluble ionic liquid (IL, the solute) and a water-structuring inorganic salt (IS, the salting-out agent). In their study at room temperature, the authors [1] used aqueous solutions of the ionic liquid 1-butyl-3-methylimidazolium chloride, $[C_4mim]Cl$, and a concentrated solution of potassium phosphate, $K_3[PO_4]$.

With the hindsight provided by such pioneering work [1], ABS formation in the context of (IL + IS) aqueous solutions is a relatively straightforward process to

J.N. Canongia Lopes (✉)
Centro de Química Estrutural, Instituto Superior Técnico, Universidade de Lisboa, 1049-001 Lisbon, Portugal

Instituto de Tecnologia Química e Biológica, Universidade Nova de Lisboa, Av. República, Ap. 127, 2780-901 Oeiras, Portugal
e-mail: jnlopes@ist.utl.pt

M.G. Freire (ed.), *Ionic-Liquid-Based Aqueous Biphasic Systems*, Green Chemistry and Sustainable Technology, DOI 10.1007/978-3-662-52875-4_2

explain. It occurs because (*i*) the salted-out salt (IL solute) has a high difficulty in forming a precipitate since the melting temperatures of ILs are lower and their solubility in water is higher when compared to inorganic salts and (*ii*) upon separation from the original solution, the IL solute will carry with it a relatively large amount of solvent (water) molecules. The use of a hydrophilic (water-miscible) ionic liquid as solute fulfills both requirements. Hydrophobic ionic liquids are not adequate for the formation of ABS because they already form two phases when mixed with water, even without the presence of a salting-out agent (IS). In this case, the water-rich phase is quite poor in ionic liquid, which means that the usefulness of an ABS for separation purposes is thus very limited. As a result, the following discussion is limited to hydrophilic ionic liquids which in fact form ABS.

2.2 ABS Depicted in Ternary Phase Diagrams

A global ternary phase diagram of a generic IL + IS + water system provides a simple image of the phase equilibria that one may find for different mixture composition ranges. It must be stressed that most literature reports ABS in orthogonal phase diagrams limited to the water-rich corner of the diagram [2]. Although this is justified because the region showing the coexistence of two aqueous phases is of main interest, one should not forget that other phenomena may occur, including the precipitation of the salting-out agent (SAP phenomena).

Figure 2.1 shows different kinds of ternary systems where water is mixed with different inorganic salts (IS), molecular solvents (MS), or ionic liquids. It must be stressed that since all solutes present in (IS + IL) aqueous systems are ionic, ion exchange upon ABS formation cannot be excluded (the speciation of ions in the two phases may differ from that of the original ions that form the IS and IL). However, since the electroneutrality of each phase must be preserved, it was shown [3] that an interpretation of the results based on phase diagrams and equilibrium binodal curves built considering the speciation of the original salts and not the concentration of individual ions is usually sufficiently accurate from an operational (if not formal) point of view.

Figure 2.1a shows the typical case of a water + IS + IS system (e.g., H_2O + NaCl + KCl) [4], where the precipitation of two mutually immiscible salts causes the existence of an extended triple-immiscibility region and where a water solution saturated in both ISs coexists with both IS precipitates.

When one of the salts is substituted by a molecular solvent that is completely miscible with water, one can have a phase diagram such as the one represented in Fig. 2.1b (H_2O + K[NO]$_3$ + C_2H_5OH) [5], which displays a one-phase region and a two-phase region consisting of an IS precipitate and a saturated (water + MS) solution.

Under some conditions, particular MSs can be salted-out from the aqueous solution and this leads to the formation of ABS. Figure 2.1c shows the (H_2O

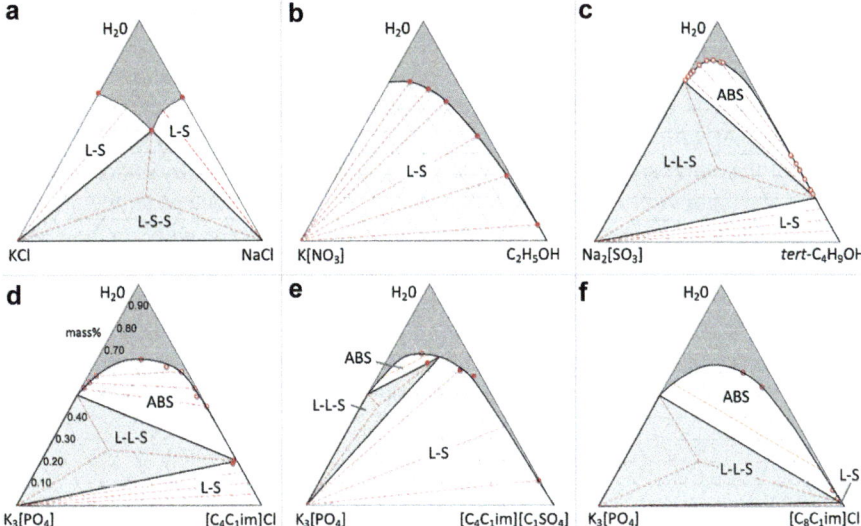

Fig. 2.1 Ternary phase diagrams depicting different situations where ABS may occur in aqueous solutions containing an inorganic salt and either a second inorganic salt (**a**), a molecular compound (**b, c**), or an ionic liquid (**d–f**). The *dark gray* areas correspond to one-phase aqueous regions, *white* areas to two-phase regions (ABS or L-S equilibria), and *light gray* areas to three-phase regions (L-S-S or L-L-S equilibria). The *red* and *orange dashed lines* correspond to selected experimental and inferred tie lines of the multiphase regions, respectively

+ $Na_2[SO]_3$ + *iso*-C_4H_9OH) system [6], where the salting-out of isobutanol from the aqueous IS solution leads to the formation of a second liquid phase that still contains a substantial amount of water. One important issue to mention at this point is the formation of a triple-immiscibility region where two IS-saturated aqueous phases are in equilibrium with an IS precipitate. In some cases the positioning of such region also allows the existence of an additional two-phase region (at the bottom part of the diagram) where the IS precipitate is in equilibrium with a single solute-rich aqueous phase – a situation where the salting-out agent becomes salted out by the presence of large amounts of solute.

The phase diagram just described is conceptually very similar to the ones obtained in (water + IS + IL) systems. In fact, Fig. 2.1c is strikingly akin to Fig. 2.1d. The latter represents the whole ternary phase diagram of the system originally studied by Gutowski et al. [1] (H_2O + $K_3[PO_4]$ + [C_4mim]Cl). Like in the case of some specific MSs, the ionic liquid acts as a species that is completely miscible with water and that under some circumstances can be salted-out by the IS. It can always be speculated however that the intricate balance found in ionic liquids between electrostatic and dispersion forces (polar heads versus alkyl side chains) has an echo on MSs such as isobutanol that also displays a balance between hydrogen bonding and hydrophobic interactions. Finally, the versatility of ILs can be further recognized by looking at the effects caused by changes in the anion or in the alkyl side chain of the IL cation; Fig. 2.1e, f show phase diagrams with distinct types of ABS formation and salting-out of the IS for two distinct types of IL [7].

The versatility of ABS formed by different IS + IL combinations has been recognized since the pioneering work of Gutowski et al. [1]. The remaining part of this chapter will discuss how the IL nature, namely, the nature of the cation, the nature of the anion and the length of the alkyl side chains generally attached to the cations, the nature of the salting-out IS, and other variables, such as temperature or pH, influences ABS formation. The discussion will focus exclusively on the water-rich tip of the phase diagram, where the most striking L-L equilibria and separations in ABS can be achieved. Nevertheless, the precipitation of the salting-out agent and the formation of a third phase occupying a significant region of the corresponding phase diagrams cannot be discarded and influence the outcome of many separation processes. Further studies are still needed in this particular area.

2.3 Nature of the IL Cation Core

As mentioned above, ABS formation in aqueous IS plus IL systems are mainly defined by salting-out effects on the later salt caused by the former one. Such effects are the result of the creation of strongly bound water–IS complexes that cause the dehydration of the IL and increase the surface tension of any cavity in aqueous media [8–10]. In other words, the addition of high-charge-density ISs to IL aqueous solutions leads to liquid–liquid demixing due to the preferential hydration of the IS over the ionic liquid, leading therefore to the salting-out of the ionic liquid and to the creation of a second IL-rich aqueous phase. Thus, ABS formation is mainly dominated by ion–ion versus ion–water interactions (and ensuing entropically driven effects) and not by any underlying water structure modifications, as classically accepted.

One of the first investigations of the effect of the IL cation nature on the formation of ABS was presented by Bridges et al. [3] in 2007. Figure 2.2a presents a summary of their data for ABS containing aqueous solutions of imidazolium-, pyridinium-, ammonium-, and phosphonium-based chloride ILs combined with K_3PO_4. The binodal curves closer to the pure water tip of the ternary diagram – especially if the diagram is represented in terms of mole fraction (*cf.* inset of Fig. 2.2a) – represent ABS containing less hydrophilic ILs. In other words, ILs containing cations whose charged moieties are less prone to interact with water have a higher capacity to phase separate and promote larger biphasic regions. The corresponding order, $[P_{4\,4\,4\,4}Cl] > [N_{4\,4\,4\,4}]Cl > [C_4py]Cl > [C_4C_1im]Cl$, can be rationalized taking into account the strong steric shielding of the charged moiety of the phosphonium and ammonium cations by four alkyl groups, a smaller degree of shielding of the pyridinium cation (the charge is mainly concentrated in the sp_2 tertiary nitrogen atom but is also partially delocalized in the rest of the aromatic ring), and the possibility of H-bonding (mainly via the acidic H2 hydrogen) in the less hindered and more charge-delocalized imidazolium cation. Later, it was

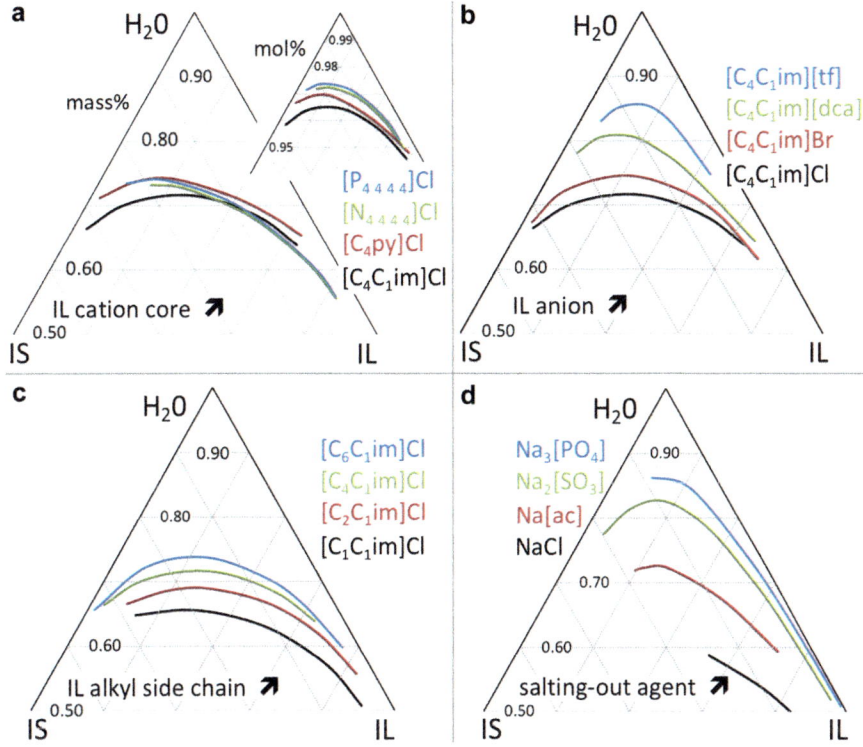

Fig. 2.2 Partial ternary phase diagrams (water tip) depicting different ABS boundary lines in aqueous solutions containing an inorganic salt (IS) and an ionic liquid (IL): (**a**) IL cation core effect [3], (**b**) IL anion nature effect[12], (**c**) IL alkyl side chain effect [14], and (**d**) IS anion salting-out effect [18]. In each case four representative systems were selected, and the corresponding trends discussed in the text are indicated by the arrow

demonstrated that the solubility of ionic liquids in water (and the concomitant effect on the size of the ABS region) is in large measure ruled by the molar volume of the ionic fluids [11]. Larger cations, such as pyridinium and piperidinium (6-atom heterocyclic compounds), are better at inducing ABS than smaller cations such as imidazolium or pyrrolidinium (containing 5-atom rings) [11]. In summary, the influence of the cation charged cores in the formation of ABS is dominated by steric and entropic effects and to a lesser extent by the possible interactions of the charged cores with water molecules.

2.4 Nature of the IL Anion

The influence of the IL anion (already illustrated by the comparison of Figs. 2.1d and 2.1e, both representing data from Najdanovic-Visak et al. [7]) was further studied in a more comprehensive and systematic way by Coutinho and co-workers

[12]. Figure 2.2b shows a summary of a selected set of their ABS formed by aqueous solutions of [C$_4$C$_1$im][X], where X represent different IL anions, and K$_3$PO$_4$.

First, it must be stressed that anion–water interactions generally define the hydrophilicity or hydrophobicity of an ionic liquid. In other words, since anions are generally more polarizable and have a more diffused valence electronic structure than cations, they are able to form (in most cases) more extended networks with water molecules. Thus, the ability of the different anions to form more or less clear ABS regions can be rationalized in terms of the relative intensity of their hydrogen-bond acceptor capacity; anions with lower hydrogen-bond basicity values (usually quantified via the use of solvatochromic probes) coordinate less with water molecules and are more easily salted-out by the IS. Such order is generally independent of the salting-out agent [13]. Unlike their cationic counterparts, IL anions affect ABS formation via direct competition with the IS for interactions with water molecules. The outcome of such direct effect is that the distinction between ABS boundaries (even in a ternary diagram based in mass fractions) caused by ILs based on different anions is more clear than the ABS effects caused by different IL cations (*cf.* Figs. 2.2a and 2.2b).

2.5 Length and Number of IL Alkyl Side Chains

Figure 2.2c shows the effect of the IL cation alkyl side chain on ABS formation in aqueous solutions of [C$_n$C$_1$im]Cl ($1 \leq n \leq 6$) and K$_3$PO$_4$ [14]. The arrow in the figure indicates that ABS formation is easier for systems containing ILs with longer alkyl side chains at the cation (and thus larger ionic liquid molar volumes). Obviously, longer alkyl side chains in the cation lead to more hydrophobic ILs and thus to poorer IL solubilities in water. Such solubility decrease is driven by entropic effects similar to the well-known hydrophobic effect frequently found in aqueous solutions of nonpolar organic compounds. For this reason, ionic liquids with longer aliphatic chains generally require less IS for salting-out and are more easily excluded from the salt-rich phase to the ionic liquid-rich phase. However, the comparison of Figs. 2.1d and 2.1f reveals a different trend: when the alkyl chains start to be larger than C6, the ABS regions become less prominent as the alkyl side chains increase. More complex trends for ABS formed by ILs with long alkyl chains were also analyzed by other authors [11, 15–17]. Such trends above the "hexyl threshold" can be rationalized in terms of the possibility of self-aggregation (proto-micelle, micelle, or more complex aggregates) of the ILs in aqueous media. In fact, the complexity of the trends is the result of two opposing driving forces: (a) the entropic-driven hydrophobic effect already present for ILs with chains smaller than C6 and (b) the inherent aptitude of ionic liquids with longer aliphatic chains to self-aggregate.

Other alkyl side chain effects on ABS formation can be achieved via the substitution of the cationic cores at different positions, the replacement of alkyl substituents by hydrogen atoms leading to the formation of protic ionic liquids, or the change of the alkyl substituents in some anions. All these were discussed in detail in a previous review paper [2].

2.6 The Nature of the IS and Its Salting-Out Strength

Obviously the nature of the IS salting-out agent will play a pivotal role in the definition of any ABS. Figure 2.2d shows schematically what happens to a given IL ($[C_4C_1im][CF_3SO_3]$) when the IS used is progressively a better salting-out agent [18]. The most important fact that must be recognized here is that the ABS formation trend follows closely the well-known Hofmeister series. Moreover, the ordering of the ions in the Hofmeister series (for cations and for anions) closely follows the order of the Gibbs free energy of hydration of the ions. However, it was already shown that a significant part of the ABS phenomena that occurs in IL + IS aqueous systems is driven by entropic contributions [18]. In other words, the correlation between ABS formation and the entropy of hydration is even stronger than its counterpart based on Gibbs energy of hydration values.

2.7 Other Effects

Two other effects must be considered when optimizing a given ABS system: the influence of pH and temperature. Most IS used to promote ABS formation yield basic aqueous solutions. This is especially true for strong salting-out agents, such as K_3PO_4. The use of neutral or acidic solutions to form IL-based ABS has been much more limited [2]. Nonetheless, the use of ABS within specific pH boundaries can be of extreme importance for the separation of specific molecules, namely, in areas more related to biochemistry or biology. More studies are definitely needed in this area. The same applies to the temperature dependence of IL-based ABS.

2.8 Conclusions

The factoring of the different effects that influence the formation of ABS in (IL + IS) aqueous solutions leads to the obvious conclusion that better separation between the two aqueous phases can be achieved using imidazolium-based ILs with alkyl side chains just below the "hexyl threshold" combined with anions that are the least hydrophilic possible (just below the IL-water immiscibility threshold) and an IS salting-out agent as strong as possible. A possible combination could be, for instance, the ($[C_6C_1im][CF_3SO_3] + K_3[PO_4] + H_2O$) system. However, this type of analysis neglects three very important issues: (*i*) ABS are not an end per se but rather a means to achieve better separation processes. This means that the nature of the two aqueous phases and their interactions with different components to be separated (partitioned) between them is as important as achieving a good separation between the aqueous phases; (*ii*) ILs are an extremely versatile class of compounds. To limit the scope of their applicability in the formation of ABS by just considering their ability for well-separated aqueous solutions is a rather poor strategy. For instance, the formation of IL aggregates in solution can have some dramatic consequences for the separation/partitioning process itself; (*iii*) the use of strong IS salting-out agents can lead to undesirable operating conditions (e.g., inadequate pH, denaturation and dehydration of biomolecules, etc.). The use of alternative ABS based on ILs (as also discussed in the following chapters) indeed offers a broad range of options.

Acknowledgments The author would like to thank Prof. Luís Paulo Rebelo for his groundbreaking contributions in the area of IL-based ABS and his permanent willingness for promoting stimulating discussions that helped to shape the whole field of IL research.

References

1. Gutowski KE, Broker GA, Willauer HD et al (2003) Controlling the aqueous miscibility of ionic liquids: aqueous biphasic systems of water-miscible ionic liquids and water-structuring salts for recycle, metathesis, and separations. J Am Chem Soc 125:6632–6633
2. Freire MG, Cláudio AFM, Araújo JMM et al (2012) Aqueous biphasic systems: a boost brought about by using ionic liquids. Chem Soc Rev 41:4966–4995
3. Bridges NJ, Gutowski KE, Rogers RD (2007) Investigation of aqueous biphasic systems formed from solutions of chaotropic salts with kosmotropic salts (salt–salt ABS). Green Chem 9:177–183
4. Chou I-M (1982) Phase relations in the system NaCl-KCl-H_2O. Part I: differential thermal analysis of the NaCl-KCl liquidas at 1 atmosphere and 500, 1000, 1500, and 2000 bars. Geochim Cosmochim Acta 46:1957–1962
5. Thompson AR, Vener RE (1948) Solubility and density isotherms. Ind Eng Chem 40:478–481
6. Lynn S, Schiozer AL, Jaecksch WL et al (1996) Recovery of anhydrous Na2SO4 from SO2-scrubbing liquor by extractive crystallization: liquid – liquid equilibria for aqueous solutions of sodium carbonate, sulfate, and/or sulfite plus acetone, 2-propanol, or tert-butyl alcohol. Ind Eng Chem Res 35:4236–4245
7. Najdanovic-Visak V, Canongia Lopes JN, Visak ZP et al (2007) Salting-out in aqueous solutions of ionic liquids and K3PO4: Aqueous Biphasic Systems and salt precipitation. Int J Mol Sci 8:736–748
8. Freire MG, Carvalho PJ, Silva AMS et al (2009) Ion specific effects on the mutual solubilities of water and hydrophobic ionic liquids. J Phys Chem B 113:202–211
9. Freire MG, Neves CMSS, Silva AMS et al (2010) 1H NMR and molecular dynamics evidence for an unexpected interaction on the origin of salting-in/salting-Out phenomena. J Phys Chem B 114:2004–2014
10. Tomé LIN, Varanda FR, Freire MG et al (2009) Towards an understanding of the mutual solubilities of water and hydrophobic ionic liquids in the presence of salts: the anion effect. Phys Chem B 113:2815–2825
11. Ventura SPM, Sousa SG, Serafim LS et al (2011) Ionic liquid based aqueous biphasic systems with controlled pH: the ionic liquid cation effect. J Chem Eng Data 56:4253–4260

12. Ventura SPM, Neves CMSS, Freire MG et al (2009) Evaluation of anion influence on the formation and extraction capacity of ionic-liquid-based aqueous biphasic systems. J Phys Chem B 113:9304–9310
13. Ventura SPM, Sousa SG, Serafim LS et al (2012) Ionic-liquid-based aqueous biphasic systems with controlled pH: the ionic liquid anion effect. J Chem Eng Data 57:507–512
14. Neves CMSS, Ventura SPM, Freire MG et al (2009) Evaluation of cation influence on the formation and extraction capability of ionic-liquid-based aqueous biphasic systems. J Phys Chem B 113:5194–5199
15. Li ZY, Pei YC, Liu L et al (2010) (Liquid + liquid) equilibria for (acetate-based ionic liquids + inorganic salts) aqueous two-phase systems. J Chem Thermodyn 42:932–937
16. Pei YC, Wang JJ, Liu L et al (2007) Liquid − liquid equilibria of aqueous biphasic systems containing selected imidazolium ionic liquids and salts. J Chem Eng Data 52:2026–2031
17. Pei YC, Wang JJ, Wu K et al (2009) Ionic liquid-based aqueous two-phase extraction of selected proteins. Sep Purif Technol 64:288–295
18. Shahriari S, Neves CMSS, Freire MG et al (2012) Role of the Hofmeister series in the formation of ionic-liquid-based aqueous biphasic systems. J Phys Chem B 116:7252–7258

Chapter 3
ABS Constituted by Ionic Liquids and Carbohydrates

André M. da Costa Lopes and Rafał Bogel-Łukasik

Abstract Aqueous biphasic systems (ABS) composed of ionic liquids (ILs) and carbohydrates are an environmentally friendly and promising technology for separation and purification purposes. This chapter discloses the key parameters and mechanisms involved in the formation of this type of ABS. Phase diagrams and the factors affecting ABS formation, such as type of IL cation and anion, nature of the carbohydrate involved, temperature dependence and physicochemical properties, namely, density and viscosity of each phase, are presented and discussed in detail. Some particular applications of carbohydrate-based ABS are already reported in the literature. These comprise the removal of phenol from wastewaters and the extraction of specific value-added biomolecules. This technology is of particular relevance for the extraction of highly hydrophobic compounds, such as β-carotene. Nevertheless, the robustness and further acceptance of IL-carbohydrate-based ABS as a separation technology is largely dependent on an efficient recovery of the value-added compounds and further IL recovery and reuse.

Keywords Carbohydrates • Biomass • Cellulose • Hemicellulose • Polysaccharide • Biomolecules • Dyes • Kamlet–Taft

3.1 Introduction

Carbohydrates are naturally occurring compounds that consist of carbon, hydrogen and oxygen with the following empirical formula: $C_n(H_2O)_n$. Nevertheless, some carbohydrates also contain nitrogen, phosphorus or sulphur. By definition,

---

A.M. da Costa Lopes
Unit of Bioenergy, National Laboratory of Energy and Geology,
Estrada do Paço do Lumiar 22, 1649-038 Lisbon, Portugal

LAQV-REQUIMTE, Departamento de Química, Faculdade de Ciências e Tecnologia,
Universidade NOVA de Lisboa, Caparica, Portugal

R. Bogel-Łukasik (✉)
Unit of Bioenergy, National Laboratory of Energy and Geology,
Estrada do Paço do Lumiar 22, 1649-038 Lisbon, Portugal
e-mail: rafal.lukasik@lneg.pt

© Springer-Verlag Berlin Heidelberg 2016
M.G. Freire (ed.), *Ionic-Liquid-Based Aqueous Biphasic Systems*, Green Chemistry and Sustainable Technology, DOI 10.1007/978-3-662-52875-4_3

carbohydrates are polyhydroxy aldehydes or ketones or substances that yield such compounds after their hydrolysis. The natural production of carbohydrates mainly arises from the energy storage of plants. During photosynthesis, plants are able to convert sunlight into chemical energy by combining carbon dioxide with water to form carbohydrates and molecular oxygen [1]. Therefore, carbohydrates as high-energy storage compounds are crucial in nature for the food chain of living beings.

The predominance of carbohydrates in nature comprises structural and protective features of the cell walls of bacteria and plants and of connective tissues of animals [2]. This is an inherent effect of the carbohydrate properties that establish strong covalent bonds (α- or β-glycosidic linkages) and non-covalent bonds (intra- and intermolecular hydrogen bonding) between carbohydrate units to form polymeric aggregations, which are ascribed as polysaccharides.

Cellulose and hemicellulose (in plants), starch (in tubers and cereals), chitin (in crustaceans and insects), glycogen (in animals) and dextran (in yeast and bacteria) are main examples of polysaccharides. Differences between each other fall in the composition of the carbohydrate structural units, type of covalent bonds, chain length and degree of branching [2]. Furthermore, polysaccharides can be subdivided into homopolysaccharides or heteropolysaccharides. Cellulose is a homopolysaccharide containing only one type of unit, namely, glucose, while hemicellulose is a heteropolysaccharide that possesses at least two different kinds of carbohydrate units, such as xylose, arabinose, mannose and others.

The depolymerization of polysaccharides leads to the release of disaccharides (two units) and monosaccharides (one unit). All can be obtained by hydrolysis of polysaccharides, excepting sucrose, which as a non-reducing sugar is not originated from any polymer. As monosaccharides, carbohydrates are important compounds for sustaining life being as one of the major susceptible energy sources consumed by living beings. Figures 3.1 and 3.2 depict the chemical structures of mono-, di-, oligo- and polysaccharides.

In addition to the structural and energy source functions of carbohydrates in nature, their diversity and abundance allow their use for different purposes. Actually, several applications of carbohydrates (in both polymeric and monomeric forms) have been approached by industry due to the green and sustainable impact that is mostly connected to their renewable character. Carbohydrates have been applied in food formulations [4], in pharmaceutical and cosmetic purposes [5, 6], in biotechnological fermentation processes [7], in the synthesis of nanocomposite materials [8] and in the production of biomedical devices [9], among others. Nevertheless, the possibility to isolate carbohydrates for transformation into fuels and chemicals has been intensively explored within the biorefinery concept [10–13]. Bioethanol is an example of a biofuel that can be generated by biotechnological conversion of monosaccharides, such as glucose. Rather than that, monosaccharides can be used in other reactions, such as in their dehydration into furfural or 5-hydromethylfurfural [11] or hydrogenation into polyols. Xylitol, sorbitol and mannitol (Fig. 3.3) can be directly obtained by the hydrogenation of xylose, glucose and fructose, respectively. Those sugar-converted compounds can be further used as "building blocks" (raw materials) to produce a large variety of fine chemicals [14].

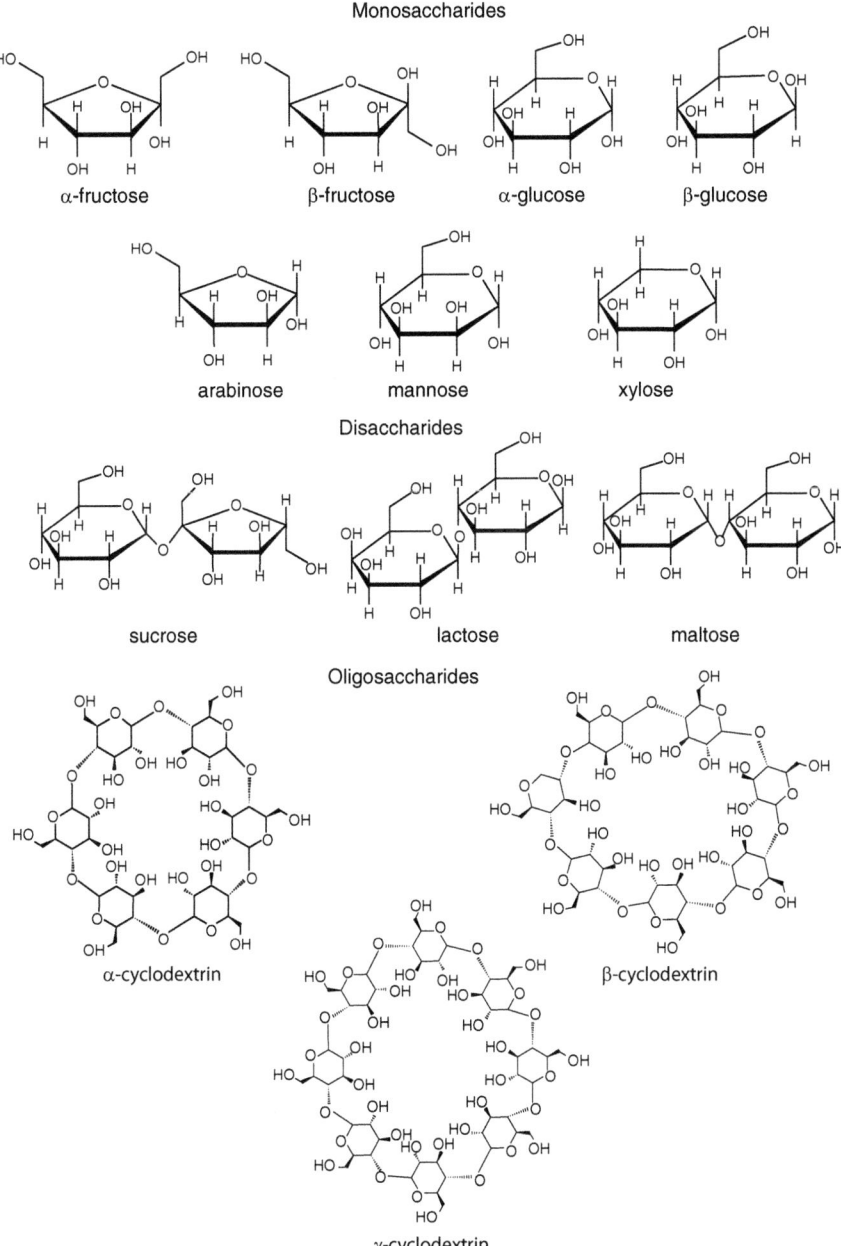

Fig. 3.1 Main examples of carbohydrate units in nature represented as mono-, di- and oligosaccharides (Reprinted with the permission from Ref. [3]. Copyright 2010 American Chemical Society)

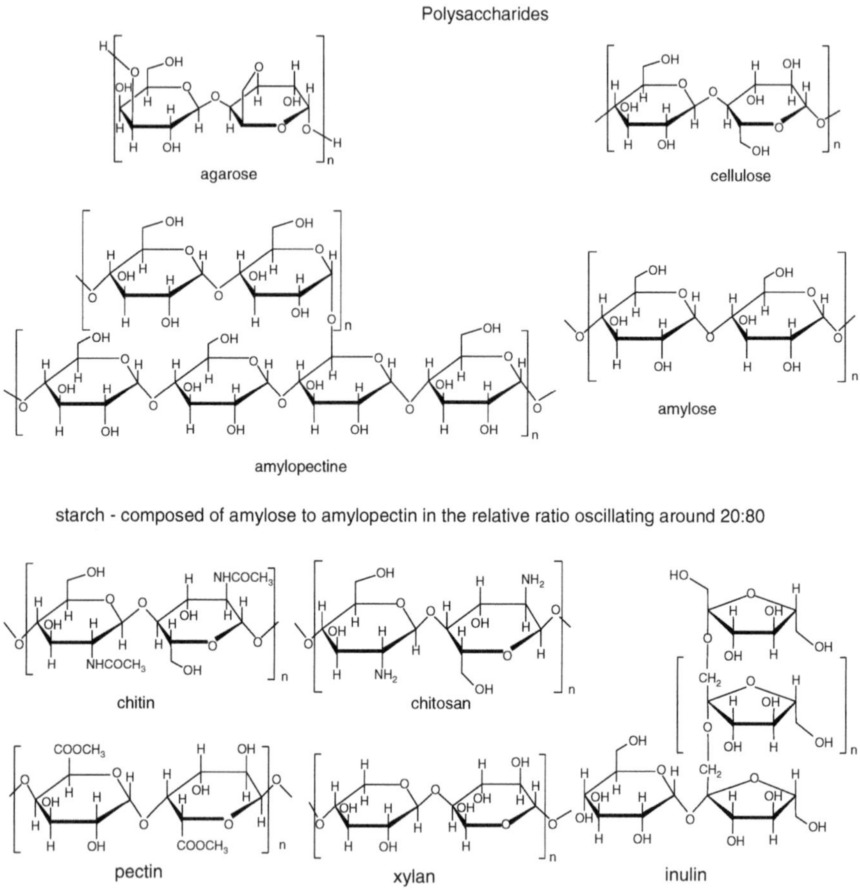

Fig. 3.2 Examples of polysaccharides (Reprinted with the permission from Ref. [3]. Copyright 2010 American Chemical Society)

One of the last disclosed applications of carbohydrates is their use in the separation and purification of biomolecules through the formation of ABS with ILs. The employment of carbohydrates as salting-out (or sugaring-out) agents for the formation of ABS with ILs is environmentally more benign in comparison to traditional polymers and salts. The employment of carbohydrates for this specific technology regarding the chemistry and physical mechanisms behind the two-phase formation is reviewed in this chapter. Their potentiality for the extraction of value-added biomolecules is discussed, and the emergent applications of this technology are presented. Table 3.1 provides a summary on ABS constituted by ILs and carbohydrates that were reported in the literature up to date.

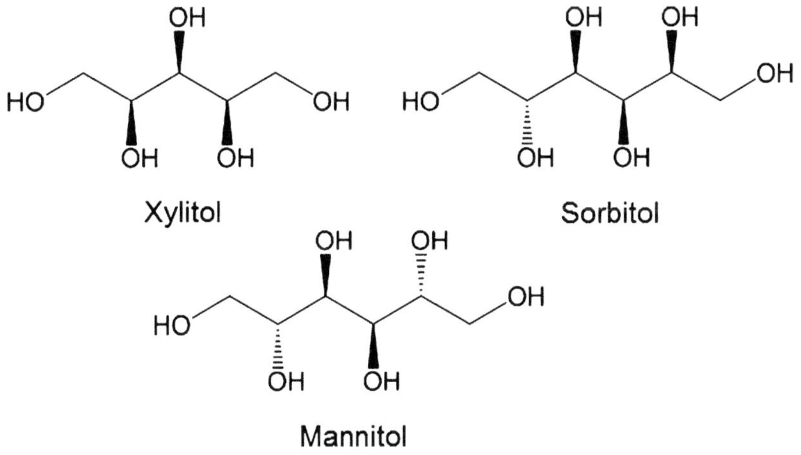

Fig. 3.3 Main examples of polyols directly obtained after hydrogenation of monosaccharides

Table 3.1 Combinations of ILs and carbohydrates investigated for ABS formation

IL	Carbohydrate
[amim]Br	Sucrose [15]
[amim]Cl	Sucrose [15]
[C_2mim][$C_4F_9SO_3$]	Maltose [16]
[C_2mpy][$C_4F_9SO_3$]	Arabinose, fructose, galactose, glucose, maltitol, maltose, mannose, sorbitol, sucrose and xylitol [16]
[C_3mim][BF_4]	Glucose [17]
[C_4mim][BF_4]	Fructose [15, 18], glucose [17, 19], maltose [20], sucrose [15, 19, 21] and xylose [19]
[C_4mim][CF_3SO_3]	Arabinose, fructose, galactose, glucose, maltitol, maltose, mannose, sorbitol, sucrose, xylitol and xylose [22]
[$C_n$mim]Cl ($n = 2$–10)	Glucose, maltose, sucrose and xylose [23]
[$C_n$mim][BF_4] ($n = 2$–10)	Glucose, maltose, sucrose and xylose [23]
[$C_n$mim][BF_4] ($n = 3$–8)	Glucose [24]
[$C_n$mim]Br ($n = 2$–10)	Glucose, maltose, sucrose and xylose [23]

3.2 Phase Behaviour

3.2.1 Methodology: Binodal Curves and Tie Lines

The most studied methodology for ABS formation and characterization is the so-called "cloud point" titration [25]. Concisely, the procedure consists of dropping-wising a carbohydrate aqueous solution repeatedly over an IL aqueous

Fig. 3.4 Ternary phase diagram of [C₄mim][BF₄] + sucrose + water showing the binodal curve (○) and determined tie lines (■) (Reprinted with permission from Ref. [19]. Copyright (2008) American Chemical Society)

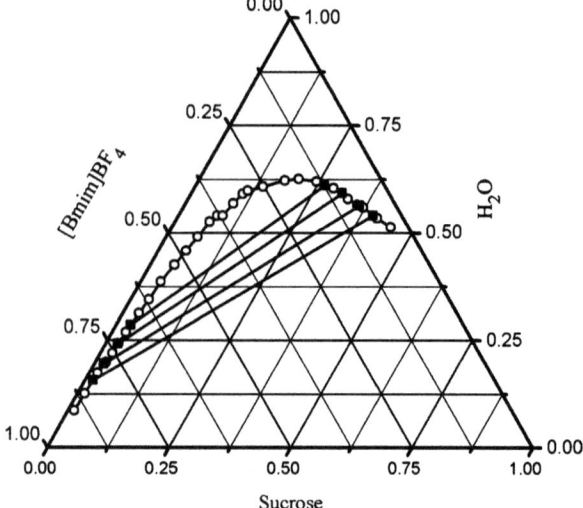

solution (or pure IL) at a given temperature until the resulting solution becomes turbid, which indicates the formation of two immiscible liquid phases. Subsequently, water is added to the system until a clear liquid monophase is observed. It is important to point out that the maximum concentration of carbohydrates in each aqueous solution is dependent on the solubility (saturation) of each individual carbohydrate in water [22].

The described methodology allows the construction of phase diagrams. After reaching the "cloud point", the carbohydrate and IL contents are determined allowing the determination of the respective binodal (solubility) curves. The binodal curves correspond to the boundaries (phase behaviour) between mono- and biphasic systems. Furthermore, the compositions of each phase correspond to the end points (nodes) of a specific tie line. The numerical length of each tie line is crucial to predict some trends in the partitioning of solutes between both phases in ABS [25]. Figure 3.4 demonstrates a particular example of the [C₄mim][BF₄] + sucrose + water ternary phase diagram, with the corresponding binodal curve and determined tie lines.

The fitting of the experimental solubility points is usually attempted by empirical equations. The empirical non-linear three-parameter equation (Eq. 3.1) developed by Merchuk and co-workers [26] is the most used in literature to fit the experimental data of ABS composed of ILs and carbohydrates, with satisfactory correlation results [15, 19, 21, 22]. Nevertheless, a five-parameter equation (Eq. 3.2) presented by Zhang and co-workers [18] also demonstrated good reliability for fitting the experimental data [17, 18, 20]:

$$IL = A \exp\left[\left(B \times CH^{0.5}\right) - \left(C - CH^{3}\right)\right] \tag{3.1}$$

$$IL = A + B \times CH^{0.5} + C \times CH + D \times CH^2 + E \times CH^3 \qquad (3.2)$$

where IL and CH are the mass fraction of the ionic liquid and carbohydrate, respectively. And A, B, C, D and E are fitting parameters obtained by least squares regression.

In respect to the tie-line fitting, empirical correlation equations given by Othmer–Tobias (Eq. 3.3) [27] and Bancroft (Eq. 3.4) [25] are usually applied:

$$\frac{1 - w_1^{IL}}{w_1^{IL}} = k_1 \left(\frac{1 - w_2^{CH}}{w_2^{CH}} \right)^n \qquad (3.3)$$

$$\frac{w_3^{CH}}{w_2^{CH}} = k_2 \left(\frac{w_3^{IL}}{w_1^{IL}} \right)^r \qquad (3.4)$$

where w_1^{IL} is the mass fraction of IL in the IL-rich phase, w_2^{CH} is the mass fraction of carbohydrate in the carbohydrate-rich phase and w_3^{CH} and w_3^{IL} are the mass fractions of water in the carbohydrate-rich phase and IL-rich phase, respectively. The adjustable parameters k_1, k_2, n, and r are obtained by least squares regression.

3.2.2 The Physicochemical Mechanism Behind the Formation of ABS

In order to explain the formation of ABS with ILs in the presence of carbohydrates, Chen et al. [17] proposed a simple mechanism which considers the IL dissociation in water. However, for this to happen, the IL should be at low concentrations, which is not always the case. As the concentration of IL increases, ions start to attract each other [28]. On the other hand, by increasing the carbohydrate concentration in water, interactions between carbohydrates are also favoured [29]. In aqueous media, water molecules start to structurally surround carbohydrates through the formation of hydration shells [17]. Based on this conjecture, with the addition of carbohydrates to a homogeneous IL solution, water molecules migrate from IL ions towards carbohydrates creating two differently structured microphases. A further increase of the IL or carbohydrate concentrations, or even both, disrupts the stability of this microemulsion leading to droplet coalescence (turbidity), which results in a two-aqueous-phase system [17].

Hence, the phase behaviour (binodal curves and tie lines) is dependent on condition variables, such as the type of both IL anion and cation, nature of carbohydrate, temperature or even physicochemical properties, like viscosity and density. In order to better understand the inherent mechanisms of ABS formation, different analyses were performed to scrutinize the effect of each variable.

3.2.2.1 Effect of ILs

The influence of ILs on the formation of ABS composed of carbohydrates was broadly studied [15, 17, 22, 23]. The ability to form ABS depends on both the IL anion and cation and seems to be a result of a salting-out ("sugaring-out") mechanism exerted by the carbohydrate. Wu et al. [15] tested 1-allyl-3-methylimidazolium bromide ([amim]Br), 1-allyl-3-methylimidazolium chloride ([amim]Cl) and 1-butyl-3-methylimidazolium tetrafluoroborate ([C_4mim][BF_4]) ILs for ABS formation with sucrose at room temperature. The obtained binodal curves are represented in Fig. 3.5. The results show that [C_4mim][BF_4] is the most efficient IL among all examined ILs, since lower mass fractions of both IL and sucrose are needed for ABS formation. The authors [15] concluded that ILs with longer alkyl side chains in the cation [$C_n$mim]$^+$ are favourable for the creation of two distinct aqueous phases. Nevertheless, [C_4mim][BF_4] cannot be considered only from the cation point of view. In fact, the IL anion exerts the primary role since [C_4mim][BF_4] is the IL that is better salted out by the carbohydrate. Theoretically, ABS formation is significantly governed by the entropy of the system. The increase of anion chaotropicity, approached mathematically by measurements of structural entropy ([BF_4]$^-$ (76.2) > Br$^-$ (61.6) > Cl$^-$ (36.6)), leads to an easier ABS formation [15]. In practice, [BF_4]$^-$ anion is less hydrated than both Cl$^-$ and Br$^-$ anions, which explains the remarkable performance of [C_4mim][BF_4] in the formation of ABS [19].

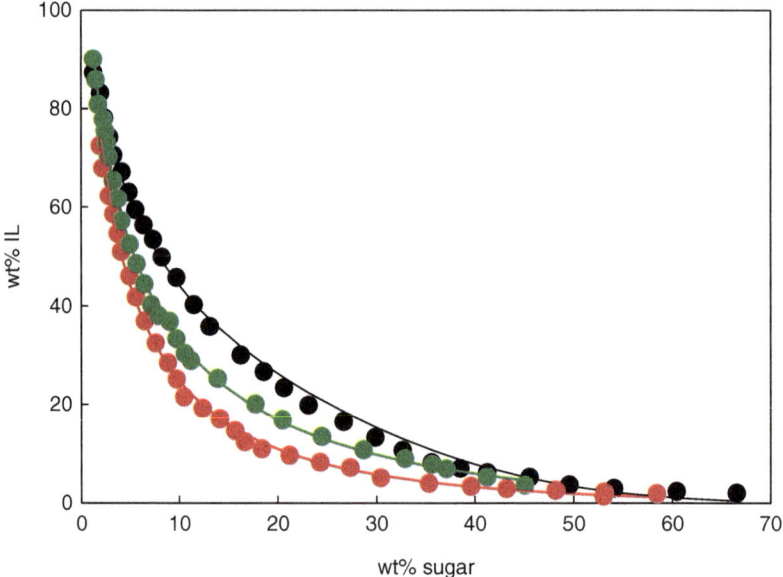

Fig. 3.5 Phase diagram for the ternary systems IL + sucrose + water in an orthogonal representation: binodal curves for [C_4mim][BF_4] (●), [amim]Br (●) and [amim]Cl (●) (Adapted with permission from Ref. [15]. Copyright (2008) American Chemical Society)

Regarding the high aptitude of [BF$_4$]-based ILs for ABS formation, Chen et al. [17, 23] further tested several ILs with different 1-alkyl-3-methylimidazolium cations. The authors [17, 23] verified the ability of hydrophilic ILs, namely, 1-propyl-3-methylimidazolium tetrafluoroborate ([C$_3$mim][BF$_4$]) and [C$_4$mim][BF$_4$], to form ABS from an initial homogeneous aqueous solution after the addition of carbohydrates. [C$_4$mim][BF$_4$] demonstrated to be more efficient in forming ABS than [C$_3$mim][BF$_4$]. ILs with longer alkyl side chains at the imidazolium were also tested; yet, they are not water miscible at room temperature. Although ABS were not formed, the authors also confirmed that the solubility of these ILs in water decreases by the addition of carbohydrates [17, 23].

The key point to produce ABS with ILs in the presence of carbohydrates comprises the competition between carbohydrates and IL ions to create hydration complexes [22]. As mentioned before, IL anions must be weakly hydrated so that carbohydrates could be more competitive to produce hydration complexes. The hydration ability of the IL anions could also be associated to the hydrogen basicity of the IL, which can be characterized by the Kamlet–Taft parameter β [30]. Therefore, low values of β for IL anions indicate a low ability to accept protons and thus a weak hydration and a high probability to form ABS when other salting-out species are added. For instance, [C$_4$mim][BF$_4$] presents a Kamlet–Taft parameter $\beta = 0.55$ [31]. Apart from [C$_4$mim][BF$_4$], Freire et al. [22] discovered other ABS composed of carbohydrates by using the IL 1-butyl-3-methylimidazolium trifluoromethanesulphonate ([C$_4$mim][CF$_3$SO$_3$]), which presents a $\beta = 0.57$ [31]. Similar to [BF$_4$]$^-$, [CF$_3$SO$_3$]$^-$ also presents a high capability to induce the creation of ABS in presence of salting-out species. Other fluorinated-based ILs, with lower β values, are also good candidates to form ABS with carbohydrates. Nonetheless, imidazolium-based ILs composed of hexafluorophosphate (PF$_6^-$) and bis((trifluoromethyl)sulphonyl)imide ((CF$_3$SO$_3$)$_2$N$^-$) are immiscible with water at room temperature avoiding thus the creation of systems of two aqueous-rich phases [32, 33].

Several other ILs were evaluated, such as [C$_4$mim]Cl, 1-hexyl-3-methylimidazolium ([C$_6$mim]Cl), 1-butyl-3-methylimidazolium methylsulphate ([C$_4$mim][MeSO$_4$]), 1-butyl-3-methylimidazolium hydrogen sulphate ([C$_4$mim][HSO$_4$]) and 1-butyl-3-methylimidazolium dicyanamide ([C$_4$mim][N(CN)$_2$]). However, these ILs were found to be unable to produce ABS with carbohydrates [22]. Furthermore, 1-alkyl-3-methylimidazolium chloride ([C$_n$mim]Cl, $n = 2-10$) and 1-alkyl-3-methylimidazolium bromide ([C$_n$mim]Br, $n = 2-10$) ILs were also tested with no ability to form ABS [23], contrarily to previous reports stating that [amim]Cl and [amim]Br are able to produce ABS with carbohydrates [15]. To clarify this contradiction, Freire et al. [25] attempted to reproduce ABS formation with [amim]Cl in presence of sucrose at room temperature, but even after 48 h, a homogeneous solution (with no two-phase creation) was always observed. This suggests that ILs containing halogen anions, which form ABS in presence of inorganic salts [34], are difficult to form ABS with carbohydrates.

Mainly imidazolium-based ILs have been examined to form ABS in presence of carbohydrates. Recently, ILs containing cholinium and pyridinium cations and either $[CF_3SO_3]^-$ or $[C_4F_9SO_3]^-$ anions have been proposed as potential phase-forming components of ABS with carbohydrates [16]. The ABS formation was achieved using $[C_2mpy][C_4F_9SO_3]$ with several carbohydrates. Actually, the obtained results demonstrated a higher propensity of ABS formation with this IL than in the case of $[C_2mim][C_4F_9SO_3]$, meaning that the 6-carbon ring of pyridinium favours ABS formation with carbohydrates when compared with the 5-carbon ring of imidazolium [16]. However, neither $[C_2mim][CF_3SO_3]$ nor $[C_2mpy][CF_3SO_3]$ was able to produce phase separation with carbohydrates. In fact, long alkyl side chains at the cation ($[C_4mim]^+$ [22]) or at the anion, $[C_4F_9SO_3]^-$ [16], are needed to produce ABS with carbohydrates. Therefore, both anion and cation influence the phase separation independently, as long as they provide sufficient hydrophobicity to the IL and a low competition with carbohydrates for water interactions. Nevertheless, the binding strength between anion and cation, the steric hindrance of both and the ability for carbohydrate dissolution should not be disregarded as factors influencing ABS formation. Concerning the cholinium-based ILs, no successful ABS was produced with carbohydrates even when this type of IL possesses the $[C_4F_9SO_3]^-$ anion. The high hydrophilicity of the cholinium cation may be the cause of such results [16].

3.2.2.2 Effect of Carbohydrates

Despite the few possibilities of ILs to generate ABS with carbohydrates, there is a great variety of carbohydrate-based salting-out agents that were explored in the literature [19, 20, 22, 23]. For instance, the binodal curves of ABS with $[C_4mim]$ $[BF_4]$ using fructose, glucose, sucrose and xylose at room temperature are shown in Fig. 3.6 [19]. The data demonstrate that in the case of sucrose, lower quantities of both phase-forming components are needed to form ABS. Wu et al. [19] mentioned that the main reason for the observed differences is the number of OH groups composing the carbohydrates. They also established the following order of carbohydrates regarding the average number of equatorial hydroxyl groups: sucrose (6.2) > glucose (4.56) > xylose (3.67) > fructose (2.56) [19]. The higher the number of equatorial hydroxyl groups, the higher the ability of a given carbohydrate to create ABS with ILs. This conclusion is in concordance with the results present in Fig. 3.6. Actually, it seems that OH groups in carbohydrates display high affinity for water favouring the formation of hydration complexes. This effect combined with the highly chaotropic nature of ILs guarantees the ABS formation [19, 22].

Freire et al. [22] further investigated the reasons behind the ABS formation with $[C_4mim][CF_3SO_3]$ and several carbohydrates at 298 K. Firstly, the authors [22] determined the phase diagrams for ABS with monosaccharides and established their capacity for liquid–liquid demixing in the following order: glucose ≈ galactose > fructose ≈ mannose > arabinose > xylose [22]. These data confirm the higher effectiveness of C6 sugars in the formation of ABS to those of C5 sugars. A higher number of hydroxyl groups also results in the formation of more hydrogen bonds with water turning them into stronger salting-out agents.

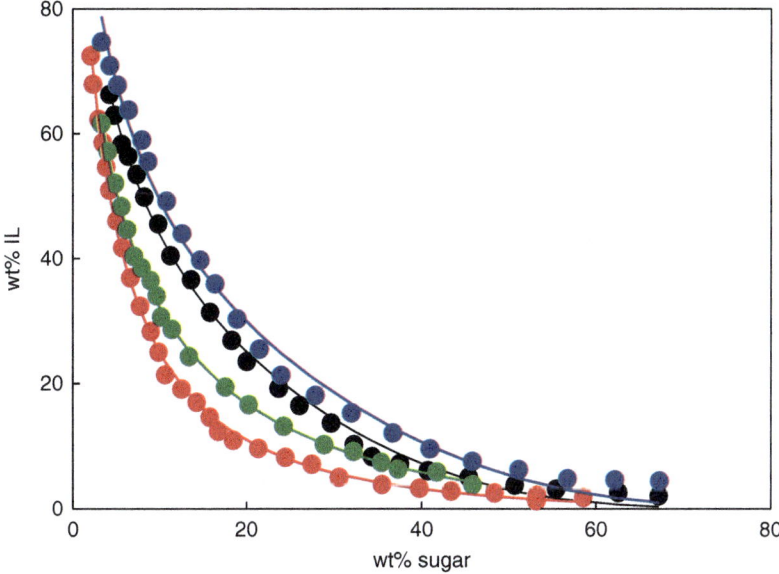

Fig. 3.6 Phase diagram for the ternary systems [C$_4$mim][BF$_4$] + carbohydrate + water in an orthogonal representation: binodal curves for sucrose (●), glucose (●), xylose (●) and fructose (●) (Adapted with permission from Ref. [19]. Copyright (2008) American Chemical Society)

Furthermore, the superior efficiency for glucose in contrast to fructose (both C6 sugars) is governed by the pyranose conformation which favours a better hydration and a higher ability to induce the creation of ABS [22].

Considering the epimers of glucose, mannose and galactose, their efficiency for ABS formation can be explained as a function of the ratio between axial and equatorial hydroxyl groups composing each pyranose ring [22]. A similar argument was given by Wu et al. [19], as it was previously discussed. Particularly, the orientations of OH groups at C2 and C4 positions were considered as key factors to define saccharides which facilitate the ABS formation. The OH groups at C2 and C4 positions in glucose are equatorial, while galactose has OH groups at C2 in equatorial and at C4 in axial positions. In the case of mannose, both OH groups are in an axial orientation. Herein, the OH groups with equatorial orientation are more easily hydrated further facilitating the phase separation. This conjecture explains the ABS formation capacity that was ordered as follows: glucose > galactose > mannose [22].

It is also important to point out in the work of Freire et al. [22] that fructose presented a higher efficiency in ABS formation than xylose. This is in contrast to what was observed by Wu et al. [19], whose results are presented in Fig. 3.6. Despite the higher number of OH groups present in fructose, the structure of xylose contains more OH groups in equatorial orientation. Therefore, a higher efficiency of ABS formation should be noticed for xylose. Nevertheless, to examine this effect, different ILs were used ([C$_4$mim][CF$_3$SO$_3$] and [C$_4$mim][BF$_4$]), and therefore different chemical environments are present that somehow may lead to divergent data between both studies [19, 22]. A possible explanation can be attempted by the xylose and fructose decyclization in presence of different ILs. Actually, Wu et al. [19, 21] mentioned that a better performance of some carbohydrates in the formation of ABS could be explained by their decyclization, and where one example can be an aqueous solution of sucrose in the presence of [C$_4$mim][BF$_4$]. In aqueous solution, sucrose is a mixture of several isomers being in equilibrium, in which cyclic conformation isomers are more stable than chain conformation isomers. The presence of [C$_4$mim][BF$_4$] in the system allows the disruption of that equilibrium favouring the formation of chain conformation isomers of sucrose. However, such supposition still requires more research to deeply infer on the different behaviours observed [19, 21].

Because the number of OH groups in carbohydrates seems to be a key factor influencing the formation of ABS with ILs, different disaccharides (sucrose and maltose) and polyols (sorbitol, xylitol and mannitol) were examined to form ABS with [C$_4$mim][CF$_3$SO$_3$] [22]. The data obtained for these systems are represented in Fig. 3.7. The main advantage coming from the use of disaccharides is the increased number of OH groups *per* mole in comparison to monosaccharides. This definitively can lead to a better performance in inducing ABS. Figure 3.7a demonstrates that maltose (two linked glucose units) presents a higher efficiency to form ABS than sucrose (one glucose unit linked to one fructose unit). This confirms again that the pyranose conformation present in maltose favours the phase separation in contrast to sucrose that is based on furanose conformation.

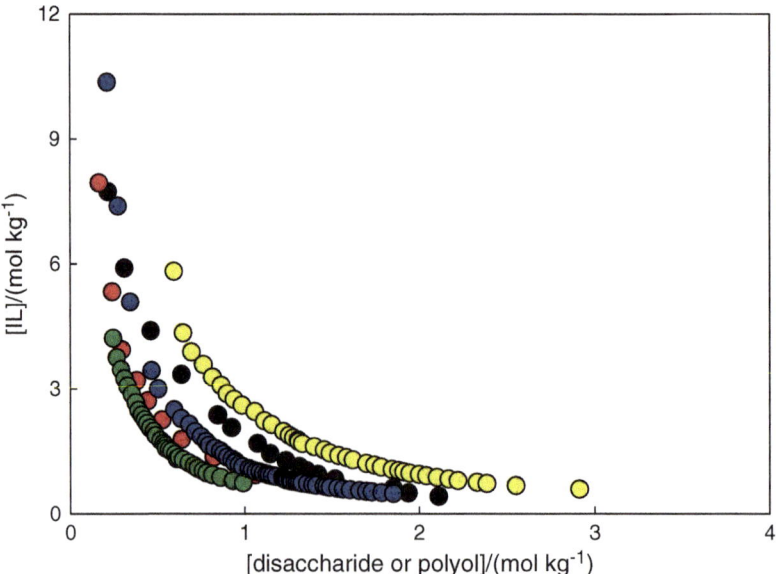

Fig. 3.7 Phase diagram for the ternary system [C$_4$mim][CF$_3$SO$_3$] + carbohydrate + water in an orthogonal representation: binodal curves for maltose (●) and sucrose (●) maltitol (●), sorbitol (●) and xylitol (●) (Reproduced from Ref. [22] by permission of The Royal Society of Chemistry)

For polyols, the increasing number of OH groups is also directly correlated to their higher efficiency in ABS formation. The efficiency pattern is as follows: maltitol > sorbitol > xylitol; it is clearly observable in Fig. 3.7b. In fact, the higher performance of polyols is a practical confirmation that the decyclization of carbohydrates enhances the formation of ABS. For instance, sorbitol displays a similar linear structure to the chain form of glucose, whereas an aldehyde (−CHO group) is replaced by a CH_2OH group. Therefore, the reduction of aldoses and ketoses into an alcohol enhances the affinity towards water improving the salting-out effect [22].

Freire et al. [22] concluded that ABS with $[C_4mim][CF_3SO_3]$ are formed according to the carbohydrates' salting-out ability. Among the examined carbohydrates, the highest ability to form ABS was exhibited by maltitol, while the remaining carbohydrates follow the following order: maltose > sorbitol > sucrose > glucose ≈ galactose > xylitol ≈ fructose ≈ mannose > arabinose > xylose. The authors [22] also found that alditols have an outstanding ability to form ABS with ILs mostly due to two main factors that are subjacent to the carbohydrates salting-out aptitude: (i) the number of OH groups and (ii) the related stereochemistry.

3.2.2.3 Effect of Temperature

Temperature is one of the major factors influencing the ABS formation ability, while an adequate selection of temperature is highly important for the application of this separation technology. The effect of temperature was evaluated in some studies demonstrating that other mechanisms in addition to those already discussed may be involved in ABS formation [17, 18, 20, 21, 23]. In respect to phase behaviour, Chen et al. [17] verified that the increase of temperature does not favour the ABS formation with ILs and carbohydrates. With the temperature decrease, the biphasic area increases as depicted in Fig. 3.8. The same trend was observed in additional studies using other carbohydrate-based ABS [18, 20, 21, 23]. This is also observed for other types of ABS, such as in polymer + polymer ABS [35]. However, the opposite effect is found in polymer + salt systems, in which a temperature increase guides to more favourable ABS formation [36].

For ABS constituted by carbohydrate and ILs, it is expected that the hydration of carbohydrates decreases for higher temperatures and simultaneously the solubility of IL in water rises. On the other hand, an increase in temperature can interfere negatively with ABS formation by the reduction of the decyclization process of carbohydrates [21]. Furthermore, because the interactions between $[C_4mim][BF_4]$ and sucrose are exothermic, an increase in temperature disfavours ABS creation [21]. Actually, in order to form ABS at higher temperature (e.g. 308 K), larger amounts of IL are needed [21].

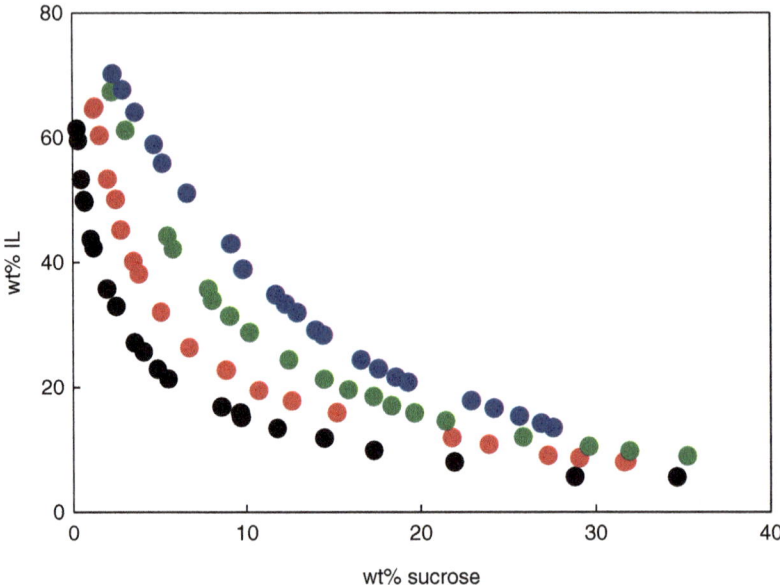

Fig. 3.8 Phase diagram for the ternary system [C$_4$mim][BF$_4$] + glucose + water in an orthogonal representation: binodal curves at 278.15 K (●), 288.15 K (●), 298.15 K (●) and 308.15 K (●) (Adapted with permission from Ref. [17]. Copyright (2010) American Chemical Society)

3.2.2.4 Effect of Physicochemical Properties

The ABS formation by an IL and a salting-out agent normally results in the formation of an upper IL-rich phase, while at the bottom, an enriched salting-out-rich phase is generated [25]. Regarding carbohydrate-based ABS formed with ILs, the addition of any carbohydrate to [C$_4$mim][BF$_4$] follows this separation pattern [15, 17–21, 23]. Nevertheless, Freire et al. [22] observed an inversion on the phases in ABS formed by [C$_4$mim][CF$_3$SO$_3$]. In this case, the IL-rich phase was found at the bottom, and the enriched carbohydrate solution corresponds to the upper phase. This interesting phenomenon is technically favourable for decantation processes [22]. The driving force provoking this inverse two-phase system corresponds to the high density of [C$_4$mim][CF$_3$SO$_3$] (1306.1 kg·m^{-3}) that is higher than that observed for the other IL – [C$_4$mim][BF$_4$] (1206.9 kg·m^{-3}) [37]. However, in systems with sucrose, an inverse phase formation occurred at a specific composition (≈40 wt% of IL and ≈25 wt% of sucrose) [22].

Freire et al. [22] further established a correlation between the potential of a specific carbohydrate to form ABS and the corresponding viscosity of the carbohydrates' aqueous solutions. The authors [22] proposed the extended Jones–Dole equation [38], which allows to find the viscosity B- and D-coefficients for the concentrated solutions of carbohydrates. Using these coefficients, it was possible to predict the potential salting-out effect of a specific carbohydrate. Following the

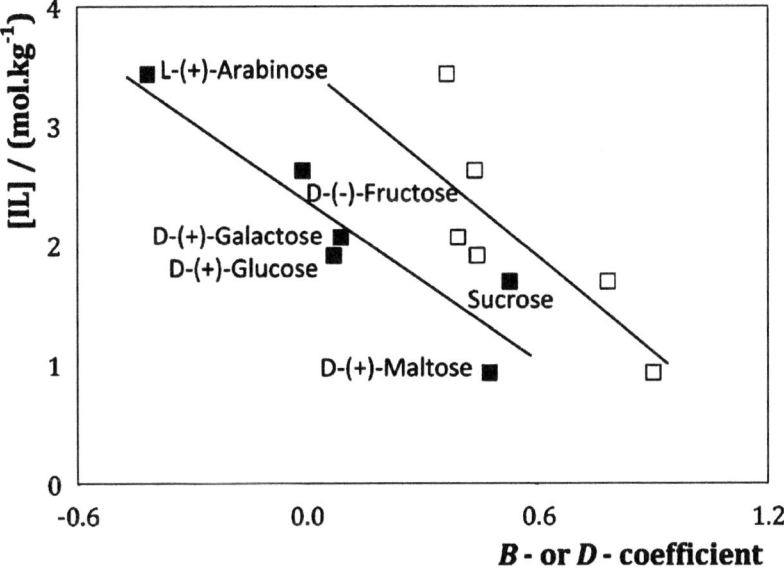

Fig. 3.9 Ionic liquid molality at which the carbohydrate molality is equal to 1.0 mol·kg^{-1} as a function of viscosity B-coefficient (□) and D-coefficient (■) of aqueous solutions of carbohydrates at 308.15 K (Reproduced from Ref. [22] by permission of The Royal Society of Chemistry)

proposed equation, higher viscosity B-coefficient values indicate a stronger ability of the carbohydrate to change the viscosity of the aqueous solution and to act as a salting-out agent, considering only water–water and water–carbohydrate interactions. In the case of the viscosity D-coefficient, it represents the strength of both carbohydrate–carbohydrate and carbohydrate–water interactions, coupled to the size and shape effects of the solute (entropic contributions). In other words, for a given carbohydrate to act as a strong salting-out agent, the favourable interactions between water and carbohydrates must overwhelm the favourable interactions between IL and water [22]. Freire et al. [22] represented the IL molality concentration found for 1.0 mol·kg^{-1} of each carbohydrate as function of viscosity B- and D-coefficients at 308.15 K (determined elsewhere [39]) as shown in Fig. 3.9. The results clearly show a trend demonstrating that higher viscosity B- and D-coefficients lead to a stronger ability of the carbohydrate to establish intermolecular interactions with water, favouring thus ABS formation [22].

In contrast to other types of ABS, such as IL + salt and typical PEG + salt or PEG + polysaccharide, [C$_4$mim][CF$_3$SO$_3$] + carbohydrates present a lower viscosity. This property favours mass transfer within the system that is crucial for the potential scale-up of the technology [22].

3.3 Applications of IL-Carbohydrate-Based ABS

The applications of ABS composed of ILs and carbohydrates have been demonstrated mainly as a separation technology. The environmentally benign nature of carbohydrates is advantageous for the separation of biomolecules. Since it is a recent technology, only few applications have been proposed up to date; however, the obtained results are very promising. Nowadays, ABS with ILs and carbohydrates were applied in phenol extraction [23, 24] and extraction of bio-based compounds [16, 22].

3.3.1 Removal of Phenol

Phenols are usually found as pollutants in wastewaters. Therefore, regarding the human health security and environmental issues, phenol removal from water is an important aspect. The traditional phenol extraction processes are liquid–liquid-based processes, and the solvents used are generally volatile organic compounds that are recognized as environmental non-friendly chemicals. Therefore, the need to find a solution to this problem drove to the employment of ABS with ILs and carbohydrates. Chen et al. [23, 24] demonstrated that ABS composed of [$C_n$mim][BF_4] ($n = 3$–8) and glucose can enhance the phenol removal in comparison to traditional extraction processes. The authors [23, 24] addressed main variables, namely, glucose and IL concentrations, initial phenol concentration, temperature and alkyl chain length of the IL cation towards the phenol extraction from water by ABS. The phenol extraction capability was evaluated as the distribution ratio (D) between the concentration of phenol in the IL-rich to that in the glucose-rich phase [24].

Regarding the concentrations of the phase-forming components, either the increase of IL or glucose amount within ABS allowed an increase in the D value, although it was found that the glucose concentration increase effect is more dominant than that displayed by the IL [23, 24]. Interestingly, with the increase of the initial concentration of phenol, the extraction efficiency into the IL-rich phase declines, resulting in lower D values, meaning that the saturation of the IL-rich phase could be reached – although no saturation values were provided by the authors [23, 24]. Therefore, ABS formed by ILs and carbohydrates could be favourable routes for the extraction of phenol at low concentrations that indeed correspond to typical concentrations found in wastewaters.

As previously mentioned, temperature has a strong influence on the ABS formation with ILs and carbohydrates [17, 18, 20, 21, 23]. The increase of temperature disfavours the ABS formation, which in the specific case of the [C_4mim][BF_4] + glucose system also acts negatively on the extraction of phenol. Furthermore, if the extraction is performed at temperatures below 278.15 K, the D values significantly increase [24]. To establish a correlation between the D value and

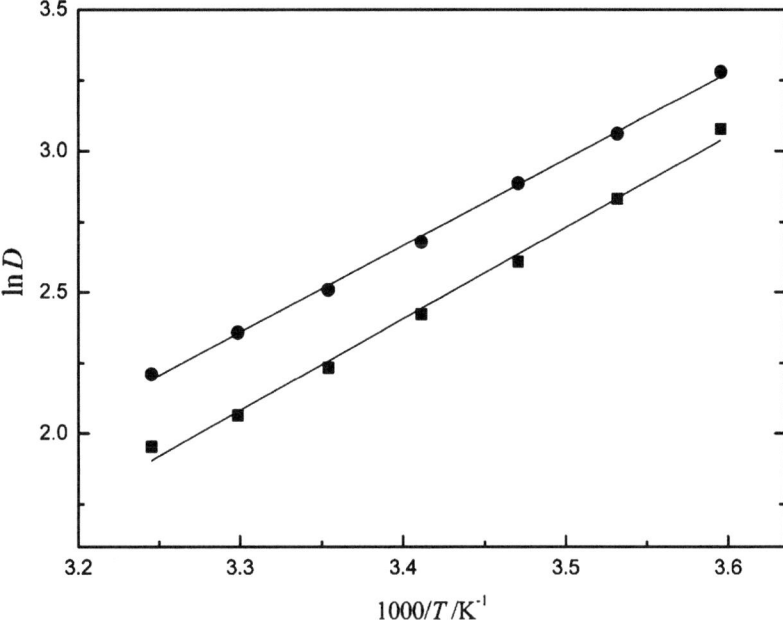

Fig. 3.10 Linear correlation between ln D and $1/T$ in the [C$_4$mim][BF$_4$] + glucose system using a glucose mass fraction of 0.1500 (■) and 0.2000 (●) (Reprinted with the permission from Ref. [24]. Copyright 2012 American Chemical Society)

temperature, Chen et al. [24] employed the van't Hoff equation and plotted the natural logarithm of the distribution ratio (ln D) versus the corresponding temperature ($1/T$) with good correlation ($R^2 > 0.9931$), as shown in Fig. 3.10.

Imidazolium-based ILs with different alkyl side chain lengths combined with [BF$_4$]$^-$ were additionally examined for phenol removal, and the results are presented in Fig. 3.11. It can be concluded that ILs with longer alkyl side chains lead to higher D values – higher efficiency to extract phenol for the IL-rich phase. For instance, a maximum D value of 78 was reached with the [C$_8$mim][BF$_4$] + glucose system [23, 24]. This trend of phenol extraction is very similar to that observed in liquid–liquid extraction with hydrophobic IL + water systems [40]. It should be pointed out that [C$_8$mim][BF$_4$] is not completely water miscible at room temperature and two aqueous-rich phases are not formed. Nevertheless, the efficiency of phenol extraction into the IL-rich phase was deduced to be related to hydrogen bonding between the hydroxyl group in phenol and the hydrogen attached to C2 carbon in the imidazolium cation [23] as well as to π–π interactions between the aromatic and imidazolium rings of phenols and ILs, respectively. In summary, the addition of carbohydrates, the increase of the length of the alkyl side chain in the imidazolium cation and the decrease of temperature improve the extraction of phenol to the IL-rich phase.

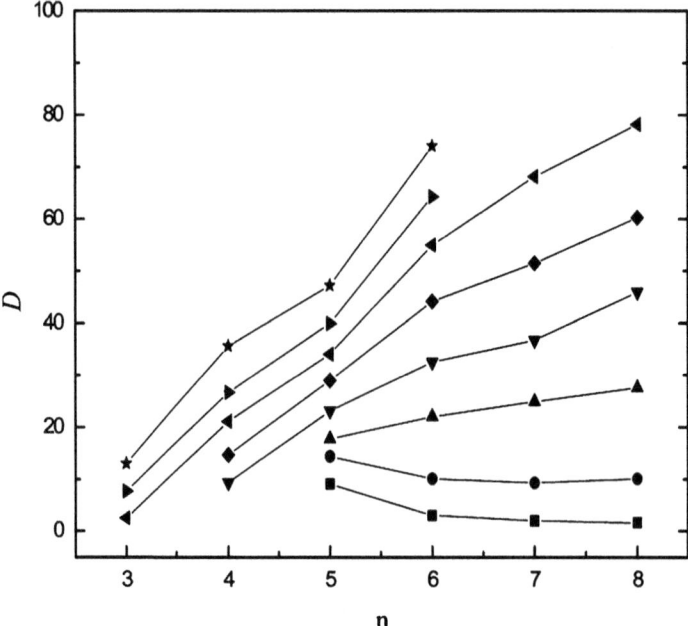

Fig. 3.11 Distribution ratio (D) of phenols obtained for ABS formed by [C$_n$mim][BF$_4$] ($n = 3$ to 8) and glucose at 298.15 K. IL mass fraction of 0.3500 (■) and glucose mass fractions of 0.0000 (●), 0.0500 (▲), 0.0999 (▼), 0.1501 (♦), 0.2000 (◄), 0.2498 (►) and 0.3500 (★) (Reprinted with the permission from Ref. [24]. Copyright 2012 American Chemical Society)

3.3.2 Extraction of Biomolecules

The fact that carbohydrates are natural organic compounds can play an important role in the extraction of added-value biomolecules using ABS. One of the major disadvantages of using, for instance, inorganic salts as salting-out agents for the extraction of biomolecules through ABS is the undesired speciation of biomolecules. It occurs mostly promoted by very high pH values afforded by the majority of inorganic salts. Therefore, the use of carbohydrates as salting-out agents is more benign to the separation process itself as well as to the final product. The partitioning of model biomolecules, such as L-tryptophan, caffeine and β-carotene, was investigated in [C$_4$mim][CF$_3$SO$_3$] + carbohydrate ABS [22]. The data obtained in this study are summarized in Fig. 3.12. The extraction efficiency of each compound was evaluated according to the partition coefficient, K, expressed as the ratio of concentration of each biomolecule X in the IL-rich phase to that in the carbohydrate-rich phase.

The partition coefficients obtained for L-tryptophan were practically the same for all performed ABS trials demonstrating that its extraction is independent of the carbohydrate employed. Furthermore, the obtained K value (≈ 1) shows that a homogeneous dispersion of L-tryptophan occurs between the two phases [22]. In

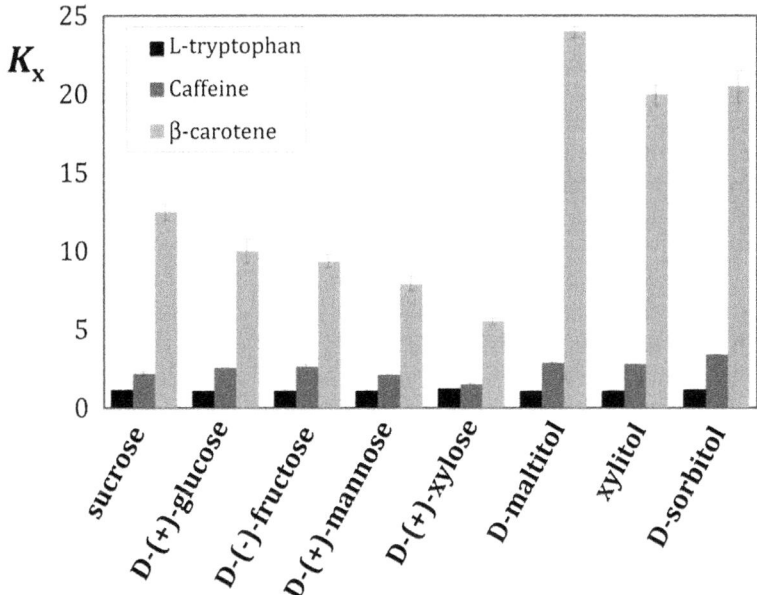

Fig. 3.12 Partition coefficients (K) of L-tryptophan, caffeine and β-carotene obtained in ABS formed by [C$_4$mim][CF$_3$SO$_3$] (≈40 wt%) and several carbohydrates (≈25 wt%) at 298.15 K (Reproduced from Ref. [22] by permission of The Royal Society of Chemistry)

respect to the caffeine biomolecule, an improvement of K value (maximum 2.68 with sorbitol) was observed in comparison to that obtained with L-tryptophan. This can be explained by the higher hydrophobicity presented by caffeine. Furthermore, polyols demonstrated the highest efficiency as salting-out agents for obtaining higher partitioning of caffeine into the IL-rich phase, due to the more extensive hydration of that type of carbohydrates as described before [22]. Still, among the studied biomolecules, β-carotene was the most efficiently extracted by [C$_4$mim] [CF$_3$SO$_3$] + carbohydrate ABS. The K value reached a maximum of 24.0 when maltitol is used as the salting-out agent, which corresponds to a 95 % of β-carotene extraction efficiency. Herein, the very high hydration aptitude of maltitol is the driving force for the partitioning of β-carotene, which is also supplemented by the higher hydrophobicity of β-carotene in comparison to other biomolecules [22]. Therefore, [C$_4$mim][CF$_3$SO$_3$] + carbohydrate ABS demonstrate to be an excellent extraction system for hydrophobic solutes. Freire et al. [22] mentioned that the partitioning of hydrophilic and aromatic solutes, such as L-tryptophan, depends on the IL nature (other rather than [C$_4$mim][CF$_3$SO$_3$]) and on the IL-amino-acid-specific interactions. Furthermore, for a middle rank polarity compound, such as caffeine, a slight dependence of the carbohydrate hydration aptitude was verified. The authors [22] finally found that the extraction of a more extreme hydrophobic solute, such as β-carotene, is successfully achieved with [C$_4$mim][CF$_3$SO$_3$] + carbohydrate ABS; however specific carbohydrates must be used.

3.3.3 Extraction of Azo Dyes

Azo dyes are one of the most used colourants in food, cosmetic, textile and pharmaceutical industries. However, processing these dyes into new products leads to losses of about 10–15 % [41]. Therefore, their recovery is highly demanded from economic and environmental reasons. ABS composed of maltose and [C$_2$mpy][C$_4$F$_9$SO$_3$] were recently examined to recover azo dyes, such as Brilliant Blue FCF (E133), Green S (E142), tartrazine (E102) and Ponceau 4R (E124), at 298 K, and the respective results are presented in Fig. 3.13 [16]. In general, almost all dyes were concentrated in the carbohydrate-rich phase, excepting Brilliant Blue FCF which preferentially migrates to IL-rich phase. The extraction of dyes into the carbohydrate-rich phase was highlighted as not usual among the existing ABS. Generally, the high salting-out effect of inorganic salts leads to the migration of the targeted compound to the IL-rich phase. In the studied IL-carbohydrate-based ABS, an inverse behaviour was observed due to the high polarity and weak sugaring-out power of carbohydrates. Therefore, polar molecules can be easily extracted to the carbohydrate-rich fraction, while hydrophobic ones, such as Brilliant Blue FCF, preferably partition to the IL-rich phase [16].

It is important to highlight that this kind of extraction could be an outstanding advantage once the desired molecules are extracted to the more benign and biocompatible phase, essentially constituted by carbohydrates. On the other hand, the concentrated IL-rich phase could be easily recycled and reused for several times without contamination of the extracted compounds [16].

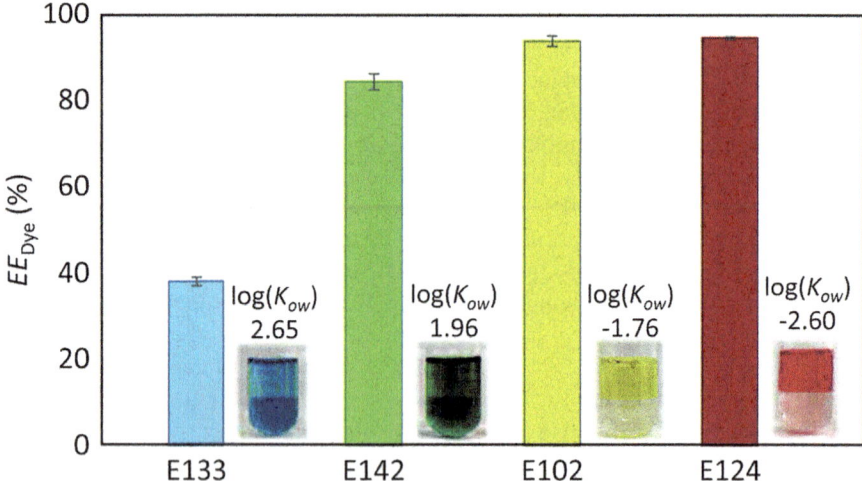

Fig. 3.13 Extraction efficiency (EE$_{Dye}$) of Brilliant Blue FCF (E133), Green S (E142), tartrazine (E102) and Ponceau 4R (E124) dyes to the carbohydrate-rich phase using ABS composed of [C$_2$mpy][C$_4$F$_9$SO$_3$] (40 wt%) + maltose (18 wt%) + H$_2$O at 298 K. The partition of each dye is represented as log(K$_{ow}$) (Reproduced from Ref. [16] by permission of The Royal Society of Chemistry)

3.4 Conclusions and Outlook

Carbohydrates as renewable, diverse and abundant raw materials could play an important role in the development of green and sustainable processes. This chapter discloses the importance of ABS composed of ILs and carbohydrates as innovative and environmentally friendly technologies for the separation and purification of several compounds. The profound study of the respective phase diagrams, including binodal curves and tie lines, has aided on the understanding of the mechanisms inherent to ABS formation (regarding the chemical interactions and physicochemical characteristics of carbohydrates, ILs and water).

The main factors affecting carbohydrate-IL-based ABS formation are the type of IL cation and anion and carbohydrate used, as well as temperature. In the case of ILs, both anion and cation play different roles within the system, but the chaotropic nature of IL anions is crucial for phase separation. IL anions must be less hydrated by water than carbohydrates. This behaviour can be predicted by the hydrogen-bond basicity of the anion, which can be given by the Kamlet–Taft parameter β (hydrogen-bond basicity). ILs with anions with lower β values, such as [C$_4$mim] [BF$_4$] and [C$_4$mim][CF$_3$SO$_3$], demonstrate an enhanced ability to form ABS with carbohydrates. Nevertheless, [C$_4$mim][CF$_3$SO$_3$] seems to be more preferable due to its high stability in aqueous media. In contrast to ILs, carbohydrates must be highly hydrated to act as salting-out agents.

A high number of OH groups and the equatorial position of those in the carbohydrate structure are driving forces to enhance ABS formation. Furthermore, pyranose conformation of hexoses favours ABS, and the decyclization process of carbohydrates to the chain form conformation is believed to enhance the phase separation. Temperature, as an external condition of the system, also interferes with the ABS formation ability. Low temperatures facilitate the ABS formation decreasing the quantities of IL and carbohydrate required for phase separation.

ABS composed of ILs and carbohydrates demonstrated to be a low viscous separation technology, which is an important advantage over other liquid–liquid-based separations. Applications on phenol extraction from water and selective extraction of biomolecules have been successfully achieved.

One of the major disadvantages of ABS with ILs and carbohydrates conveys on the limited number of ILs able to induce phase separation with carbohydrates, as a result of the weak salting-out effect of carbohydrates in comparison to others (for instance, inorganic salts). Actually, most hydrophilic ILs are not able to produce ABS with carbohydrates. Only ILs with moderate polarity and hydrophobicity capable to mix with water are the target ones to apply in this type of concept.

In order to attain sustainability of ABS processes, the efficient IL recycling should be investigated. Few studies have been performed regarding this subject [15, 19], but definitely the IL recovery after ABS formation needs to be attempted to develop cost-effective processes. Therefore, the IL recovery and their further performance in extraction must be addressed in more detail in the near future.

One of the important steps for the nearest future is the need to develop new applications of ABS with ILs and carbohydrates, so that the technology will be recognized as robust and promising. Furthermore, the green and environmental character of this technology could be a great advantage and an alternative to conventional separation processes. For instance, the extraction of hydrophobic compounds could be a feasible opportunity for the application of this technology in biorefineries [42, 43] or other bio-based industries.

Acknowledgments This work was supported by the Fundação para a Ciência e Tecnologia (FCT, Portugal) through strategic project UId/QUI/5006/2013, and grants SFRH/BD/90282/2012 (AMdCL) and IF/00424/2013 (RBL).

References

1. Robyt JF (1998) Essentials of carbohydrate chemistry. Springer, New York
2. Nelson DL, Cox MM (2013) Lehninger principles of biochemistry. Freeman, W.H, London
3. Zakrzewska ME, Bogel-Lukasik E, Bogel-Lukasik R (2010) Solubility of carbohydrates in ionic liquids. Energy Fuel 24:737–745
4. Fathi M, Martin A, McClements DJ (2014) Nanoencapsulation of food ingredients using carbohydrate based delivery systems. Trends Food Sci Technol 39:18–39
5. Ahmed A, Adel M, Karimi P, Peidayesh M (2014) Pharmaceutical, cosmeceutical, and traditional applications of marine carbohydrates. In: Kim S-K (ed) Advances in food and nutrition research, vol 73. Elsevier, London, pp 197–220
6. Laurienzo P (2010) Marine polysaccharides in pharmaceutical applications: an overview. Mar Drugs 8:2435–2465
7. Peters D (2006) Carbohydrates for fermentation. Biotechnol J 1:806–814
8. Siró I, Plackett D (2010) Microfibrillated cellulose and new nanocomposite materials: a review. Cellulose 17:459–494
9. Jayakumar R, Prabaharan M, Nair SV, Tokura S, Tamura H, Selvamurugan N (2010) Novel carboxymethyl derivatives of chitin and chitosan materials and their biomedical applications. Prog Mater Sci 55:675–709
10. Bozell JJ, Petersen GR (2010) Technology development for the production of biobased products from biorefinery carbohydrates-the US Department of Energy's "Top 10" revisited. Green Chem 12:539–554
11. Zakrzewska ME, Bogel-Lukasik E, Bogel-Lukasik R (2011) Ionic liquid-mediated formation of 5-hydroxymethylfurfural-A promising biomass-derived building block. Chem Rev 111:397–417
12. Morais ARC, Mata AC, Bogel-Lukasik R (2014) Integrated conversion of agroindustrial residue with high pressure CO_2 within the biorefinery concept. Green Chem 16:4312–4322
13. Kamm B, Kamm M (2007) Biorefineries – multi product processes. Adv Biochem Eng Biotechnol 105:175–204
14. Climent MJ, Corma A, Iborra S (2011) Converting carbohydrates to bulk chemicals and fine chemicals over heterogeneous catalysts. Green Chem 13:520–540
15. Wu B, Zhang YM, Wang HP (2008) Aqueous biphasic systems of hydrophilic ionic liquids plus sucrose for separation. J Chem Eng Data 53:983–985
16. Ferreira AM, Esteves PDO, Boal-Palheiros I, Pereiro AB, Rebelo LPN, Freire MG (2016) Enhanced tunability afforded by aqueous biphasic systems formed by fluorinated ionic liquids and carbohydrates. Green Chem 18:1070

17. Chen Y, Zhang S (2010) Phase behavior of (1-alkyl-3-methyl imidazolium tetrafluoroborate +6-(hydroxymethyl) oxane-2, 3, 4, 5-tetrol+water). J Chem Eng Data 55:278–282
18. Zhang YQ, Zhang SJ, Chen YH, Zhang JM (2007) Aqueous biphasic systems composed of ionic liquid and fructose. Fluid Phase Equilib 257:173–176
19. Wu B, Zhang Y, Wang H (2008) Phase behavior for ternary systems composed of ionic liquid + saccharides + water. J Phys Chem B 112:6426–6429
20. Chen Y, Meng Y, Zhang S, Zhang Y, Liu X, Yang J (2010) Liquid – liquid equilibria of aqueous biphasic systems composed of 1-butyl-3-methyl imidazolium tetrafluoroborate + sucrose/maltose + water. J Chem Eng Data 55:3612–3616
21. Wu B, Zhang Y, Wang H, Yang L (2008) Temperature dependence of phase behavior for ternary systems composed of ionic liquid + sucrose + water. J Phys Chem B 112:13163–13165
22. Freire MG, Louros CLS, Rebelo LPN, Coutinho JAP (2011) Aqueous biphasic systems composed of a water-stable ionic liquid plus carbohydrates and their applications. Green Chem 13:1536–1545
23. Chen YH, Wang YG, Cheng QY, Liu XL, Zhang SJ (2009) Carbohydrates-tailored phase tunable systems composed of ionic liquids and water. J Chem Thermodyn 41:1056–1059
24. Chen Y, Meng Y, Yang J, Li H, Liu X (2012) Phenol distribution behavior in aqueous biphasic systems composed of ionic liquids–carbohydrate–water. J Chem Eng Data 57:1910–1914
25. Freire MG, Claudio AFM, Araujo JMM, Coutinho JAP, Marrucho IM, Lopes JNC, Rebelo LPN (2012) Aqueous biphasic systems: a boost brought about by using ionic liquids. Chem Soc Rev 41:4966–4995
26. Kaul A, Pereira RA, Asenjo JA, Merchuk JC (1995) Kinetics of phase separation for polyethylene glycol-phosphate two-phase systems. Biotechnol Bioeng 48:246–256
27. Othmer DF, Tobias PE (1942) Liquid-liquid extraction data-toluene and acetaldehyde systems. Ind Eng Chem 34:690–692
28. Katayanagi H, Nishikawa K, Shimozaki H, Miki K, Westh P, Koga Y (2004) Mixing schemes in ionic liquid-H_2O systems: a thermodynamic study. J Phys Chem B 108:19451–19457
29. Mason PE, Neilson GW, Enderby JE, Saboungi ML, Brady JW (2005) Structure of aqueous glucose solutions as determined by neutron diffraction with isotopic substitution experiments and molecular dynamics calculations. J Phys Chem B 109:13104–13111
30. Kamlet MJ, Taft RW (1976) Solvatochromic comparison method.1. Beta-scale of solvent Hydrogen-bond acceptor (Hba) basicities. J Am Chem Soc 98:377–383
31. Lungwitz R, Friedrich M, Linert W, Spange S (2008) New aspects on the hydrogen bond donor (HBD) strength of 1-butyl-3-methylimidazolium room temperature ionic liquids. New J Chem 32:1493–1499
32. Freire MG, Carvalho PJ, Gardas RL, Marrucho IM, Santos LM, Coutinho JA (2008) Mutual solubilities of water and the $[C_nmim][Tf2N]$ hydrophobic ionic liquids. J Phys Chem B 112:1604–1610
33. Freire MG, Neves CM, Carvalho PJ, Gardas RL, Fernandes AM, Marrucho IM, Santos LM, Coutinho JA (2007) Mutual solubilities of water and hydrophobic ionic liquids. J Phys Chem B 111:13082–13089
34. Deng Y, Long T, Zhang D, Chen J, Gan S (2009) Phase diagram of [Amim] Cl+salt aqueous biphasic systems and its application for [Amim] Cl recovery. J Chem Eng Data 54:2470–2473
35. Forciniti D, Hall C, Kula M-R (1991) Influence of polymer molecular weight and temperature on phase composition in aqueous two-phase systems. Fluid Phase Equilib 61:243–262
36. Murugesan T, Perumalsamy M (2005) Liquid-liquid equilibria of poly(ethylene glycol) 2000 + sodium citrate plus water at (25, 30, 35, 40, and 45) degrees C. J Chem Eng Data 50:1392–1395
37. Gardas RL, Freire MG, Carvalho PJ, Marrucho IM, Fonseca IMA, Ferreira AGM, Coutinho JAP (2007) High-pressure densities and derived thermodynamic properties of imidazolium-based ionic liquids. J Chem Eng Data 52:80–88
38. Martinus N, Crawford D, Sinclair D, Vincent CA (1977) The extended Jones-Dole equation. Electrochim Acta 22:1183–1187

39. Dey PC, Motin MA, Biswas TK, Huque EM (2003) Apparent molar volume and viscosity studies on some carbohydrates in solutions. Monatsh Chem 134:797–809
40. Fan J, Fan YC, Pei YC, Wu K, Wang JJ, Fan MH (2008) Solvent extraction of selected endocrine-disrupting phenols using ionic liquids. Sep Purif Technol 61:324–331
41. Kanetake WT, Sasaki M, Goto M (2007) Decomposition of a lignin model compound under hydrothermal conditions. Chem Eng Technol 30:1113–1122
42. Morais ARC, Bogel-Lukasik R (2013) Green chemistry and the biorefinery concept. Sustain Chem Process 1:18
43. Kamm B, Kamm M (2004) Principles of biorefineries. Appl Microbiol Biotechnol 64:137–145

Chapter 4
ABS Composed of Ionic Liquids and Polymers

Rahmat Sadeghi

Abstract Polymer–ionic liquid aqueous biphasic systems (polymer–IL ABS) can be formed by the combination of distinct pairs of polymers and ionic liquids (ILs) in aqueous media. These ABS have a series of advantages relative to the conventional polymer–polymer or polymer–salt ABS. Depending on the structural features of polymers and ILs, both species seem to display the ability to act as salting-out agents. Based on a compilation and analysis of the data hitherto reported, the main issues which govern the phase behaviour of these systems are here discussed. In this respect, the effects of the chemical structure and molecular weight of polymer and IL and temperature on the liquid–liquid equilibria are addressed. The molecular-level mechanisms behind the formation of these ABS are further highlighted and discussed based on liquid–liquid and vapour–liquid equilibria behaviour of ternary polymer–IL–water systems. It is shown that the salting-in/salting-out effects in these systems are controlled by a complex interplay of polymer–IL, polymer–water and IL–water interactions.

Keywords Aqueous biphasic systems • Ionic liquids • Polymers • Salting-in/salting-out • Phase diagrams • Isopiestic • Binodal curves

4.1 Introduction

The oldest types of aqueous biphasic systems (ABS) are polymer–salt and polymer–polymer ABS. In these systems, the more hydrophobic solute is usually salted out by the more hydrophilic one, and the ability for ABS formation increases as the difference between the hydrophobic properties of the two solutes increases [1, 2]. The hydrophobicity of ionic liquids (ILs) can be tailored by a proper manipulation of the cation/anion design and their combinations, and therefore, by virtue of their tailoring ability, ILs practically cover the whole hydrophilicity/hydrophobicity range. In typical polymer–salt ABS, both the polymer and salt can be replaced by a particular IL to form a new IL–polymer or IL–salt ABS. In

---

R. Sadeghi (✉)
Department of Chemistry, University of Kurdistan, Sanandaj, Iran
e-mail: rahsadeghi@yahoo.com; rsadeghi@uok.ac.ir

© Springer-Verlag Berlin Heidelberg 2016
M.G. Freire (ed.), *Ionic-Liquid-Based Aqueous Biphasic Systems*, Green Chemistry and Sustainable Technology, DOI 10.1007/978-3-662-52875-4_4

IL–salt ABS, the IL (more hydrophobic solute) is salted out by the salt (more hydrophilic solute) to form an ABS with an upper IL-rich phase and a lower salt-rich phase. However, in the case of IL–polymer ABS, depending on the hydrophobic properties of the IL and polymer, either the salting-out of a hydrophobic polymer by a hydrophilic IL or the salting-out of a hydrophobic IL by a hydrophilic polymer can occur leading to the formation of ABS. Furthermore, for polymer–IL ABS, depending on the type of IL, both "upper polymer-rich and lower IL-rich phases" and "upper IL-rich and lower polymer-rich phases" can be formed [3]. Another surprising behaviour of these systems is that, similar to polymer–polymer ABS, polymer–IL ABS can be created either with polymer–IL mixtures of limited solubility or by completely miscible pairs.

The thermodynamic properties of aqueous solutions of ILs depend on their chemical structures can be similar to those of either aqueous polymer solutions or aqueous salt solutions [4]. As an example, in Fig. 4.1, the comparison between the water activity data of the $[C_4mim]Br + H_2O$, $[C_4mim][HSO_4] + H_2O$, $NaCl + H_2O$ and $PEG200 + H_2O$ systems is presented at $T = 298.15$ K. As can be seen, at the same molality, the values of the water activity of NaCl aqueous solutions are closer to those of $[C_4mim][HSO_4]$ aqueous solutions than to those of $[C_4mim]Br$ aqueous solutions. However, the values of the water activity for the $PEG200 + H_2O$ system are closer to those of the $[C_4mim]Br + H_2O$ system [2–4].

The anion HSO_4^- can establish hydrogen bonds with water molecules and it hydrates more water molecules than the anion Br^-. Therefore, it is expected a

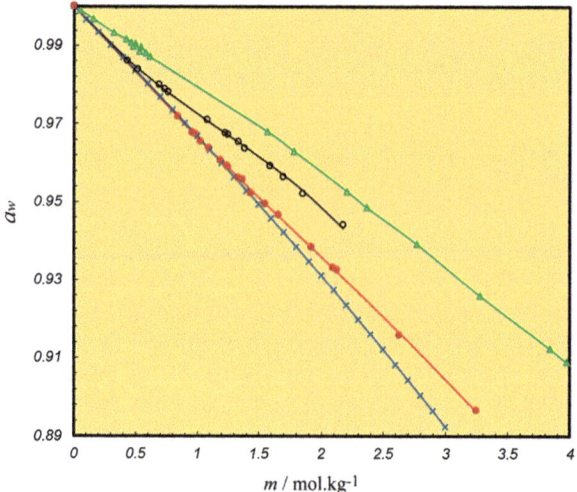

Fig. 4.1 Water activity, a_w, of binary aqueous solutions against molality, m, at 298.15 K: ○, $[C_4mim]Br + H_2O$ [1]; ●, $[C_4mim][HSO_4] + H_2O$ [1]; ×, $NaCl + H_2O$ [5]; △, $PEG200 + H_2O$ [6]; lines indicate Pitzer's model [7]

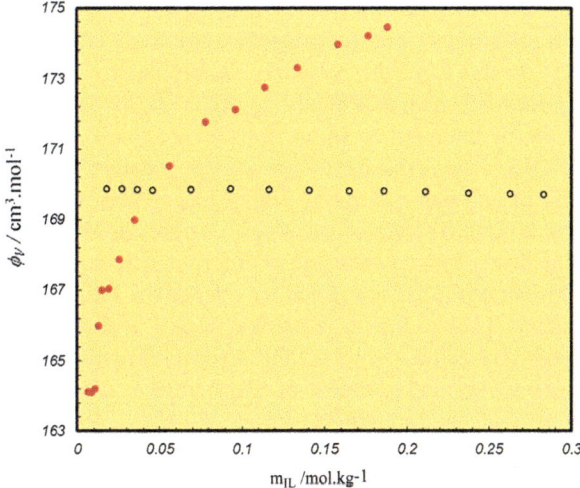

Fig. 4.2 Apparent molar volume, ϕ_V, of [C$_4$mim]Br and [C$_4$mim][HSO$_4$] in aqueous solutions against molality of IL, m_{IL}, at 298.15 K: ○, [C$_4$mim]Br + H$_2$O [8]; ●, [C$_4$mim][HSO$_4$] + H$_2$O [1]

greater depression of the vapour pressure of [C$_4$mim][HSO$_4$] solutions than those of [C$_4$mim]Br solutions. Furthermore, protonation/deprotonation equilibria associated with HSO$_4^-$ lead to the production of SO$_4^{2-}$ which strongly hydrates in aqueous solutions. The IL molality dependence of the water activity or vapour pressure depression follows the order [C$_4$mim][HSO$_4$] > [C$_4$mim]Br, which implies that the HSO$_4^-$–water interactions are stronger than Br$^-$–water interactions. The behaviour of the investigated ILs can be further compared if the plots of the apparent molar volume, ϕ_V, as a function of IL molality are considered (Fig. 4.2).

As can be depicted from Fig. 4.2, similar to simple electrolytes [9–12], the apparent molar volumes of [C$_4$mim][HSO$_4$] in water increase by increasing the IL molality. For low concentrations of IL, the small volume is attributed to the strong attractive interactions due to the hydration of ions. By increasing the IL concentration, the ion–ion interactions increase, and the positive initial slope of ϕ_V against IL concentration is attributed to these interactions. However, the values of ϕ_V of [C$_4$mim]Br in water slightly decrease by increasing the IL molality. This behaviour is similar to that observed in aqueous polymer solutions [13]. Therefore, it can be concluded that the properties of the ILs [C$_4$mim][HSO$_4$] and [C$_4$mim]Br, respectively, are similar to simple electrolytes and water-soluble polymers. In the case of polymer–salt ABS, the polymer can be replaced by [C$_4$mim]Br to form a new IL–salt ABS (such as [C$_4$mim]Br-Na$_3$ citrate ABS) [8, 14], and the salt can be replaced by [C$_4$mim][HSO$_4$] to form a new IL–polymer ABS (such as PPG–[C$_4$mim][HSO$_4$] ABS) [1]. Since both ILs and water-soluble polymers have the flexibility to be designed either as hydrophobic or hydrophilic solutes, it is possible to obtain two types of polymer–IL ABS. The first type can be formed by mixing one hydrophilic IL and one hydrophobic polymer in an aqueous solutions (such as PPG-[C$_4$mim][HSO$_4$] ABS) [1]. In this case, the polymer is salted out by the IL. In the second type of polymer–IL ABS, such as PEG–[P$_{4444}$]Cl ABS, the IL (more hydrophobic solute) is salted out by the polymer (more hydrophilic solute) [15].

IL–polymer ABS have a series of advantages relative to the conventional polymer–polymer or polymer–salt ABS. While the polymer–polymer or polymer–salt ABS display a limited range of polarities or hydrogen-bond basicity (and/or hydrogen-bond acidity) of their coexisting phases, polarities of coexisting phases of polymer–IL ABS can easily be controlled by selecting a specific IL with a proper hydrophilic character (due to the large number of ILs available). In other words, polymer–IL-based ABS exhibit a much wider hydrophilic–hydrophobic range than conventional systems reported to date [16]. Furthermore, salt crystallization problems can be avoided when employing an IL which melts below room temperature. Even in the case of an IL melting above room temperature, the presence of water will largely preclude its solidification, that is, the saturation levels of ILs in aqueous solutions are usually significantly higher than those observed with common salts. In addition, ILs can optionally be designed to have a low corrosive character relative to the inorganic salts, which largely hinder their use in industrial processes [15]. Although there is no experimental support for the related claims, the idea of polymer–IL ABS was present in some patented examples in the mid-2000s [17, 18]. Approximately at the same time, the changes of the fluid phase behaviour of PEG aqueous solutions by the addition of different ILs were reported [19, 20]. Recently, liquid–liquid equilibria phase diagrams were reported for several PPG-IL, PEG-IL and EO$_{10}$PO$_{90}$-IL ABS. Table 4.1 summarizes the combinations of ILs and distinct polymers for the formation of ABS studied hitherto.

Although the binodal curves have been reported for all polymer–IL ABS given in Table 4.1, less information has been reported in the literature about the tie-lines (compositions of the two phases in equilibrium) of these systems. Table 4.2 summarizes the polymer–IL ABS for which the tie-lines have been reported in the literature.

For the PPG400–ILs, PEG4000–ILs and PEG600–[Ch][L-ma], [Ch][Fum] and [Ch][Suc] ABS at 298.15 K, the upper and lower phases are polymer rich and IL rich, respectively. However, in the case of PEG600–[Ch][Ox], [Ch][Mal] and [Ch][Glu] ABS, the upper and lower phases are IL rich and polymer rich, respectively. As can be seen, the more hydrophobic polymers such as PPG or high molecular weight of PEG tend to enrich the upper phase.

In this chapter, the mechanisms for polymer–IL ABS formation will be discussed and then the effects of different parameters, such as temperature, type and molecular weight of polymer and IL on the liquid–liquid equilibria phase diagram of these systems will be presented.

4 ABS Composed of Ionic Liquids and Polymers

Table 4.1 ABS composed of ILs and polymers reported in the literature

Polymer	Ionic liquid
PEG2000 [15]	[im]Cl, [mim]Cl, [C_1mim]Cl, [C_2mim]Cl, [C_4mim]Cl, [C_6mim]Cl, [C_8mim]Cl, [C_4mim][CF_3SO_3], [C_4mim]Br, [C_4mim][CH_3SO_3], [C_4mim][CH_3CO_2], [C_2mim]Cl, [C_2mim][HSO_4], [C_2mim][CH_3SO_3], [C_2mim][CH_3CO_2], [C_2mim][$(CH_3)_2PO_4$], [C_4mpy]Cl, [C_4mpyr]Cl, [C_4mpip]Cl, [P_{4444}]Cl, [amim]Cl, [HOC_2mim]Cl,
PEG1000 [15], PEG3400 [15], PEG4000 [15]	[C_4mim]Cl
PPG400 [1], PPG725 [1]	[C_4mim][HSO_4]
PPG400 [21]	[amim]Cl, [C_4mim][CH_3CO_2], [C_4mim]Cl
PPG1000 [21]	[amim]Cl, [C_4mim]Cl
PPG400 [22]	[C_2mim]Br, [C_4mim]Br
PPG400 [23]	Tri-cholinium citrate ([Ch]$_3$[Cit]), di-cholinium oxalate ([Ch]$_2$[Ox]), cholinium glycolate ([Ch][Gly]), cholinium lactate ([Ch][Lac]), cholinium butyrate ([Ch][But]), cholinium formate ([Ch][For]), cholinium propionate ([Ch][Pro]), cholinium acetate ([Ch][Ac])
PPG400 [24], PPG1000 [24], $EO_{10}PO_{90}$ [24]	[Ch][Pro], cholinium chloride ([Ch]Cl), [Ch][Lac], [Ch][Gly]
PEG600 [25]	Cholinium oxalate ([Ch][Ox]), cholinium malonate ([Ch][Mal]), cholinium succinate ([Ch][Suc]), cholinium glutarate ([Ch][Glu]), cholinium L-malate ([Ch][L-ma]), cholinium fumarate ([Ch][Fum])
PEG4000 [25]	[Ch][Mal], [Ch][Glu], [Ch][$_L$-ma], [Ch][Fum], [Ch][Suc], [Ch][DHcit]
PEG400 [26], PEG600 [26], PEG1000 [26]	[Ch][Lac], [Ch][Gly], [Ch][Ac], cholinium dihydrogencitrate ([Ch][DHcit]), cholinium bicarbonate ([Ch][Bic]), cholinium dihydrogenphosphate ([Ch][DHph]), cholinium bitartrate ([Ch][Bit]), [Ch]Cl
PEG1500 [27]	[C_4mim]Cl
PEG1500 [16]	[C_4mpip]Cl, [C_4mpyrr]Cl, [C_2mim]Cl, [C_4mim]Cl

Table 4.2 Polymer–IL ABS for which the tie-lines at 298.15 K have been reported in the literature

Polymer	Ionic liquid
PPG400 [23]	[Ch][Pro], [Ch][Gly]
PPG400 [24]	[Ch][Pro], [Ch][Gly], [Ch][Lac]
PPG400 [22]	[C_2mim]Br
PPG400 [21]	[amim]Cl, [C_4mim]Cl, [C_4mim][Ac]
PEG600 [25]	[Ch][Ox], [Ch][Mal], [Ch][Suc], [Ch][Glu], [Ch][L-ma], [Ch][Fum]
PEG4000 [25]	[Ch][Mal], [Ch][Glu], [Ch][$_L$-ma], [Ch][Fum], [Ch][Suc], [Ch][DHCit]

4.2 Mechanisms of Phase Separation in Polymer–IL ABS

The phase behaviour in polymer–IL ABS is not only dominated by the ability of the solutes to interact with water, but also it is a direct consequence of the favourable (or non-favourable) interactions that occur between the polymer and IL in a large extent. Both factors affect the water activity of the systems, and by studying the vapour–liquid equilibria behaviour of polymer–IL–water systems, valuable information can be derived. In fact, for the understanding of interactions in liquids, the activities of different components are of great interest. They are the most relevant thermodynamic reference data, and they are often the starting point of any modelling. Furthermore, water is the main component in ABS and the dissolved components have a high affinity for it. Thus, water activity in these systems can provide useful information about the salting-in/salting-out mechanisms produced by the addition of salting-out or salting-in inducing ILs to polymers aqueous solutions. The isopiestic method is the most accurate and a simple experimental technique available for measuring the solvent activity of solutions. It is based on the phenomenon that different solutions, when connected through the vapour space, approach equilibrium by transferring solvent mass by distillation. Equilibrium has been established once the temperature, pressure and solvent chemical potential are uniform throughout the system, provided that no concentration gradients exist in the same liquid phase.

For ternary aqueous electrolyte A + electrolyte B solutions under isopiestic equilibrium, the following empirical linear isopiestic relation (Zdanovskii–Stokes–Robinson rule) was proposed by Zdanovskii [28]:

$$\frac{m_A}{m_A^0} + \frac{m_B}{m_B^0} = 1 \, (a_w = \text{constant}) \tag{4.1}$$

where m_i is the molality of solute i in the ternary solution and m_i^o is the molality of solute i in the binary solution of equal a_w.

Stokes and Robinson [29] theoretically derived this equation for isopiestic mixed non-electrolyte aqueous solutions from the semi-ideal hydration model. According to Zdanovskii's rule, binary aqueous solutions having equal water activity, when mixed in any proportion, produce a ternary aqueous solution with the same water activity. In other words, many solutions with equal water activity exhibit no net effective interaction when mixed, that is, changes of hydration between the dissolved components on mixing are apparently absent. Interactions between the solutes and the solvent occur and can be important in binary solutions, but interactions between the solutes in mixed solutions are not evident, and the behaviour is termed semi-ideal. In fact, the thermodynamic behaviour of the mixed solution according to equation (1) is as simple as that of an ideal solution, that is, the constituent binary solutions mixed ideally under isopiestic equilibrium. According to Eq. 4.1, the plot of m_A against m_B is linear with a negative slope. Studying the vapour–liquid equilibria behaviour of different ternary aqueous solute A + solute B

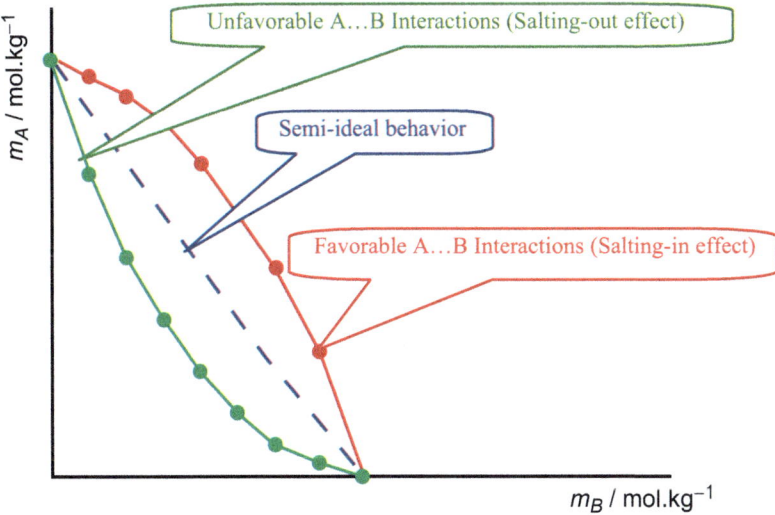

Fig. 4.3 The constant water activity lines for different solutions under isopiestic equilibrium

solutions [1, 2, 14, 30], which were or were not capable of inducing phase separation, showed that there is a relation between the salting-out or salting-in effects of solute A on the solute B aqueous solutions and their vapour–liquid equilibria behaviour. This relation is summarized in Fig. 4.3. As can be seen, in the case of the salting-out effect, the constant water activity lines have a concave slope, but in the case of the salting-in effect, the constant water activity lines have a convex slope.

For completely miscible nonvolatile solute A + nonvolatile solute B + water solutions (salting-in effect), as a consequence of preferential A–B interactions, the interaction of each solute with water becomes less favourable in the presence of the other solute. Therefore, water molecules are allowed to relax to the bulk state and more free water molecules would be available in respect to the semi-ideal behaviour in which the solute–water interactions in the ternary solution are the same as those in the binary solutions. In this context, the concentrations of solutes A and B in a ternary solution which is in isopiestic equilibrium with certain binary solute A + water and solute B + water solutions are larger than those expected in the case of semi-ideal solution, and then, these systems show positive deviations from Eq. 4.1. In these systems, the activity coefficient of solute A in aqueous solution is lowered by the presence of B. This is an example of "salting-in" of A by B in which increases on the mutual solubilities of A–water are observed. On the other hand, in the case of aqueous solute A + solute B systems which form ABS (salting-out effect), because of the unfavourable A–B interactions (relative to A– or B–water interactions), the solute A–water interaction becomes more stable in the presence of solute B, and therefore, in these ternary systems, less free water molecules would be available in respect to the semi-ideal behaviour, and then these systems show

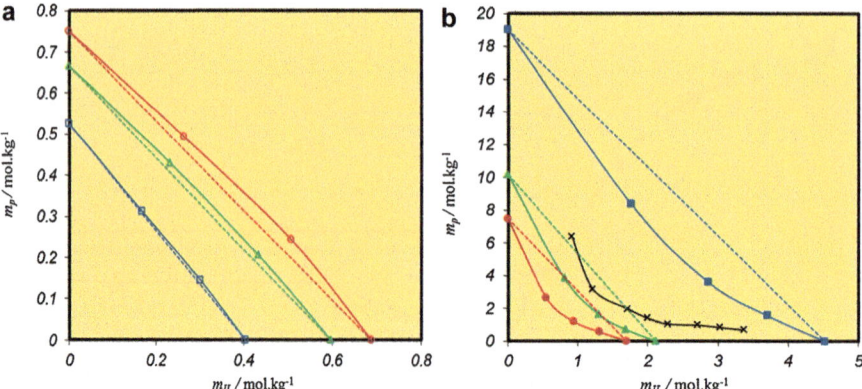

Fig. 4.4 (a) Molality of PEG400, m_p, against molality of [C$_4$mim]Br, m_{IL}, for the constant water activity lines of PEG400 (p) + [C$_4$mim]Br (IL) + H$_2$O (w) system at 298.15 K: ○, $a_w = 0.9800$; △, $a_w = 0.9829$; □, $a_w = 0.9885$. (b) Molality of PPG400, m_p, against molality of [C$_4$mim][HSO$_4$], m_{IL}, for the constant water activity lines of PPG400 (p) + [C$_4$mim][HSO$_4$] (IL) + H$_2$O (w) system at 298.15 K: ■, $a_w = 0.87712$; ▲, $a_w = 0.9326$; ●, $a_w = 0.9461$; ×, binodal curve, calculated by Eq. 4.1 (Reprinted with the permission from Ref. [1]. Copyright 2013 American Chemical Society)

negative deviations from Eq. 4.1. Since the association of solute A with solute B is a highly unfavourable process, they exclude themselves from the vicinity of each other due to their preferential hydration. By increasing the concentration of the solutes, this exclusion will increase, and ultimately, the system could reach a state where, for entropic reasons, phase formation would become favourable. As a result, these systems separate into an A-rich aqueous phase and a B-rich aqueous phase. Formation of aqueous two-phase systems can only involve partial dehydration of solutes, where both solutes A and B remain strongly associated with water but exclude each other by separating into two phases. As an example, the experimental constant water activity lines of PEG400 + [C$_4$mim]Br + H$_2$O (not capable of inducing phase separation) and PPG400 + [C$_4$mim][HSO$_4$] + H$_2$O (capable of inducing phase separation) systems along with the results of Eq. 4.1 are shown in Fig. 4.4. Ternary aqueous mixtures of PEG400 + [C$_4$mim]Br and PPG400 + [C$_4$mim][HSO$_4$], respectively, show positive and negative deviations from Eq. 4.1. Aqueous solutions of PPG400 + [C$_4$mim][HSO$_4$] form ABS and for which the binodal curve is also plotted in Fig. 4.4.

Recently, Tomé et al. [27] showed that the experimental ternary phase diagrams of the system composed of PEG1500, [C$_4$mim]Cl and water, at 323.15 K and 333.15 K, are of type 0—characterized by an immiscibility gap in the ternary region, while all the binary PEG–IL mixtures are fully miscible. The authors [27] demonstrated that, unlike what happens in IL–salt-based ABS, the formation of PEG–IL-based ABS is controlled by the IL anion solvation by water which leads to the destruction of the hydrogen bonds established between the IL anion and the hydroxyl groups of the polymer. The authors [27] concluded that ABS formation in

PEG–IL systems does not result from a salting-out effect of the polymer by the IL or of the IL over the polymer; instead, the PEG–IL-based ABS formation is a result of what is labelled as a "washing-out" phenomenon. More recently, in order to further delve into the interactions that govern the mutual solubilities between ILs and PEGs and the formation of PEG–IL-based ABS, the same authors [31] performed ^{1}H NMR spectroscopy in combination with classical molecular dynamics (MD) simulations for binary mixtures of TEG and $[C_n\text{mim}]$Cl ILs and for the corresponding ternary TEG/$[C_n\text{mim}]$Cl/water solutions at 298.15 K. The MD results for binary TEG + IL systems showed that the hydrogen bonding between Cl^- and the –OH group of TEG is the main interaction controlling the mutual solubilities between the IL and the polymer. Additionally, the formation of IL–PEG-based ABS is controlled by a competition between water and chloride to interact with the hydroxyl group of TEG. Introduction of water into the PEG–IL mixtures leads to the complete replacement of the hydrogen bond established between the chloride anion and the TEG hydroxyl group by Cl^-–water hydrogen bonding. As a result, the formation of PEG/IL-based ABS becomes controlled by the hydrophobic interactions occurring between the cation alkyl chain and the less polar moieties of the polymer.

4.3 Effect of Different Parameters on the Phase Behaviour of Polymer–IL ABS

Generally, polymer–IL ABS can be classified into two types. In the first type of ABS, such as PPG–$[C_4\text{mim}]$Cl ABS, a hydrophobic polymer is salted out by a hydrophilic IL in aqueous solution. The most part of polymer–IL ABS reported up to date belong to this category. However, in the second type of polymer–IL ABS, such as PEG–$[C_4\text{mpip}]$Cl ABS, the salting-out of a hydrophobic IL by a hydrophilic polymer seems to occur in aqueous solutions. In the case of the former, the ability to induce ABS increases as the hydrophilicity of the IL and/or hydrophobicity of the polymer is increased. However, in the case of the latter, the ability to induce ABS increases as the hydrophilicity of the polymer and/or hydrophobicity of the IL is increased. As a result, in addition to the IL–water and polymer–water interactions (governed by the hydrophobicity of solutes and that rule the phase behaviour), polymer–IL interactions cannot be discarded and could additionally control the phase diagram behaviour. In general, the larger the immiscibility between the IL and polymer, the greater the ability of the IL to induce the polymer separation in aqueous media. In summary, the salting-in/salting-out effects in these systems are controlled by a complex interplay of polymer–IL, polymer–water and IL–water interactions. These interactions can be influenced by the structure of polymer and IL, molecular weight of polymer and temperature. The effect of these parameters on the liquid–liquid equilibria phase behaviour of polymer–IL ABS is discussed in detail below.

4.3.1 Effect of the Polymer

Among all the water-soluble polymers available, ABS formed by ILs were only studied with PPG and PEG. In the case of hydrophilic IL + hydrophobic polymer + water systems, in which the polymer is salted out by the IL, for the same polymer molecular weight, PPG is much more likely to undergo phase separation in aqueous media than PEG. In Fig. 4.5, the binodal curve of PPG1000 + [C$_4$mim]Cl ABS [21] is compared to that of PEG1000 + [C$_4$mim]Cl ABS [15] at 298.15 K. Figure 4.5 also shows the binodal curves of PEG400, PEG600, PEG1000 and PPG400 + [Ch]Cl + water ABS [24, 26] at 298.15 K. It should be mentioned that the solubility curves are presented according to the number of moles of solute in 1 kg of water. As can be seen, for a common IL and polymers with the same molar mass, the polymer concentration required for the formation of two phases increases in the order PEG >> PPG, which implies that the ability of polymers to induce phase separation follows the trend PPG >> PEG. PPG contains a longer hydrocarbon chain, while the side-chain methyl groups in PPG also hinder hydrogen bonding between water molecules and the ether oxygen atoms. Thus, PPG is more easily salted out by a salting-out-inducing species and more easily forms ABS (compared to PEG). This trend also agrees with the affinity/miscibility patterns observed in polymer–IL binary systems in the sense that PPG is less soluble in ILs than PEG [32].

According to Fig. 4.6, the negative deviation of the constant water activity lines from Eq. 4.1 for the systems containing [C$_4$mim]Br and PPG400 is larger than that containing [C$_4$mim]Br and PEG400. In fact, more hydrophobic polymers are more easily excluded from the near-surface region of the ions. Furthermore, the side-chain methyl groups in PPG hinder the PPG–ion interactions, and then the salting-in effect of ions on PPG400 is smaller than those on PEG400. Both factors lead to a

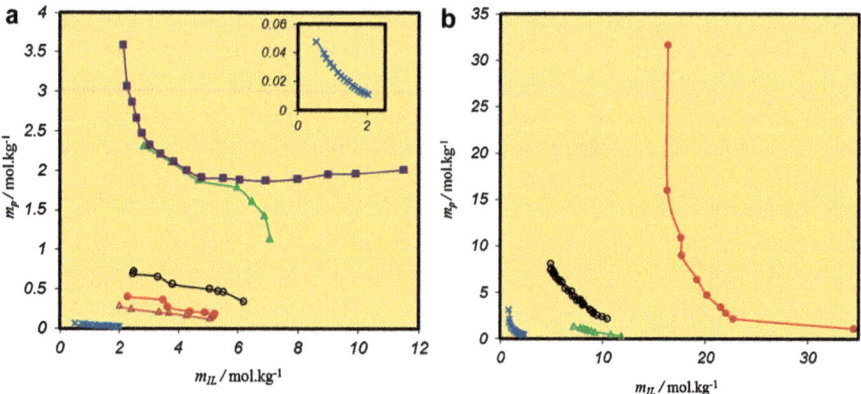

Fig. 4.5 (**a**) Binodal curves for polymer (p) + [C$_4$mim]Cl (IL) + H$_2$O (w) systems at 298.15 K: ■, PPG400 [21]; ▲, PEG1000 [15]; ○, PEG2000 [15]; ●, PEG3400 [15]; △, PEG4000 [15]; ×, PPG1000 [21]. (**b**) Binodal curves for polymer (p) + [Ch]Cl (IL) + H$_2$O (w) systems at 298.15 K: ●, PEG400 [26]; ○, PEG600 [26]; △, PEG1000 [26]; ×, PPG400 [24]

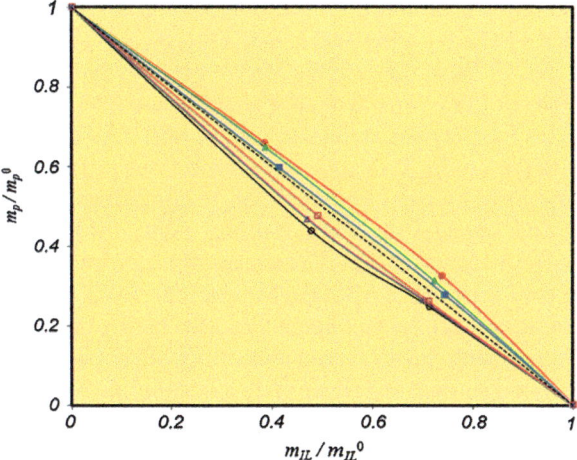

Fig. 4.6 Plot of m_p/m_p^0 against m_{IL}/m_{IL}^0 for constant water activity lines of PPG400 (p) + [C$_4$mim]Br (IL) + H$_2$O (w) (*empty symbols*) and PEG400 (p) + [C$_4$mim]Br (IL) + H$_2$O (w) (*filled symbols*) systems at 298.15 K: ● and ○, $a_w = 0.9800$; ▲ and △, $a_w = 0.9829$; ■ and □, $a_w = 0.9885$, calculated by Eq. 4.1 (Reprinted with the permission from Ref. [1]. Copyright 2013 American Chemical Society)

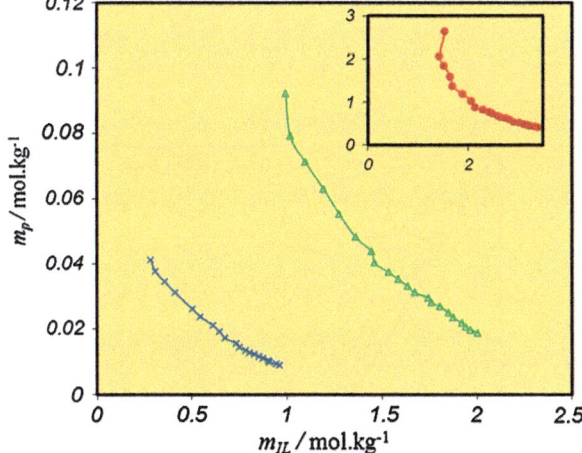

Fig. 4.7 Binodal curves of polymer (p) + [Ch]Cl (IL) + H$_2$O (w) ABS at 288.15 K: ●, PPG400; △, PPG1000; ×, EO$_{10}$PO$_{90}$ (Reprinted from Ref. [24]. Copyright 2013, with permission from Elsevier)

larger negative deviation of constant water activity lines from the linear isopiestic relation.

The binodal curves determined at 288.15 K for ABS formed by polymers (PPG400, PPG1000 and EO$_{10}$PO$_{90}$) + [Ch]Cl + H$_2$O are given in Fig. 4.7. As can be seen, the phase-forming ability of the polymers in cholinium IL-based ABS follows the order: PPG400 < PPG1000 < EO$_{10}$PO$_{90}$.

Figures 4.5 and 4.7 show that, similarly to polymer–polymer and polymer–salt ABS, for both PPG or PEG + imidazolium-based ILs and PPG or PEG + cholinium-based ILs ABS, as the polymer molecular weight is increased, the binodal curves become closer to the origin. It is known that the closer to the origin the binodal curve is, the lower the polymer concentration required for the formation of two phases and then the stronger the phase-forming ability of the polymer to create ABS. Hydrophobicity increases with the increase of the polymer molecular weight, thus facilitating the formation of ABS. Higher molecular weight polymers present a lower affinity for water and are preferentially salted out by the IL. This behaviour might also be related with the decreased miscibility of polymers with ILs due to a decrease of the hydrogen-bonding donor sites *per* molecule (i.e. the terminal –OH groups on each polymer molecule) [26]. Rodriguez et al. [3] demonstrated that an increase in the molecular weight of the polymer leads to larger immiscibility gaps with chloride-based ILs. Hence, higher molecular weight polymers are less soluble in ILs and are more readily separated from aqueous media. These trends also agree with the negative deviation of constant water activity lines from the linear isopiestic relation observed in polymer–IL–water systems. As shown in Fig. 4.8, the negative deviations of constant water activity lines from Eq. 4.1 for the PPG + [C$_4$mim] [HSO$_4$] + H$_2$O system increase as the PPG molecular weight increases.

Regarding the effect of the polymer molecular weight on the compositions of the coexisting phases, in Fig. 4.9, the complete phase diagrams are plotted for ABS formed by PEG(600 and 4000) + IL ([Ch][Fum] and [Ch][Suc]) + water at

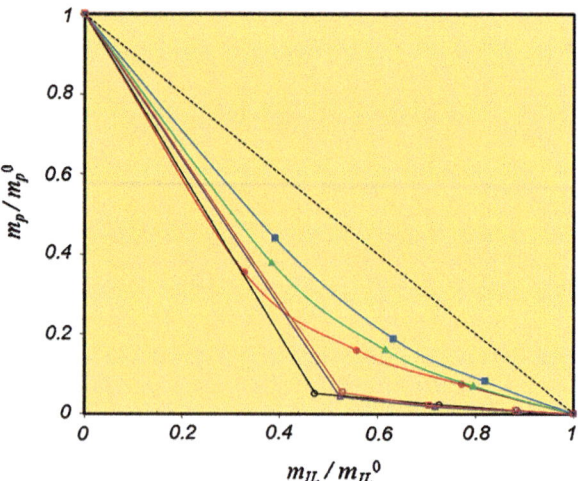

Fig. 4.8 Plot of m_p/m_p^0 against m_{IL}/m_{IL}^0 for constant water activity lines of PPG725 (p) + [C$_4$mim] [HSO$_4$] (IL) + H$_2$O (w) (*empty symbols*) and PPG400 (p) + [C$_4$mim][HSO$_4$] (IL) + H$_2$O (w) (*filled symbols*) systems at 298.15 K: ○, a_w = 0.9926; △, a_w = 0.9894; □, a_w = 0.9873; ●, a_w = 0.9461; ▲, a_w = 0.9326; ■, a_w = 0.8712, calculated by Eq. 4.1 (Reprinted with the permission from Ref. [1]. Copyright 2013 American Chemical Society)

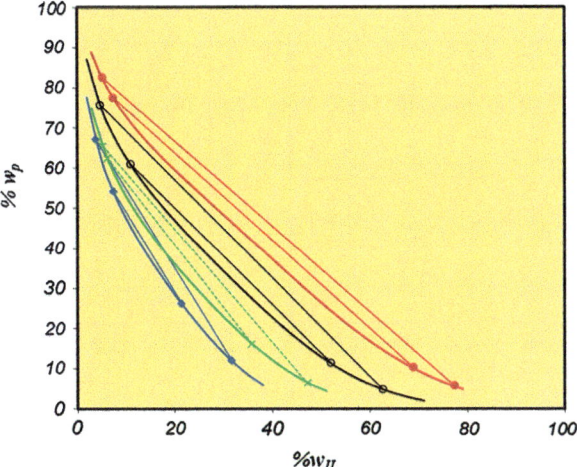

Fig. 4.9 Binodal curves and tie-lines for several PEG+IL ABS at 298.15 K: —●—, PEG600 +[Ch][Suc]; ···×···, PEG4000+[Ch][Suc]; —○—, PEG600+[Ch][Fum]; ▲, PEG4000+[Ch][Fum] (Reprinted with the permission from Ref. [25]. Copyright 2014 American Chemical Society)

298.15 K. For PEG–cholinium-based ILs ABS, the slope and length of the tie-lines, respectively, increase and decrease with an increase on the PEG molecular weight.

4.3.2 Effect of Temperature

Two types of temperature dependency behaviour of binodal curves have been observed for polymer–IL ABS. For the first type, an increase in temperature decreases the immiscibility region (Fig. 4.10). The examples for these systems are PEG2000+[C_2mim]Cl [15], PEG2000+[C_2mim][CH_3CO_2] [15] and PPG400 +[C_2mim]Br [22] ABS. However, for the PPG400+[C_4mim]Br [22], PPG400 +([Ch][Pro], [Ch][Gly], [Ch][Lac] and [Ch]Cl) [23, 24] and PEG600+([Ch]Cl and [Ch][DHph]) [26] ABS, an increase in temperature enhances the immiscibility region (Fig. 4.10).

The reduction of the immiscibility regime with an increase in temperature within the same polymer–IL ABS is in close agreement with that observed in ABS composed of ILs and inorganic salts, carbohydrates or amino acids [33]. This trend is a reflection of the hydrophilic nature of ILs and their capacity to create water–ion complexes and then is a consequence of the typical behaviour from the upper critical solution temperature (UCST) observed in binary mixtures composed of ILs and water [15, 34]. The larger differences among the binodal curves at various temperatures for the PEG2000+[C_2mim][CH_3CO_2] ABS relative to the PEG2000+[C_2mim]Cl ABS are a direct result of the increased salting-out aptitude

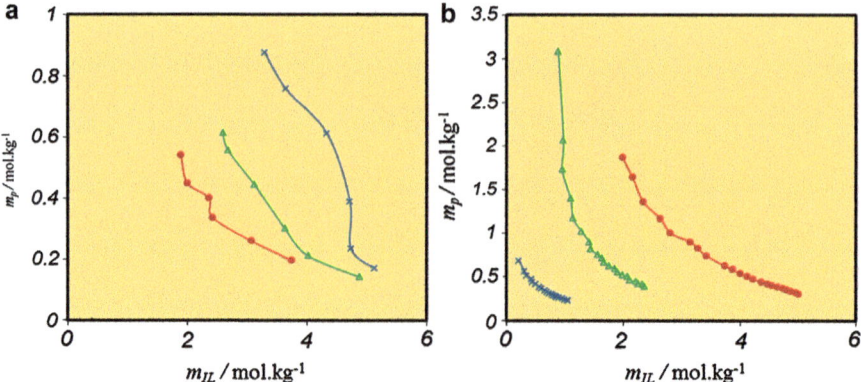

Fig. 4.10 (a) Binodal curves for PEG2000 (p) + [C$_2$mim][CH$_3$CO$_2$] (IL) + H$_2$O (w) ABS at: •, 298.15 K; △, 308.15 K; ×, 323.15 K. Reproduced from Ref. [15] by permission of John Wiley & Sons Ltd. (b) Binodal curves for PPG400 (p) + [Ch]Cl (IL) + H$_2$O (w) ABS at: •, 278.15 K; △, 298.15 K; ×, 318.15 K (Reprinted from Ref. [24]. Copyright 2013, with permission from Elsevier)

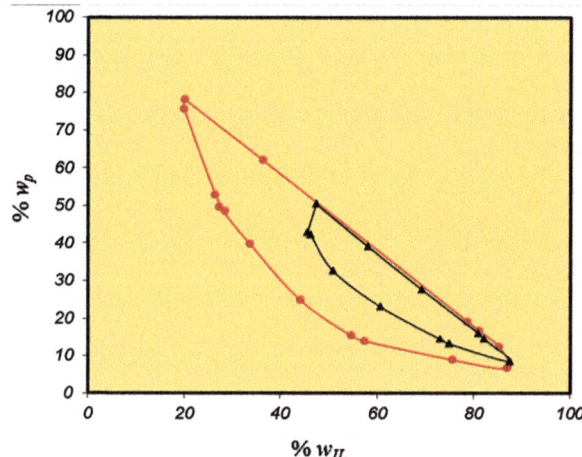

Fig. 4.11 Experimental phase diagrams for the ABS composed of PEG1500 and [C$_4$mim]Cl at: •, 323.15 K; ▲, 333.15 K (Reproduced from Ref. [27] by permission of The Royal Society of Chemistry)

of the acetate anion, which is reproduced in a more pronounced dependence on temperature [15]. Similarly (as shown in Fig. 4.11) the immiscibility loop of PEG1500-[C$_4$mim]Cl-water ABS [27] becomes smaller by increasing temperature, and therefore lower temperatures are favourable for the two-phase formation.

The later temperature dependency behaviour of binodal curves (the reduction of the miscibility region with an increase in temperature) is in close agreement with that observed for some polymer + salt ABS.[1] In these systems, the influence of temperature on the liquid–liquid demixing process seems to be more dominant by the hydrogen-bonding interactions between the polymer and water and, as usually explained, based on the polymer–water lower critical solution temperature (LCST)-type behaviour [23, 26]. For systems dominated by hydrogen bonding, an increase in temperature leads to a decrease in the interaction strengths and consequently to an easier liquid–liquid demixing. It has been shown that an increase in temperature decreases the binary cholinium-based ILs–PEG hydrogen-bonding interactions and facilitates the creation of ABS [26].

As mentioned above, the relative differences in the ABS-forming capability of the PPG400 + [C_2mim]Br and PPG400 + [C_4mim]Br [22] are suppressed as the temperature is increased. [C_4mim]Br, because of its higher hydrophobicity, may produce [C_4mim]Br–PPG hydrophobic interactions *via* its longer cation alkyl side chain which may cause the salting-in effects and then compensate the demixing capability in the ABS. It has been found that, at room temperature, an apparently homogeneous binary PPG400–[C_4mim]Br system splits into two immiscible aqueous phases when some water is added to the system [22] (similar to the PEG1500-[C_4mim]Cl system [27]). In this case, when the temperature is raised, the PPG becomes more hydrophobic and it can be more easily salted out to form ABS. Furthermore, the temperature rise can also decrease the PPG400–[C_4mim]Br interactions.

4.3.3 Effect of the Ionic Liquid

It has been discussed that, in some cases, ILs with a better ability to form ion–water complexes (more hydrophilic ILs) are more effective in promoting polymer–IL ABS. On the other hand, for other ABS, the higher the affinity for water and/or the hydrophilic nature of the IL, the less effective is such an IL in promoting phase separation. It should be kept in mind that these trends seem to be affected by the polymer–IL interactions. The effect of ILs on the phase behaviour of polymer–IL ABS is discussed below on the basis of their chemical structure.

---

[1]In the case of polymer-salt ABS, two types of temperature dependency behaviour of binodal curves have been observed. For some polymer-salt ABS such as PPG-salts [2], PEGDME2000-(NH4)2HPO4, PEG6000-Na3Cit, PVP-Na3Cit, etc., an increase in temperature enhances the immiscibility region in the whole polymer and salt concentration range. However, for some else polymer-salt ABS, such as PEG400-Na2CO3 [2], PEG1000-Na2HPO4, PEGDME250-Na2CO3, PVP-Na2HPO4, etc., the crossing of binodal curves at different temperatures has been observed. In the polymer-rich region an increase in temperature caused the expansion of the one-phase area, while for the salt-rich region the expansion of the two-phase area occurs with an increase in temperature

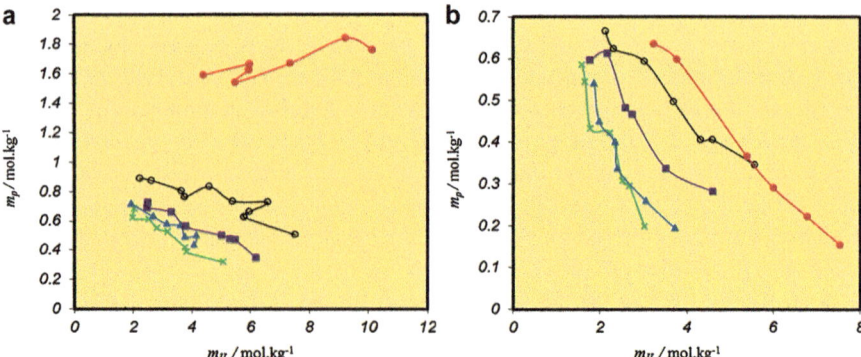

Fig. 4.12 (a) Binodal curves for PEG2000 (p) + [C$_4$mim][X] (IL) + H$_2$O (w) system at 298.15 K for X = ×, [CH$_3$CO$_2$]$^-$; ▲, [CH$_3$SO$_3$]$^-$; ■, Cl$^-$; ○, Br$^-$; ●, [CF$_3$SO$_3$]$^-$. (b) Binodal curves for PEG2000 (p) + [C$_2$mim][X] (IL) + H$_2$O (w) system at 298.15 K for X = ×, [(CH$_3$)$_2$PO$_4$]$^-$; ▲, [CH$_3$CO$_2$]$^-$; ■, [CH$_3$SO$_3$]$^-$; ○, [HSO$_4$]$^-$; ●, Cl$^-$ (Reproduced from Ref. [15] by permission of John Wiley & Sons Ltd)

4.3.3.1 Effect of the IL Anion

The binodal curves of several PEG2000–[C$_2$mim][X] and PEG2000–[C$_4$mim][X] ABS [15] determined at 298.15 K are shown in Fig. 4.12. The ability of anions of imidazolium-based ILs to induce PEG-based ABS, as measured by the minimum required combined concentrations, follows the trend: [(CH$_3$)$_2$PO$_4$]$^-$ > [CH$_3$CO$_2$]$^-$ > [CH$_3$SO$_3$]$^-$ > [HSO$_4$]$^-$ > Cl$^-$ > Br$^-$ > [CF$_3$SO$_3$]$^-$ (which is the opposite order to that observed in K$_3$PO$_4$–IL ABS [35]).

In fact, in the case of IL–inorganic salts ABS, the IL (the more hydrophobic solute) is salted out by the inorganic ions, and therefore, the ability to induce ABS increases with the decrease of the affinity of the IL for water. However, for PEG–[C$_n$mim][X] ABS, the PEG (the more hydrophobic solute) can be salted out by [C$_n$mim][X] (if it is a more hydrophilic solute). As a result, more hydrophilic ILs can be more effective in promoting polymer–IL ABS. The salting-out aptitude of an ion can be directly related to its Gibbs free energy of hydration. The acetate anion (ΔG_{hyd} = 373 kJ.mol^{-1}) is a stronger salting-out species and more prone to induce liquid–liquid demixing (of PEG from aqueous media) than, for example, the chloride or bromide anions (ΔG_{hyd} = 347 and 321 kJ.mol^{-1}, respectively) [15, 36]. The trends observed in Fig. 4.12 correlate well with the phase diagrams of binary poly(ethyl glycidyl ether)/imidazolium-based ILs [37]. It has been shown that [37] fluorinated anions and/or anions with lower hydrogen-bond basicity are more miscible with the investigated polymers than anions with a higher aptitude to create ion–water complexes, such as acetate. As a consequence of this lower affinity between the acetate anion and polymers, the acetate-based IL is more efficient for the separation of the PEG aqueous phase than, for example, the [CF$_3$SO$_3$]-based fluid. Usually, strongly basic anions tightly interact with imidazolium cations and

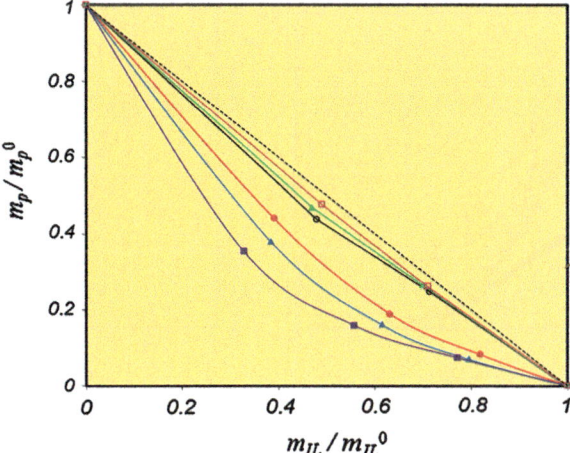

Fig. 4.13 Plot of m_p/m_p^0 against m_{IL}/m_{IL}^0 for constant water activity lines of PPG400 (p) + [C$_4$mim][HSO$_4$] (IL) + H$_2$O (w) (*filled symbols*) and PPG400 (p) + [C$_4$mim]Br (IL) + H$_2$O (w) (*empty symbols*) systems at 298.15 K: ○, a_w = 0.9800; ▲, a_w = 0.9829; □, a_w = 0.9885; ●, a_w = 0.8712; ▲, a_w = 0.9326; ■, a_w = 0.9461, calculated by Eq. 4.1 (Reprinted with the permission from Ref. [1]. Copyright 2013 American Chemical Society)

interrupt the hydrogen-bonding interactions between the PEG polymer and the aromatic cation of the IL, thereby lowering their miscibility [15].

The changes in the fluid phase behaviour of PEG35000 aqueous solutions by the addition of ionic liquids ([C$_2$mim]Cl and [C$_2$mim][C$_2$H$_5$OSO$_3$]) have been also evaluated through the determination of the critical solution temperature shifts [19]. It was found that a small salting-in effect (by an increase of the cloud point temperature) occurs when [C$_2$mim][C$_2$H$_5$OSO$_3$] is added to PEG aqueous solutions. However, at higher concentrations of added salt, the effect is reversed and a salting-out effect (drop in the cloud point temperature) takes over. On the other hand, the addition of [C$_2$mim]Cl produces a salting-out effect in the whole IL concentration range. Therefore, the salting-out ability of these anions follows the order Cl$^-$ > [C$_2$H$_5$OSO$_3$]$^-$.

As can be seen from Fig. 4.13, in the same line as the discussion previously presented, the negative deviation of constant water activity lines from Eq. 4.1 for the systems containing [C$_4$mim][HSO$_4$] and PPG400 is larger than that containing [C$_4$mim]Br and PPG400. This trend also agrees with the higher salting-out ability of [C$_4$mim][HSO$_4$] compared to [C$_4$mim]Br.

As mentioned above, the anion HSO$_4^-$ can establish strong hydrogen bonds with water molecules and can be expected to be more effective than Br$^-$ in the salting-out of polyethers. Furthermore, the double-charged anion SO$_4^{2-}$, produced from the protonation/deprotonation equilibria associated with HSO$_4^-$ in aqueous media, has a marked salting-out effect on polymers. Therefore, the negative deviation of constant water activity lines from Eq. 4.1 for the systems containing [C$_4$mim][HSO$_4$] and PPG is larger than that containing [C$_4$mim]Br and PPG.

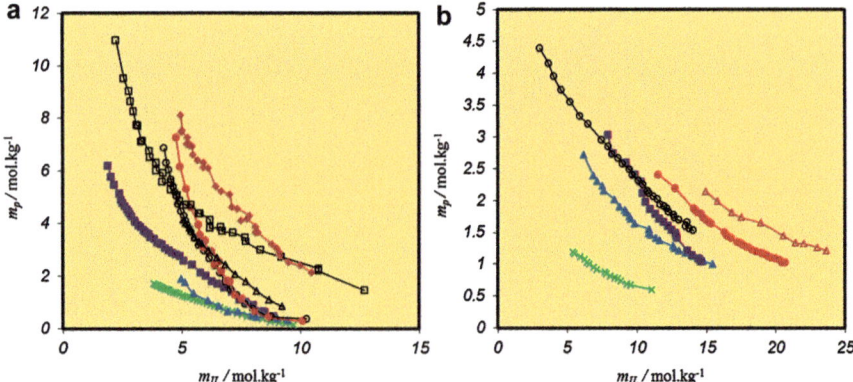

Fig. 4.14 (**a**) Binodal curves for PEG600 (p) + [Ch][X] (IL) + H$_2$O (w) systems at 298.15 K for X = ×, [DHph]$^-$; ▲, [Bit]$^-$; ■, [Bic]$^-$; ○, [Ac]$^-$; ●, [Lac]$^-$; Δ, [Gly]$^-$; □, [DHcit]$^-$; ♦, Cl$^-$. Reproduced from Ref. [26] by permission of The Royal Society of Chemistry. (**b**) Binodal curves for PEG600 (p) + [Ch][X] (IL) + H$_2$O (w) systems at 298.15 K for X = ×, [Fum]$^-$; ▲, [L-ma]$^-$; ■, [Glu]$^-$; ○, [Suc]$^-$; ●, [Ox]$^-$; Δ, [Mal]$^-$ (Reprinted with the permission from Ref. [25]. Copyright 2014 American Chemical Society)

Figure 4.14 presents a comparison between the binodal curves for ABS formed by PEG600 and cholinium-based ILs at 298.15 K.

As can be seen from Fig. 4.14, the ability of the investigated ILs ([Ch][X]) to promote the formation of ABS with PEG follows the trend: [Ch][DHph] > [Ch][Bit] > [Ch][Bic] > [Ch][DHcit] ≈ [Ch][Ac] ≈ [Ch][Lac] ≈ [Ch][Gly] > [Ch]Cl. The chemical structure of the studied cholinium-based salt anions and respective properties are given in Table 4.3. The salts [Ch][DHcit], [Ch][Bic], [Ch][DHph], [Ch][Bit] and [Ch]Cl present high melting temperatures, ranging from 103 to 302 °C. [Ch][Ac] and [Ch][Gly] present a melting point of 85 °C and 38 °C, respectively, whereas [Ch][But], [Ch][Pro] and [Ch][Lac] are liquid at room temperature [26]. The room temperature liquid salts, [Ch][But], [Ch][Pro] and [Ch][Lac], are completely soluble in both PEG and water. Neither [Ch][But] nor [Ch][Pro] induces ABS with any of PEG400, PEG600 and PEG1000, and [Ch][Lac] only forms ABS with the higher molecular weight (more hydrophobic polymers) PEG600 and PEG1000. All other salts, which have lower solubilities in PEG, form biphasic systems with all three PEGs. In general, a higher affinity between the PEG and these ILs leads to a lower ability for phase separation in the presence of water. The analysis of the data in Fig. 4.14 shows that there are two different shapes of binodal curves: one for the salts with relatively low melting points or ILs ([Ch][Lac], [Ch][Gly] and [Ch][Ac]) and another for the salts with higher melting points. For the first class of cholinium-based species, whose binodal curves present a more intensive dependence on the amount of salt added, the abilities of the ILs to promote an ABS are similar, as are their solubilities with PEG and water (Table 4.3). Here, the balance of interactions between the ions, PEG and water are fairly close due to the similar H-bonding abilities of the [Gly]$^-$, [Ac]$^-$ and [Lac]$^-$ anions. On the other

Table 4.3 Chemical structure of the studied salts and ILs and respective properties

Name	Anion	m.p./°C	Solubility in PEG600 (mol.kg^{-1})	Solubility in water (mol.kg^{-1})	Log k_{ow}[a]	Anion polar surface[b]/(Å^2)	pKa of the conjugated acid of the anion
[Ch][But]		Liquid at RT	Totally soluble	Totally soluble	Not available	40.13	4.83
[Ch][Pro]		Liquid at RT	Totally soluble	Totally soluble	Not available	40.13	4.87
[Ch][Lac]		Liquid at RT	Totally soluble	Totally soluble	−3.70	60.36	3.08
[Ch][Ac]		85	4.473	20.871	−4.66	40.13	4.47
[Ch][Gly]		38	3.221	Totally soluble	−1.20	60.36	3.83
[Ch][DHcit]		103–107	<0.005	3.295	−1.32	134.96	3.13
[Ch][Bic]		114–115	Not available	24.215	−3.70	60.36	6.37

(continued)

Table 4.3 (continued)

Name	Anion	m.p./°C	Solubility in PEG600 (mol.kg^{-1})	Solubility in water (mol.kg^{-1})	Log k_{ow}[a]	Anion polar surface[b]/(A^2)	pKa of the conjugated acid of the anion
[Ch] [DHph]	HO–P(=O)(OH)–O$^-$	119	<0.005	13.450	−3.70	90.40	2.12
[Ch] [Bit]	$^-$O–C(=O)–CH(OH)–CH(OH)–C(=O)–OH	149.5	<0.005	4.824	−1.43	82.06	3.22
[Ch]Cl	Cl$^-$	302	0.260	21.121	−3.70	–	<1

Reproduced from Ref. [26] by permission of The Royal Society of Chemistry
[a]The log k_{ow} values are a measure of the differential solubility of a particular solute between octanol and water
[b]The anion polar surface value is the surface sum over all polar atoms, primarily oxygen and nitrogen, and also includes their attached hydrogens

hand, the trend in the ability to induce ABS formation by the higher melting salts is as follows: [Ch][DHph] > [Ch][Bit] > [Ch][Bic] > [Ch][DHcit] > [Ch]Cl, following the increase in their anions polar surface and decrease in lipophilicity reflected by octanol–water partition coefficients (log k_{ow}). This fact suggests that the behaviour of the higher melting cholinium-based salts is governed mainly by solvation in water (where a higher affinity for water implies a higher ability to promote phase separation) [26]. This behaviour is in close agreement with that previously observed in conventional PEG–crystalline salt ABS, in which the anions with a higher charge density are more able to create ion–water complexes and more intensive repulsive interactions with the ether oxygens of PEG [2]. There is, however, an exception with [Ch][DHcit]. Although this anion should be a stronger salting-out agent, it shows a reduction in the two-phase region when compared with the binodal curves of the [DHph]-, [Bit]- and [Bic]-based salts. The [DHcit]$^-$ anion is based on citric acid, which has a high ability to act as a H-bond and/or a H-acceptor. In addition, according to the respective crystal structure [38], citrate anions can exhibit intramolecular hydrogen bonds between the hydroxyl hydrogen atoms and one of the oxygens of the central carboxyl group. Such self-aggregation would induce a more hydrophobic character to the anion and decrease its interaction with water and consequently the respective salting-out ability. In this case, the anion–anion interactions may have a larger influence on ABS formation than in the other studied salts [26].

The estimated partial molar excess enthalpies, h_i^E, by COSMO-RS (COnductor-like Screening MOdel for Real Solvents) of PEG600 in the ternary system show that favourable interactions occur between PEG600 and the anion, in which hydrogen-bonding contributes highly to the exothermic behaviour of the system. The hydrogen bond strength of PEG600–[Ch]$^+$ salt can be ranked as: [DHph]$^-$ (the lowest negative h_i^E values) < [Bit]$^-$ < [Bic]$^-$ < [Lac]$^-$ ≈ [Gly]$^-$ ≈ [Ac]$^-$ < [DHCit]$^-$ ≈ Cl$^-$ (the highest negative h_i^E values). This sequence closely follows the trend observed in the ability to induce the ABS formation. These results show that ILs with stronger interactions with PEG are less likely to phase separate in aqueous environments and *vice versa* [26].

Figure 4.14 shows the ability of the cholinium dicarboxylate ILs to promote a biphasic system with PEG600 and where the following trend was observed: [Fum]$^-$ > [L-ma]$^-$ > [Glu]$^-$ > [Suc]$^-$ > [Ox]$^-$ > [Mal]$^-$. The chemical structures of these anions are given in Fig. 4.15. As can be seen, the longer the alkyl chain length of dicarboxylated anion is, the larger the immiscibility region. In other words, the salting-out ability of [Ch][X] increases by increasing the alkyl chain length of the dicarboxylated anion [25].

The measured water activities of binary [Ch][X] + water solutions [25] show that the increase in the alkyl chain length of the anion leads to an increase in the hydrophilicity of [Ch][X], with [Ch][Glu] being the most hydrophilic and [Ch][Ox] the most hydrophobic. This fact can probably be explained through the existence of intramolecular interactions between the two carboxylic groups of the anion, which are very close in these molecules, leading to an increase in hydrophobicity. ILs with smaller alkyl chains at the anion, such as [Ch][Ox] and [Ch][Mal],

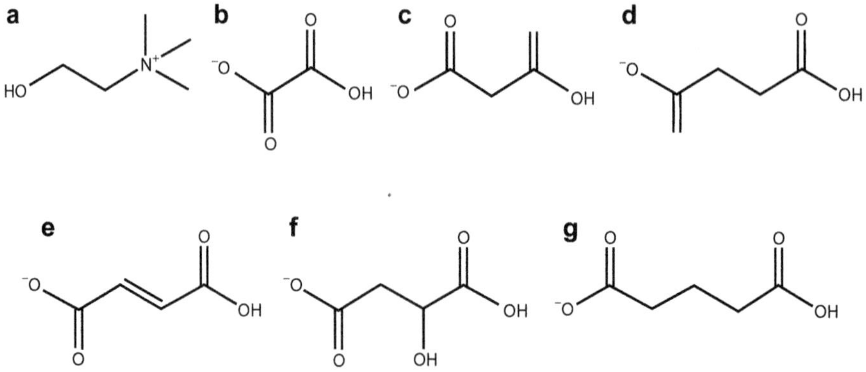

Fig. 4.15 Chemical structures of the cation and anions: (**a**) cholinium ([Ch]$^+$), (**b**) oxalate ([Ox]$^-$), (**c**) malonate ([Mal]$^-$), (**d**) succinate ([Suc]$^-$), (**e**) fumarate ([Fum]$^-$), (**f**) L-malate ([L-ma]$^-$), (**g**) glutarate ([Glu]$^-$)

display similar behaviours, with a somewhat smaller biphasic region than what could be expected from the results of the longer alkyl chain ILs, namely, [Ch][Glu] and [Ch][Suc]. The lower aptitude to create complexes with water of [Ch][Ox] can be attributed to the existence of intramolecular interactions between the two carboxylic groups of the anion. Consequently, the IL displaying the largest biphasic region is the most hydrophilic one, and therefore the ability to induce ABS increases with the increase in the affinity of the IL for water.

In order to address the effect of substituent groups in the anion alkyl chains, the binodal curves for ABS composed of PEG600 and the ILs [Ch][Suc] (anion alkyl chain of four carbons without a substituent group), [Ch][L-ma] (anion alkyl chain of four carbons with an hydroxyl substituent group), [Ch][Bit] (anion alkyl chain of four carbons with two hydroxyl substituent groups) and [Ch][Fum] (anion alkyl chain of four carbons with a double bond) are compared in Fig. 4.16. As can be seen, the ability of these compounds to induce ABS with PEG600 follows the trend: [Ch][Bit] ≈ [Ch][Fum] > [Ch][L-ma] > [Ch][Suc]. Thus, the immiscibility region increases with the presence of hydroxyl substituent groups.

The effect of the anion structure of the IL on the binodal curves of the PPG400 + cholinium-based ILs ABS at 298.15 K is depicted in Fig. 4.17. It has been shown that the phase-forming abilities of ILs to form ABS with PPG400 decrease in the order [Ch]$_3$[Cit] > [Ch]$_2$[Ox] > [Ch][Gly] > [Ch][Lac] ≈ [Ch][Pro] ≈ [Ch][Ac] > [Ch][For] > [Ch][But] > [Ch]Cl [23, 24]. [Cit]$^{3-}$ and [Ox]$^{2-}$, with higher charge valence, have stronger hydration capacities than those with lower charge valence and therefore are more prone to induce liquid–liquid demixing of PPG from aqueous media than, for example, [Ac]$^-$ and [For]$^-$ anions.

It was also found [23] that the addition of either [Ch][Lac], [Ch][Pro] or [Ch][Ac] decreases the cloud point temperature of aqueous PPG400 solutions. For a given cholinium-based IL, the cloud point temperature of the aqueous PPG400 + IL solutions decreases with increasing the concentration of the polymer, and the effect of the ILs on the cloud point temperature of the aqueous PPG400 is in agreement with the order of their phase-forming ability.

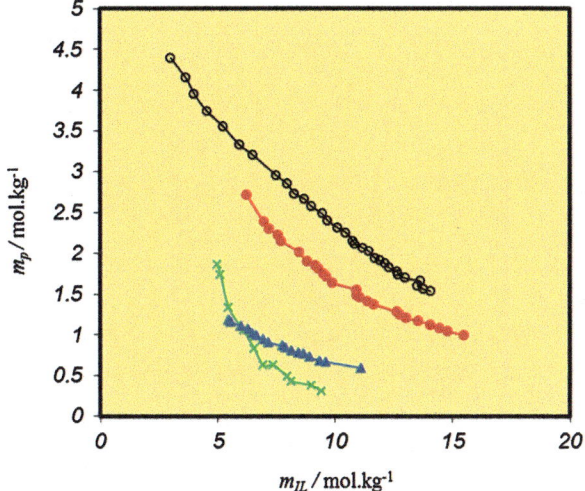

Fig. 4.16 Binodal curves for PEG600 (p) + [Ch][X] (IL) + H$_2$O (w) systems at 298.15 K for X=: ×, [Bit]$^-$ [26]; ▲, [Fum]$^-$ [25]; ●, [L-ma]$^-$ [25]; ○, [Suc]$^-$ [25]

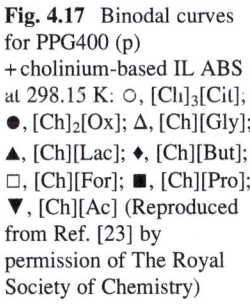

Fig. 4.17 Binodal curves for PPG400 (p) + cholinium-based IL ABS at 298.15 K: ○, [Ch]$_3$[Cit]; ●, [Ch]$_2$[Ox]; Δ, [Ch][Gly]; ▲, [Ch][Lac]; ♦, [Ch][But]; □, [Ch][For]; ■, [Ch][Pro]; ▼, [Ch][Ac] (Reproduced from Ref. [23] by permission of The Royal Society of Chemistry)

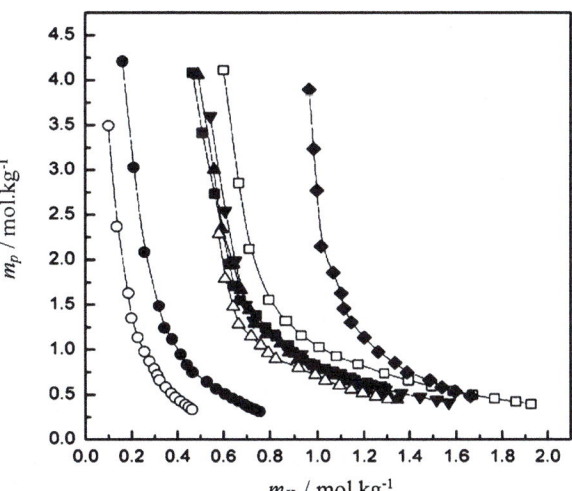

4.3.3.2 Effect of the IL Cation

In order to analyse the effect of the cationic core of ILs on their ability to induce ABS, the binodal curves of several PEG1500/IL ABS at 323.15 K are compared in Fig. 4.18.

Figure 4.18 reveals that the ability of the ILs to induce a PEG1500-based ABS follows the trend: [C$_4$mim]Cl < [C$_4$mpyr]Cl < [C$_4$mpip]Cl. A similar behaviour was obtained for PEG2000/IL ABS at 298.15 K [15]. This order correlates well

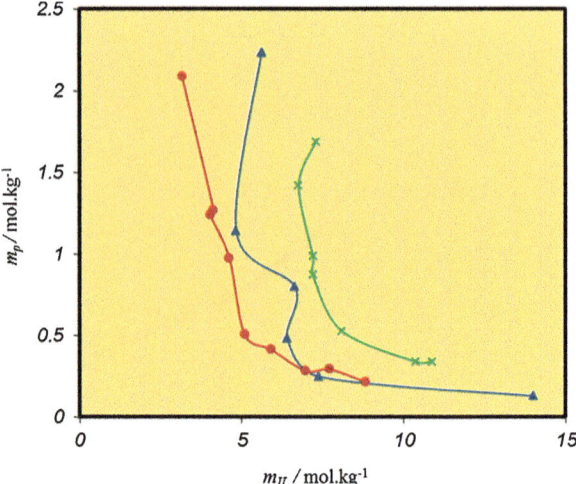

Fig. 4.18 Binodal curves for PEG1500 (p) + IL (IL) + H$_2$O (w) systems at 323.15 K for: ×, [C$_4$mim]Cl; ▲, [C$_4$mpyr]Cl; ●, [C$_4$mpip]Cl (Reproduced from Ref. [16] by permission of The Royal Society of Chemistry)

with the relative hydrophobic nature of the IL, with the most hydrophobic species showing cloud point at lower concentrations of IL. In fact, the imidazolium-based compounds exhibit stronger interactions with water as a result of their aromatic character. On the other hand, piperidinium and pyrrolidinium are five- and six-sided saturated rings, respectively, and are thus less hydrophilic ILs, with a lower propensity for interaction with water [16, 34]. Ubiquitously, this cationic pattern is in good agreement with the results obtained with ABS of ILs in the presence of inorganic salts [33]. Hence, the cation trend here observed is in stark contradiction to that obtained for the anions discussed above. When dealing with more hydrophobic ILs, the ABS formation is, as for the salt + IL systems, a direct result of the PEG salting-out aptitude over the moderate hydrophobic ILs considered. PEG seems to be preferentially hydrated, and, thus, tends to salt out ILs. As mentioned above, the interactions between PEGs and ILs, and how they are affected by water, cannot be neglected and have a significant impact on their phase separation. PEG is a polyether and, thus, hydrogen-bonding interactions are expected to occur with the chloride anion as well as with aromatic cations. The presence of πelectrons leads to strong interactions with PEG and thus explains the low ability of imidazolium-based ILs to form ABS with PEG1500 [16]. Indeed, results on the phase behaviour of binary ILs–polyether compounds proved that polymers are more soluble in ILs containing aromatic cations [37].

The length of the cationic alkyl substituent of ILs is one of the major structural features that allows the hydrophobicity of the IL to be tailored. Regarding the effect of the alkyl chain length of the cation of ILs, the binodal curves of several PEG2000–IL and PPG400–IL ABS are compared in Fig. 4.19.

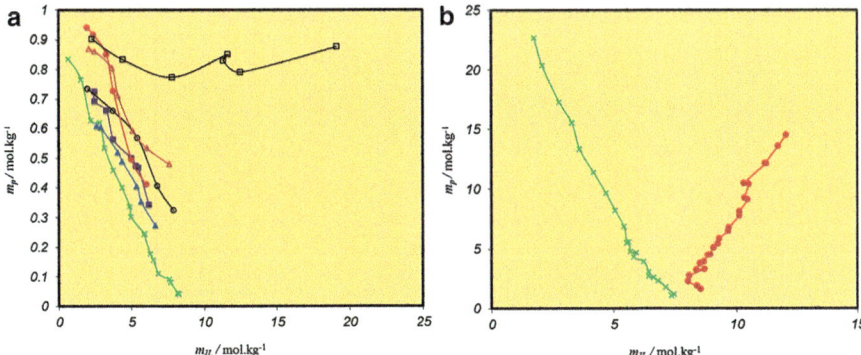

Fig. 4.19 (a) Binodal curves for PEG2000 (p) + IL (IL) + H$_2$O (w) systems at 298.15 K: ×, [C$_2$mim]Cl; ▲, [C$_1$mim]Cl; ■, [C$_4$mim]Cl; ○, [mim]Cl; ●, [C$_8$mim]Cl; ∆; [C$_6$mim]Cl; □, [im]Cl. Reproduced from Ref. [15] by permission of John Wiley & Sons Ltd. (b) Binodal curves for PPG400 (p) + IL (IL) + H$_2$O (w) system at 298.15 K: ×, [C$_2$mim]Br; ●, [C$_4$mim]Br (Reproduced from Ref. [22] by permission of John Wiley & Sons Ltd)

For low IL concentrations, the aptitude of the cation of 1-alkyl-3-methylimidazolium-based ILs to induce polymer-based ABS decreases in the following order: [C$_2$mim]$^+$ ≈ [C$_1$mim]$^+$ > [C$_4$mim]$^+$ > [mim]$^+$ > [C$_6$mim]$^+$ ≈ [C$_8$mim]$^+$ > [im]$^+$. However, for the high concentrations of IL, this trend is [C$_2$mim]$^+$ ≈ [C$_1$mim]$^+$ > [C$_4$mim]$^+$ > [C$_8$mim]$^+$ > [mim]$^+$ > [C$_6$mim]$^+$ > [im]$^+$. Moreover, the effect of the alkyl chain length of the cation of ILs on the phase behaviour of PPG–IL ABS is more significant than that of the PEG–IL ABS. ILs with shorter aliphatic chains enhance the hydrophilic character of the IL and, therefore, increase their affinity for water and increase the ABS promoting ability, which is the opposite order to that observed for the K$_3$PO$_4$–IL ABS [35]. In addition to the solute–water interactions, the interactions between the PEG polymer and ILs can also play an important role in this respect. As previously reported, an increase in the alkyl side chain of the cation enhances the solubility between the imidazolium-based ILs and the PEG polymer [3, 32], thus leading to systems that require more IL to undergo liquid–liquid demixing in aqueous media.

A similar trend has been observed in the case of the change in the cloud point temperature of PEG35000 aqueous solutions by the addition of ILs [19]. It was found that the addition of ILs with long alkyl chains improves the solubility of PEG in water (salting-in effect), whereas the impact of short-chain ILs is usually the contrary (salting-out effect). It should be reminded that imidazolium-based ILs with long alkyl side chains ($n \geq 6$) form self-assembled structures (micelles) in aqueous solution above a critical concentration [39], which may decrease their ability to interact with water or PEG, and therefore, more complex phenomena are involved.

The binodal curves of PEG2000–[amim]Cl, PEG2000–[HOC$_2$mim]Cl, PEG2000–[C$_2$mim]Cl and PEG2000–[C$_4$mim]Cl ABS are depicted in Fig. 4.20 in order to investigate the effect of the functionalization of the alkyl substituents in the cation, in particular, the influence of the presence of a double bond or a hydroxyl

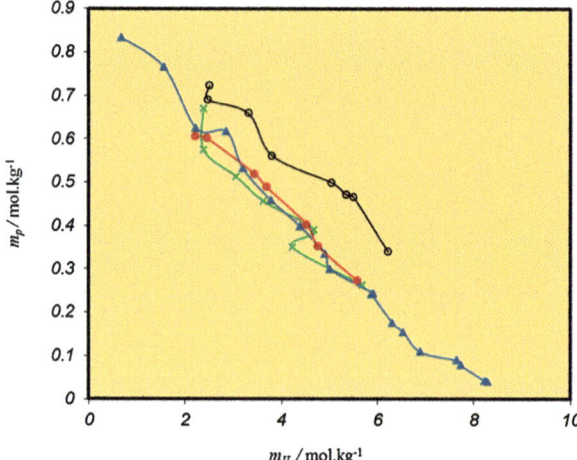

Fig. 4.20 Binodal curves of PEG2000 (p) + IL (IL) + H$_2$O (w) system at 298.15 K: ×, [HOC$_2$mim]Cl; ▲, [C$_2$mim]Cl; ●, [amim]Cl; ○, [C$_4$mim]Cl (Reproduced from Ref. [15] by permission of John Wiley & Sons Ltd)

group in the alkyl side chain of the imidazolium cation. As can be seen, the binodal curves for [amim]Cl, [HOC$_2$mim]Cl and [C$_2$mim]Cl/PEG2000 ABS are very close to each other, meaning that these substituents in the ILs considered display a marginal effect.

4.4 Conclusions and Outlook

Two types of polymer–IL ABS have been investigated in the literature: PEG–IL and PPG–IL ABS. In the case of PEG–IL ABS, the hydrophobicity of PEG and ILs is close to each other, and therefore both PEG and ILs (depending on the cation or anion of the IL) can act as a salting-out agent. Furthermore, besides the IL–water and PEG–water interactions that govern the phase behaviour of PEG + IL + water systems, the PEG–IL interactions could additionally control the phase diagrams of these systems. It seems that in the formation of a PEG–IL-based ABS, a complex interplay of different factors takes place leading to non-regular trends of the ABS promoting ability in respect to varying the structure of polymer and IL, molecular weight of polymer and temperature. However, in the case of PPG–ILs ABS, PPG is always salted out by the IL in aqueous solution, and the ability to induce ABS increases as the hydrophilicity of the IL and/or molecular weight of PPG is increased, and therefore, the phase diagrams of PPG/ILs ABS follow regular trends.

References

1. Moradian T, Sadeghi R (2013) Isopiestic investigations of the interactions of water-soluble polymers with imidazolium-based ionic liquids in aqueous solutions. J Phys Chem B 117:7710–7717
2. Sadeghi R, Jahani F (2012) Salting-in and salting-out of water-soluble polymers in aqueous salt solutions. J Phys Chem B 116:5234–5241
3. Rodriguez H, Francisco M, Rahman M, Sun N, Rogers RD (2009) Biphasic liquid mixtures of ionic liquids and polyethylene glycols. Phys Chem Chem Phys 11:10916–10922
4. Sadeghi R, Shekaari H, Hosseini R (2009) Effect of alkyl chain length and temperature on the thermodynamic properties of ionic liquids 1-alkyl-3-methylimidazolium bromide in aqueous and non-aqueous solutions at different temperatures. J Chem Thermodyn 41:273–289
5. Clarke ECW, Glew DN (1985) Evaluation of the thermodynamic functions for aqueous sodium chloride from equilibrium and calorimetric measurement below 154 °C. J Phys Chem Ref Data 14:489–610
6. Sadeghi R, Shahebrahimi Y (2011) Vapor-liquid equilibria of aqueous polymer solutions from vapor pressure osmometry and isopiestic measurements. J Chem Eng Data 56:789–799
7. Pitzer KS, Mayorga G (1973) Thermodynamics of electrolytes. II. Activity and osmotic coefficients for strong electrolytes with one or both ions univalent. J Phys Chem 77:2300–2308
8. Sadeghi R, Golabiazar R, Shekaari H (2010) The salting-out effect and phase separation in aqueous solutions of tri-sodium citrate and 1-butyl-3-methylimidazolium bromide. J Chem Thermodyn 42:441–453
9. Sadeghi R, Ziamajidi F (2007) Apparent molar volume and isentropic compressibility of trisodium citrate in water and in aqueous solutions of polyvinylpyrrolidone at T = (283.15 to 308.15) K. J Chem Eng Data 52:1037–1044
10. Sadeghi R, Ziamajidi F (2007) Thermodynamic properties of tripotassium citrate in water and in aqueous solutions of polypropylene oxide 400 over a range of temperatures. J Chem Eng Data 52:1753–1759
11. Millero FJ (1971) Molal volumes of electrolytes. Chem Rev 71:147–176
12. Millero FJ, Knox JH (1973) Apparent molal volumes of aqueous NaF, Na_2SO_4, KCl, K_2SO_4, $MgCl_2$, and $MgSO_4$ solutions at 0° and 50°C. J Phys Chem 18:407–411
13. Sadeghi R, Hosseini R, Jamehbozorg B (2008) Effect of sodium phosphate salts on the thermodynamic properties of aqueous solutions of poly(ethylene oxide) 6000 at different temperatures. J Chem Thermodyn 40:1364–1377
14. Sadeghi R, Mostafa B, Parsi E, Shahebrahimi Y (2010) Toward an understanding of the salting-out effects in aqueous ionic liquid solutions: vapor-liquid equilibria, liquid-liquid equilibria, volumetric, compressibility, and conductivity behavior. J Phys Chem B 114:16528–16541
15. Freire MG, Pereira JFB, Francisco M, Rodriguez H, Rebelo LPN, Rogers RD, Coutinho JAP (2012) Insight into the interactions that control the phase behaviour of new aqueous biphasic systems composed of polyethylene glycol polymers and ionic liquids. Chem Eur J 18:1831–1839
16. Pereira JFB, Rebelo LPN, Rogers RD, Coutinho JAP, Freire MG (2013) Combining ionic liquids and polyethylene glycols to boost the hydrophobic-hydrophilic range of aqueous biphasic systems. Phys Chem Chem Phys 15:19580–19583
17. Upfal J, MacFarlane DR, Forsyth SA (2005) Solvents for use in the treatment of lignin-containing materials. WO 2005/017252 A1
18. Dreyer S, Kragl U (2007) Verfahren zur extraktion von biomolekulen. DE102007001347 A1
19. Visak ZP, Lopes JNC, Rebelo LPN (2007) Ionic liquids in polyethylene glycol aqueous solutions: salting-in and salting-out effects. Monatsh Chem 138:1153–1157
20. Canongia Lopes JN, Rebelo LPN (2007) From aqueous biphasic system formation to salting agent precipitation. Chim Oggi 25:37–39

21. Wu C, Wang J, Pei Y, Wang H, Li Z (2010) Salting-out effect of ionic liquids on poly (propylene glycol) (PPG): formation of PPG+ionic liquid aqueous two-phase systems. J Chem Eng Data 55:5004–5008
22. Zafarani-Moattar MT, Hamzehzadeh S, Nasiri S (2012) A new aqueous biphasic system containing polypropylene glycol and a water-miscible ionic liquid. Biotechnol Prog 28:146–156
23. Li Z, Liu X, Pei Y, Wang J, He M (2012) Design of environmentally friendly ionic liquid aqueous two-phase systems for the efficient and high activity extraction of proteins. Green Chem 14:2941–2950
24. Liu X, Li Z, Pei Y, Wang H, Wang J (2013) Liquid-liquid equilibria for cholinium-based ionic liquids+polymers aqueous two-phase systems. J Chem Thermodyn 60:1–8
25. Mourao T, Tome LC, Florindo C, Rebelo LPN, Marrucho IM (2014) Understanding the role of cholinium carboxylate ionic liquids in peg-based aqueous biphasic systems. ACS Sustain Chem Eng 2:2426–2434
26. Pereira JFB, Kurnia KA, Cojocaru A, Gurau G, Rebelo LPN, Rogers RD, Freire MG, Coutinho JAP (2014) Molecular interactions in aqueous biphasic systems composed of polyethylene glycol and crystalline vs liquid cholinium-based salts. Phys Chem Chem Phys 16:5723–5731
27. Tome LIN, Pereira JFB, Rogers RD, Freire MG, Gomes JRB, Coutinho JAP (2014) "Washing-out" ionic liquids from polyethylene glycol to form aqueous biphasic systems. Phys Chem Chem Phys 16:2271–2274
28. Zdanovskii AB (1936) Regularities in the property variations of mixed solutions. Tr Solyanoi Lab Akad Nauk SSSR 6:5–70
29. Stokes RH, Robinson RA (1966) Interactions in aqueous nonelectrolyte solutions. I. Solute-solvent equilibria. J Phys Chem 70:2126–2131
30. Sadeghi R, Hamidi B, Ebrahimi N (2014) Investigation of amino acid-polymer aqueous biphasic systems. J Phys Chem B 118:10285–10296
31. Tome LIN, Pereira JFB, Rogers RD, Freire MG, Gomes JRB, Coutinho JAP (2014) Evidence for the interactions occurring between ionic liquids and tetraethylene glycol in binary mixtures and aqueous biphasic systems. J Phys Chem B 118:4615–4629
32. Rodriguez H, Rogers RD (2010) Liquid mixtures of ionic liquids and polymers as solvent systems. Fluid Phase Equilib 294:7–14
33. Freire MG, Claudio AFM, Araujo JMM, Coutinho JAP, Marrucho IM, Lopes JNC, Rebelo LPN (2012) Aqueous biphasic systems: a boost brought about by using ionic liquids. Chem Soc Rev 41:4966–4995
34. Freire MG, Neves CMSS, Carvalho PJ, Gardas RL, Fernandes AM, Marrucho IM, Santos LMNBF, Coutinho JAP (2007) Mutual solubilities of water and hydrophobic ionic liquids. J Phys Chem B 111:13082–13089
35. Ventura SPM, Neves CMSS, Freire MG, Marrucho IM, Oliveira J, Coutinho JAP (2009) Evaluation of anion influence on the formation and extraction capacity of ionic-liquid-based aqueous biphasic systems. J Phys Chem B 113:9304–9310
36. Marcus Y (1997) Ion properties. Marcel Dekker, New York
37. Kodama K, Tsuda R, Niitsuma K, Tamura T, Ueki T, Kokubo H, Watanabe M (2011) Structural effects of polyethers and ionic liquids in their binary mixtures on lower critical solution temperature liquid-liquid phase separation. Polym J 43:242–248
38. Glusker JP (1980) Citrate conformation and chelation: enzymatic applications. Acc Chem Res 13:345–352
39. Blesic M, Marques MH, Plechkova NV, Seddon KR, Rebelo LPN, Lopes A (2007) Self-aggregation of ionic liquids: micelle formation in aqueous solution. Green Chem 9:481–490

Chapter 5
Extraction of Amino Acids with ABS

Mohammed Taghi Zafarani-Moattar and Sholeh Hamzehzadeh

Abstract Among the several applications reported for ionic-liquid-based aqueous biphasic systems (IL-based ABS), these systems have shown to be highly effective in the extraction of amino acids. In this chapter, the extraction potential for amino acids of IL-based ABS composed of hydrophilic ILs and different salting-out/-in agents, such as inorganic and organic salts, carbohydrates, and polypropylene glycol (PPG) 400, is reviewed. A brief overview on factors affecting the extraction potential of ABS, such as the chemical structures of the ILs cations and/or anions and the salting-out effects, is also provided. The impact of the charge and the structural/physicochemical properties of the amino acids on their partition coefficients are also discussed. Finally, the effect of the addition of various imidazolium-based ILs to conventional polyethylene glycol (PEG)-salt ABS in what concerns their extraction capability for amino acids is reviewed and discussed.

Keywords Ionic liquids • Aqueous biphasic systems • Extraction • Amino acids

5.1 Introduction

Extraction by means of aqueous biphasic systems (ABS) is known as a promising tool in biotechnology for the separation and purification of value-added bio-products. ABS consist of two aqueous-rich phases that are generally formed when a mixture of two water-soluble, but mutually incompatible polymers, a polymer and a salt, or two surfactants exceeds specific threshold concentrations in aqueous media [1–5]. ABS are environmentally safe and benign since volatile organic solvents are not employed in the whole process. ABS provide a technically simple, easily scalable, energy-efficient, mild and nontoxic separation technique for product recovery in biotechnology [5]. Separation of biological molecules and particles using ABS was initiated more than half a century ago by P.-Å. Albertsson [6] and later proved to be of immense utility in chemical as well as in biochemical and cell biological basic and applied research. Today, aqueous biphasic extraction is established as an economical and efficient downstream processing method in biotechnology.

Amino acids are very important bio-products with extensive industrial applications that can be produced by biological or biochemical processes. The conventional methods for the separation of amino acids include ionic exchange, reversed micelle methods [7], and liquid membrane extraction processes [8]. However, organic solvents are applied in the majority of these processes with all their inherent problems, such as high volatility, high flammability, and toxicity to humans and microorganisms [9], so their recovery via aqueous two-phase extraction from bio-reaction media may represent realistic alternatives to more traditional methods. Furthermore, the study of the partition behavior of amino acids in ABS is of academic and practical importance for the separation of proteins, since amino acid residues determine the protein surface properties. Therefore, further understanding of the driving forces for the partitioning of more complex proteins in a given ABS can be obtained from the study of the distribution behavior of single amino acids. Amino acids are simpler than the larger protein molecules and only differ from each other by one characteristic group to which clear physical and chemical properties can be assigned [4].

Apart from the conventional polymer-based ABS, ionic liquid-based (IL-based) ABS have been shown to be outstanding platforms for the extraction of amino acids [10–17]. The main interest on ILs is due to their unique characteristics, among them a negligible vapor pressure, non-flammability, a large liquid temperature range, high thermal and chemical stabilities, and strong solvation abilities [18]. Although polymers also exhibit most of these properties, of particular interest on ABS composed of ILs is the potential of tuning the ABS physicochemical properties by judiciously selecting the ILs cations and/or anions, thus making possible the manipulation of the properties of the extractive phases for achieving enhanced yield of product recovery [10–12, 16]. The results found in the literature for the extraction of amino acids using IL-based ABS are listed in Table 5.1.

The efficient application of ABS for preparative purposes, e.g., for separation, isolation, or purification of a product, requires the knowledge of specific mechanism behind the product partitioning between the two phases. In general, partitioning between the two phases of ABS is a complex phenomenon guided mainly by the interaction of the partitioned solute and the phase components, e.g., through hydrogen bonds, van der Waals, hydrophobic, and electrostatic interactions, steric and conformational effects, etc. The net effect of these interactions is likely to be different in the two phases, and the solute will partition preferentially into one phase [1, 3]. As a result, solute partitioning is dependent on the properties of the solute as well as on the two aqueous phases properties and polarities.

5 Extraction of Amino Acids with ABS

Table 5.1 IL-based ABS reported in the literature for the extraction of amino acids

System		
Component 1	Component 2	Ref
L-Tryptophan		
Imidazolium chloride ([im]Cl)	K_3PO_4	[10]
1-Methylimidazolium chloride ([C_1im]Cl)	K_3PO_4	[10]
1-Ethyl-3-methylimidazolium chloride ([C_2C_1im]Cl)	K_3PO_4	[10, 11]
1-Butyl-3-methylimidazolium chloride ([C_4C_1im]Cl)	K_3PO_4	[10, 11]
1-Allyl-3-methylimidazolium chloride ([aC_1im]Cl)	K_3PO_4	[10]
1-Hydroxyethyl-3-methylimidazolium chloride ([OHC_2C_1im]Cl)	K_3PO_4	[10]
1-Benzyl-3-methylimidazolium chloride ([$C_7H_7C_1$im]Cl)	K_3PO_4	[10]
1-Ethyl-3-methylimidazolium acetate ([C_2C_1im][C_1CO_2])	K_3PO_4	[11]
1-Ethyl-3-methylimidazolium methylsulfate ([C_2C_1im][C_1SO_4])	K_3PO_4	[11]
1-Ethyl-3-methylimidazolium ethylsulfate ([C_2C_1im][C_2SO_4])	K_3PO_4	[11]
1-Ethyl-3-methylimidazolium triflate ([C_2C_1im][CF_3SO_3])	K_3PO_4	[11]
1-Butyl-3-methylimidazolium bromide ([C_4C_1im]Br)	K_3PO_4	[11]
1-Butyl-3-methylimidazolium methylsulfonate ([C_4C_1im][C_1SO_3])	K_3PO_4	[11]
1-Butyl-3-methylimidazolium dicyanamide ([C_4C_1im][N(CN)$_2$])	K_3PO_4	[11]
1-Butyl-3-methylimidazolium trifluoroacetate ([C_4C_1im][TFA])	K_3PO_4	[11]
1-Butyl-3-methylimidazolium triflate ([C_4C_1im][CF_3SO_3])	K_3PO_4	[11]
1-Butyl-3-methylimidazolium acetate ([C_4C_1im][C_1CO_2])	K_3PO_4	[12]
1-Hexyl-3-methylimidazolium acetate ([C_6C_1im][C_1CO_2])	K_3PO_4	[12]
1-Octyl-3-methylimidazolium acetate ([C_8C_1im][C_1CO_2])	K_3PO_4	[12]
Methyl(tributyl) phosphonium methylsulfate ([P_{1444}][C_1SO_4])	K_3PO_4	[13]
1-Butyl-3-methylimidazolium triflate ([C_4C_1im][CF_3SO_3])	Sucrose	[14]
1-Butyl-3-methylimidazolium triflate ([C_4C_1im][CF_3SO_3])	D-(+)-glucose	[14]
1-Butyl-3-methylimidazolium triflate ([C_4C_1im][CF_3SO_3])	D-(−)-fructose	[14]
1-Butyl-3-methylimidazolium triflate ([C_4C_1im][CF_3SO_3])	D-(+)-mannose	[14]
1-Butyl-3-methylimidazolium triflate ([C_4C_1im][CF_3SO_3])	D-(+)-xylose	[14]
1-Butyl-3-methylimidazolium triflate ([C_4C_1im][CF_3SO_3])	D-maltitol	[14]
1-Butyl-3-methylimidazolium triflate ([C_4C_1im][CF_3SO_3])	Xylitol	[14]
1-Butyl-3-methylimidazolium triflate ([C_4C_1im][CF_3SO_3])	D-sorbitol	[14]
1-Butyl-3-methylimidazolium bromide ([C_4C_1im]Br)	$K_3C_6H_5O_7$	[15]
1-Butyl-3-methylimidazolium triflate ([C_4C_1im][CF_3SO_3])	$K_3C_6H_5O_7$	[16]
1-Butyl-3-methylimidazolium thiocyanate ([C_4C_1im][SCN])	$K_3C_6H_5O_7$	[16]
1-Butyl-3-methylimidazolium dicyanamide ([C_4C_1im][N(CN)$_2$])	$K_3C_6H_5O_7$	[16]
1-Butyl-3-methylimidazolium chloride ([C_4C_1im]Cl)	$K_3C_6H_5O_7$	[16]
1-Butyl-1-methylpiperidinium chloride ([C_4C_1pip]Cl)	$K_3C_6H_5O_7$	[16]
1-Butyl-1-methylpyrrolidinium chloride [C_4C_1pyr]Cl	$K_3C_6H_5O_7$	[16]
(Tetrabutyl)phosphonium chloride ([P_{4444}]Cl)	$K_3C_6H_5O_7$	[16]
(Tetrabutyl)ammonium chloride ([N_{4444}]Cl)	$K_3C_6H_5O_7$	[16]
1-Ethyl-3-methylimidazolium bromide ([C_2C_1im]Br)	PPG 400	[17]

(continued)

Table 5.1 (continued)

System		
Component 1	Component 2	Ref
L-Phenylalanine		
1-Butyl-3-methylimidazolium bromide ([C$_4$C$_1$im]Br)	K$_3$C$_6$H$_5$O$_7$	[15]
L-Tyrosine		
1-Butyl-3-methylimidazolium bromide ([C$_4$C$_1$im]Br)	K$_3$C$_6$H$_5$O$_7$	[15]
L-Leusine		
1-Butyl-3-methylimidazolium bromide ([C$_4$C$_1$im]Br)	K$_3$C$_6$H$_5$O$_7$	[15]
L-Valine		
1-Butyl-3-methylimidazolium bromide ([C$_4$C$_1$im]Br)	K$_3$C$_6$H$_5$O$_7$	[15]

Achieving successful ABS separation depends on the ability to manipulate phase properties to obtain the appropriate partition coefficients and selectivity for the target biomolecule. There are several approaches to manipulate a particular solute partitioning, namely, by (1) adjusting the system by applying different salts and/or ILs controlling the solute affinity for one of two aqueous phases; (2) changing the system concentration of salt and/or IL; and (3) introducing additional cosolvents, antisolvents, or amphiphilic structures to the system.

In this chapter, the extraction potential for amino acids of IL-based ABS composed of hydrophilic ILs and different salting-out and salting-in agents, such as inorganic and organic salts, carbohydrates, and polypropylene glycol (PPG) 400, is reviewed. We present here a brief review on factors affecting the extraction potential of ABS such as the chemical structures of the IL cations and/or anions and the salting-out/-in effects induced by different species. The effects of the charge and the structural/physicochemical properties of the amino acids on their partition coefficients are also discussed. We then review the effect of the addition of various imidazolium-based ILs (in small quantities, as adjuvants) to conventional polyethylene glycol (PEG)-salt ABS on the extraction capability for amino acids.

In the following sections, the extraction aptitude for amino acids of IL-based ABS is evaluated through partition coefficient (K) values defined as

$$K_{\text{amino acid}} = \frac{[\text{amino acid}]_{\text{IL}}}{[\text{amino acid}]_{\text{salting-out/-in agent}}} \quad (5.1)$$

which is defined by the ratio of the concentration of amino acid in the aqueous top "IL"-rich phase to that in the aqueous bottom "salting-out/-in agent"-rich phase. The higher the $K_{\text{amino acid}}$, the higher the tendency for the amino acid to migrate to the IL-rich phase.

5.2 Extraction of Amino Acids with ABS Composed of Ionic Liquids and Salts

Rogers and coworkers [19] were the first to show that aqueous solutions of hydrophilic ILs can form biphasic systems with concentrated aqueous solutions of water-structuring salts. After this pioneer work, IL-based ABS have been explored as novel liquid-liquid partitioning systems for amino acids. In this context, systematic studies of the extractive potential for amino acids of ABS composed of hydrophilic imidazolium-based ILs, and the inorganic salt K_3PO_4 (the strongest salting-out agent evaluated) have been conducted to examine the effect of the chemical structures of IL cations and/or anions [10–12]. In regard to the extraction capability for amino acids of IL-based ABS with ILs not based on imidazolium cations, only one work with ABS composed of $[P_{1444}][C_1SO_4]$ and K_3PO_4 has been found in the literature, reported by Louros et al. [13]. In these studies, the extraction capacity of IL-salt ABS for amino acids was evaluated using the essential amino acid L-tryptophan (Trp) [10–13]. The results found in the literature for the extraction of Trp using ABS composed of ILs and salts are summarized in Table 5.2.

To gain an insight into the driving forces of amino acid partitioning in ABS, we have studied the extraction capability of ABS composed of $[C_4C_1im]Br$ and potassium citrate, as "greener" alternatives to the previously studied systems [10–13], for five model amino acids: L-tryptophan (Trp), L-phenylalanine (Phe), L-tyrosine (Tyr), L-leucine (Leu), and L-valine (Val) [15]. Citrate-based salts are biodegradable and nontoxic and can be discharged into biological wastewater treatment plants. In this study, the influences of aqueous medium pH and the amino acids' structural/physicochemical properties have been evaluated at $T = 298.15$ K and different phase compositions. Moreover, in this work, a model was established to describe the partition coefficients of three amino acids, Trp, Phe, and Val, and employed to predict the partition coefficients of two other amino acids, Tyr and Leu. The work of Freire et al. [16] has provided new information regarding the effects of the IL cations and/or anions on the partitioning of amino acids, by studying a large array of IL-based ABS, making use of potassium citrate.

5.2.1 Influence of the Ionic Liquid

The efficient application of ILs as extraction media in ABS requires the knowledge of the factors governing the partitioning of biomolecules between the two equilibrated aqueous-rich phases. The partition coefficients of biomolecules in ABS are dependent on hydrophobic interactions, electrostatic forces, molecular size, solubility, and affinity for both phases, and their magnitudes further depend on the two-phase compositions and on the nature of the biomolecules [1]. Among the different interactions between molecules, hydrophobic interactions play a key role

Table 5.2 Partition coefficients of L-tryptophan (K_{Trp}) in IL-salt ABS reported in the literature

IL	Salt	IL (wt.%)	Salt (wt.%)	K_{Trp}	Ref
[aC$_1$im]Cl	K$_3$PO$_4$	26.91	15.91	124 ± 5	[10]
[C$_7$H$_7$C$_1$im]Cl	K$_3$PO$_4$	25.11	18.30	78.4 ± 0.5	[10]
[OHC$_2$C$_1$im]Cl	K$_3$PO$_4$	40.57	16.05	73.1 ± 0.8	[10]
[C$_2$C$_1$im]Cl	K$_3$PO$_4$	25.90	14.90	59.2 ± 0.4	[10, 11]
[C$_2$C$_1$im][CF$_3$SO$_3$]	K$_3$PO$_4$	24.87	16.35	17.5 ± 0.8	[11]
[C$_6$C$_1$im][C$_1$CO$_2$]	K$_3$PO$_4$	25.00[a]	15.00[a]	25.9 ± 0.7	[12]
[C$_4$C$_1$im][C$_1$CO$_2$]	K$_3$PO$_4$	25.00[a]	15.00[a]	23.3 ± 0.8	[12]
[C$_2$C$_1$im][C$_1$CO$_2$]	K$_3$PO$_4$	24.94	14.96	16.4 ± 0.8	[11]
[C$_8$C$_1$im][C$_1$CO$_2$]	K$_3$PO$_4$	25.00[a]	15.00[a]	5.2 ± 0.2	[12]
[C$_2$C$_1$im][C$_2$SO$_4$]	K$_3$PO$_4$	24.97	15.06	5.96 ± 0.5	[11]
[C$_2$C$_1$im][C$_1$SO$_4$]	K$_3$PO$_4$	24.98	14.94	4.47 ± 0.9	[11]
[C$_4$C$_1$im][N(CN)$_2$]	K$_3$PO$_4$	24.40	16.82	45.1 ± 0.9	[11]
[C$_4$C$_1$im]Cl	K$_3$PO$_4$	25.35	15.97	36.6 ± 0.6	[10, 11]
[C$_4$C$_1$im][TFA]	K$_3$PO$_4$	25.13	15.22	36.1 ± 0.7	[11]
[C$_4$C$_1$im]Br	K$_3$PO$_4$	25.82	15.17	35.6 ± 0.8	[11]
[C$_4$C$_1$im][CF$_3$SO$_3$]	K$_3$PO$_4$	25.08	16.18	16.6 ± 0.6	[11]
[C$_4$C$_1$im][C$_1$SO$_3$]	K$_3$PO$_4$	25.03	14.96	10.4 ± 0.4	[11]
[C$_1$im]Cl	K$_3$PO$_4$	15.36	30.53	21.3 ± 0.3	[10]
[im]Cl	K$_3$PO$_4$	15.24	32.80	14.2 ± 0.4	[10]
[P$_{1444}$][C$_1$SO$_4$]	K$_3$PO$_4$	22.95	10.66	9.0 ± 0.1	[13]
[N$_{4444}$]Cl	K$_3$C$_6$H$_5$O$_7$	39.87	20.21	67.35	[16]
[P$_{4444}$]Cl,	K$_3$C$_6$H$_5$O$_7$	39.92	20.06	46.85	[16]
[C$_4$C$_1$pyr]Cl	K$_3$C$_6$H$_5$O$_7$	40.00	20.06	12.57	[16]
[C$_4$C$_1$pip]Cl	K$_3$C$_6$H$_5$O$_7$	40.00	20.06	12.19	[16]
[C$_4$C$_1$im]Cl	K$_3$C$_6$H$_5$O$_7$	39.74	20.35	9.39	[16]
[C$_4$C$_1$im][N(CN)$_2$]	K$_3$C$_6$H$_5$O$_7$	39.92	19.98	13.20	[16]
[C$_4$C$_1$im][SCN]	K$_3$C$_6$H$_5$O$_7$	39.94	20.02	8.37	[16]
[C$_4$C$_1$im][CF$_3$SO$_3$]	K$_3$C$_6$H$_5$O$_7$	39.82	20.41	2.83	[16]

[a]For the ABS formation, 0.5 g IL, 0.3 g K$_3$PO$_4$, and 1.2 mL H$_2$O were added [12]

to control the partition coefficients of biomolecules in ABS. Therefore, the relative hydrophobicity of ILs can provide some insights regarding their ability to extract amino acids.

The first investigation regarding the effect of ILs on the extractive potential of ABS for amino acids was conducted by Coutinho and coworkers in 2009 [10]. In this study, different hydrophilic imidazolium chloride-based ILs and the inorganic salt potassium phosphate, K$_3$PO$_4$, were evaluated as phase-forming components of ABS and then used for the extraction of the amino acid L-tryptophan (as a model amino acid). The main objective of this study was to evaluate the influence of different characteristics of the IL cation's structure on the partitioning of amino acids such as the alkyl side chain length, the number of alkyl groups present in the cation, and the presence of double bonds, aromatic rings, or hydroxyl groups on the

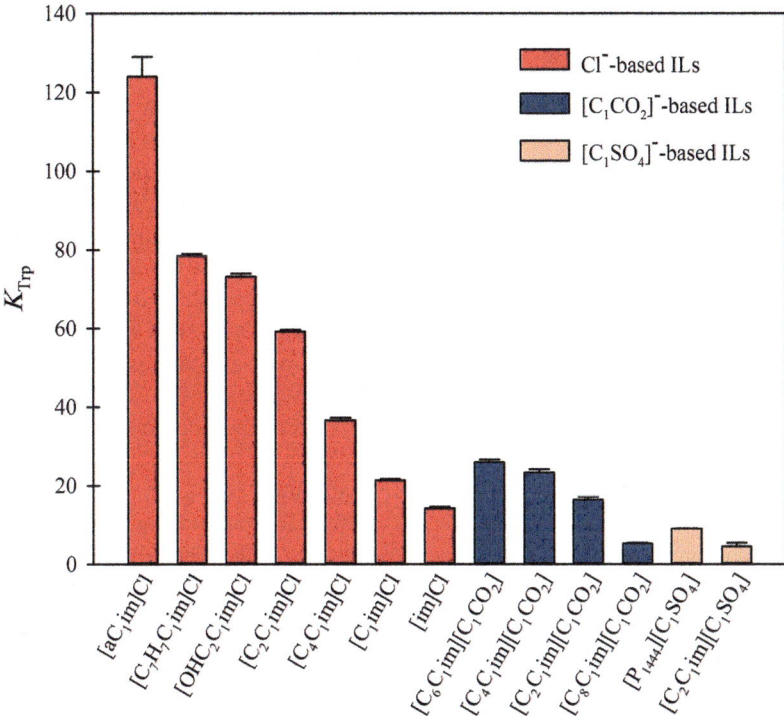

Fig. 5.1 Influence of the IL cation on the partition coefficients of L-tryptophan (K_{Trp}) within ABS composed of ILs and K_3PO_4 at $T = 298$ K [9–13]

alkyl side chain [10]. Table 5.2 presents the mixture compositions employed in this study. The graphical representation of the partition coefficients obtained for L-tryptophan is presented in Fig. 5.1. For a common mixture composition, the partition coefficients of L-tryptophan decrease in the following order of employed imidazolium chloride-based ILs: [aC$_1$im]Cl > [C$_7$H$_7$C$_1$im]Cl > [OHC$_2$C$_1$im]Cl > [C$_2$C$_1$im]Cl > [C$_4$C$_1$im]Cl > [C$_1$im]Cl > [im]Cl [10].

The obtained results demonstrate that the highest partition coefficient is obtained with [aC$_1$im]Cl followed by [C$_7$H$_7$C$_1$im]Cl and [OHC$_2$C$_1$im]Cl, indicating that the presence of double bonds, aromatic rings, or hydroxyl groups in the imidazolium side alkyl chain increases substantially the partition coefficients of L-tryptophan. Regarding the disubstituted imidazolium-based ILs investigated, increasing the IL cation alkyl chain decreases the L-tryptophan partition coefficients. Also, lower L-tryptophan partition coefficients are obtained for the mono- and unsubstituted imidazolium-based ILs due to a decrease of the hydrophobic interactions between the amino acid and the IL [10]. Thus, the partition behavior of L-tryptophan in the studied IL-based ABS seems to be governed by the $\pi \cdots \pi$ stacking, hydrogen-bonding, and hydrophobic interactions.

In regard to the influence of IL anion on the partition coefficients of L-tryptophan, Coutinho and coworkers [11] used the $[C_2C_1im]^+$- and $[C_4C_1im]^+$-based ILs with a large set of different anions to form ABS with K_3PO_4. The partition coefficients and the initial mixture compositions employed for the amino acid partitioning are collected in Table 5.2. The partition coefficients for L-tryptophan obtained in this study are depicted in Fig. 5.2. The partition coefficients of L-tryptophan decrease in the following order: $[C_2C_1im]Cl > [C_2C_1im][CF_3SO_3] \cong [C_2C_1im][C_1CO_2] > [C_2C_1im][C_2SO_4] \cong [C_2C_1im][C_1SO_4]$ and $[C_4C_1im][N(CN)_2] > [C_4C_1im]Cl \cong [C_4C_1im][TFA] \cong [C_4C_1im]Br > [C_4C_1im][CF_3SO_3] > [C_4C_1im][CH_3SO_3]$. In general, it was found that L-tryptophan partitions preferentially into the IL-rich phases composed of halogenated ions, such as Cl^- or Br^-, or of the most hydrophobic anions, that is, anions with higher hydrogen bonding accepting strength such as $[N(CN)_2]^-$ and fluoride-based anions. The authors [11] also concluded that the trend obtained for the effect of the IL anion on the partition coefficients closely follows the Hofmeister series.

A further inspection of Fig. 5.2 indicates that for the 1-alkyl-3-methylimidazolium $[CF_3SO_3]^-$-based ILs, the effect of increasing the cation alkyl side chain length from 2 to 4 is less significant than that observed for imidazolium

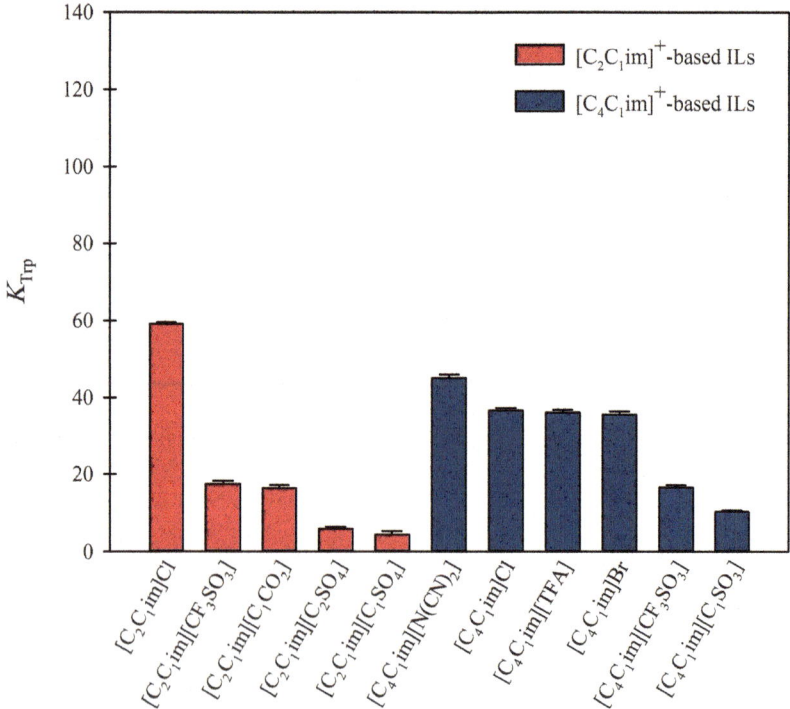

Fig. 5.2 Influence of the IL anion on the partition coefficients of L-tryptophan (K_{Trp}) within ABS composed of ILs and K_3PO_4 at $T = 298$ K (Reproduced from Ref. [9] by permission of The Royal Society of Chemistry)

chloride-based fluids [10, 11]. Still regarding the influence of the IL cation alkyl side chain length on the partitioning behavior of amino acids, Li and coworkers [12] reported the extractive capacity of $\{[C_nC_1im][C_1CO_2]$ ($n=4, 6, 8) + K_3PO_4\}$ ABS for L-tryptophan. The authors [12] demonstrated that the trend for the L-tryptophan partitioning in imidazolium acetate-based ILs is also different of that in imidazolium chloride-based ILs and follows the order: $[C_6C_1im][C_1CO_2] > [C_4C_1im][C_1CO_2] > [C_2C_1im][C_1CO_2] > [C_8C_1im][C_1CO_2]$ (*see* Fig. 5.1) [10, 12]. The partition coefficients for L-tryptophan in $\{[C_nC_1im][C_1CO_2]$ ($n=4, 6, 8) + K_3PO_4\}$ ABS along with the initial mixture compositions employed for partitioning experiments are presented in Table 5.2.

Due to the different behavior displayed by the imidazolium-based ILs with different anions, it can be concluded that the increase in the IL cation alkyl side chain length can either increase or decrease the L-tryptophan partition coefficients. Moreover, the anomalous behavior observed for $[C_8C_1im][C_1CO_2]$ shows that the inherent aptitude of ILs to self-aggregate in aqueous media interferes with their ability to extract amino acids. Self-aggregation does not exist in systems containing ILs with alkyl lengths below hexyl but can occur in aqueous solutions of $[C_nC_1im]^+$- based ILs with n > 8 (after the corresponding critical micelle concentration values are attained) [20].

Louros and coworkers [13] have assessed the ability of phosphonium-based ILs to extract amino acids with reference to the extraction capacity for L-tryptophan of the ABS composed of $[P_{1444}][C_1SO_4]$ and K_3PO_4. The obtained result is presented in Table 5.2. The value of the partition coefficient obtained in this study for L-tryptophan ($K_{Trp}=9.00$) is larger than that observed in the ABS formed by the respective imidazolium-based IL, $[C_2C_1im][C_1SO_4]$, and K_3PO_4 ($K_{Trp}=4.47$) [11]. It must be noted that alkylphosphonium-based ILs are, in general, less dense than water – a fact that can be highly beneficial in product work-up steps for decanting aqueous streams – whereas imidazolium-based ILs are usually denser than water [21]. Phosphonium-based ILs are also thermally more stable and have no acidic protons, making them more stable than imidazolium-based compounds under nucleophilic and basic conditions [22]. Consequently, some of these inherent characteristics of phosphonium-based ILs can be valuable for specific applications.

Since citrate-based salts are biodegradable and nontoxic, Freire and coworkers [16] used potassium citrate ($K_3C_6H_5O_7$, a common biodegradable organic salt) with aqueous solutions of a large array of ILs to form ABS. The authors also evaluated the extractive capacity for amino acids of $\{IL + K_3C_6H_5O_7\}$ ABS using the amino acid L-tryptophan, aimed at exploring their applicability in the biotechnology field [16]. The results obtained in this study for the L-tryptophan partition coefficients are summarized in Table 5.2. The graphical representation of the L-tryptophan partitioning into Cl^--based and $[C_4C_1im]^+$-based ABS is shown in Fig. 5.3. In general, for a common initial mixture composition, the partition coefficient of L-tryptophan decreases in the following order of ILs: $[N_{4444}]Cl > [P_{4444}]Cl > [C_4C_1im][N(CN)_2] > [C_4C_1pyr]Cl \cong [C_4C_1pip]Cl > [C_4C_1im]Cl > [C_4C_1im][SCN] > [C_4C_1im][CF_3SO_3]$. The results obtained indicate that besides the π···π stacking, the hydrophobic interactions, and the possibility of the amino acid-IL

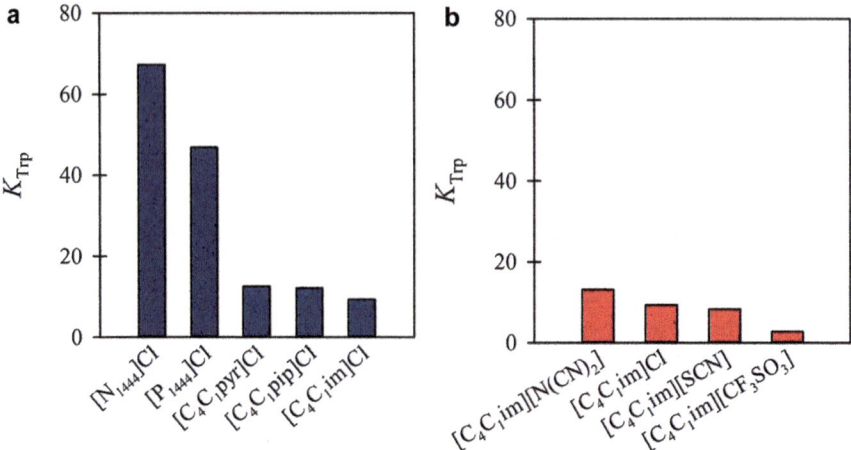

Fig. 5.3 Influence of the IL (**a**) cation and (**b**) anion on the partition coefficients of L-tryptophan (K_{Trp}) in ABS composed of ILs and $K_3C_6H_5O_7$ at $T = 298$ K [16]

hydrogen-bonding interactions [10, 11], the partition coefficient of L-tryptophan depends on a delicate balance between the amount of charged species and the composition of the coexisting phases [16].

Figure 5.3a depicts the partition coefficients of L-tryptophan for the different Cl^--based ILs. As can be seen, the partitioning of the amino acid into the aqueous IL-rich phase decreases in the following order: $[N_{4444}]Cl > [P_{4444}]Cl > [C_4C_1pyr]Cl \cong [C_4C_1pip]Cl > [C_4C_1im]Cl$. This trend also closely correlates with the tie-line length (TLL) corresponding to the common initial ternary composition used for L-tryptophan partitioning in the employed IL-based ABS [16]. TLL is a numerical indicator of the composition difference between the two phases and is generally used to correlate trends in the partitioning of solutes between both phases [9]. Using the distribution ratios obtained for a series of short-chain alcohols in $[C_4C_1im]Cl/K_3PO_4$ ABS at various TLL, Rogers and coworkers [19] showed that the IL-rich phase becomes increasingly hydrophobic as the divergence between the two phases increases, i.e., with an increase in the TLL.

As shown in Fig. 5.3b, for the $[C_4C_1im]^+$-based ILs, the partitioning of L-tryptophan for the IL-rich phase follows the order: $[C_4C_1im][N(CN)_2] > [C_4C_1im]Cl > [C_4C_1im][SCN] > [C_4C_1im][CF_3SO_3]$. This trend is similar to that observed in ABS composed of similar ILs and K_3PO_4 [11]. Moreover, although the partition coefficients of L-tryptophan in ABS composed of diverse ILs and $K_3C_6H_5O_7$ are lower than those obtained using K_3PO_4 [11, 16], the partition coefficients obtained with the citrate-based salt are higher than those observed with ABS formed by $[C_4C_1im][CF_3SO_3]$ and a large variety of carbohydrates [14].

Generally, the results reported in the literature regarding the extraction aptitude for L-tryptophan of ABS composed of hydrophilic ILs and inorganic or organic salts demonstrate that in all the investigated systems [10–13, 16] the amino acid preferentially partitions for the IL-rich phase. Furthermore, the partition coefficients obtained for L-tryptophan using these systems are substantially higher than those observed in conventional polymer-polysaccharide ($K_{Trp} = 1$) [23], polymer-inorganic salts ($K_{Trp} = 1$–7) [24], or water-immiscible ILs two-phase extractions [25]. These studies further confirm that the physicochemical properties of ABS extractive phases can be controlled, aiming at increasing the extraction capability of ABS, by properly manipulating the IL cation/anion design and their combinations. Overall, IL-based ABS are a novel option for the purification and separation of amino acids with much larger partition coefficients than conventional ABS.

5.2.2 Influence of the Salt

Freire and coworkers [16] found that the partition coefficients obtained for L-tryptophan using ABS composed of ILs and potassium citrate are lower than those obtained previously [11] using potassium phosphate. The respective comparison is given in Fig. 5.4.

It is known that an increase of the salt water-structuring nature causes the water to stay preferably in the salt-rich phase, leading to an increase in the concentration of IL at the IL-rich phase and, consequently, to an enhanced phase separation in IL-based ABS. The IL-rich phase becomes more hydrophobic, increasing therefore the extraction capacity of ABS for the hydrophobic L-tryptophan. Apart from its influence on the hydrophobic character of the IL-rich phase, the phase-forming salt

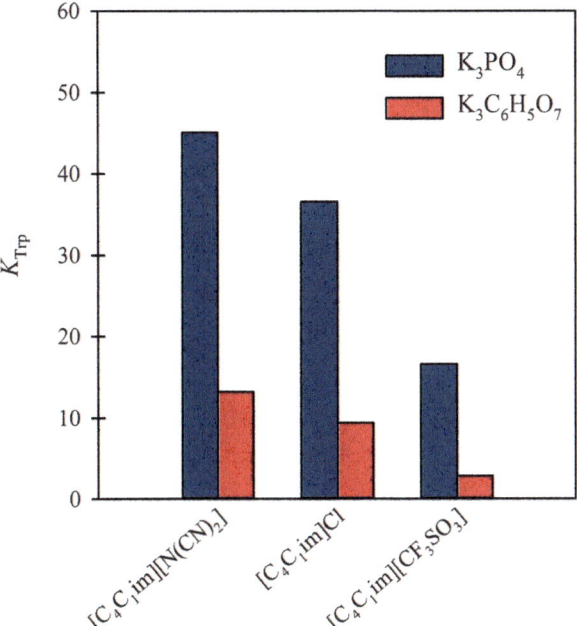

Fig. 5.4 Influence of the salt on the partition coefficients of L-tryptophan (K_{Trp}) in IL-salt ABS at $T = 298$ K [11, 16]

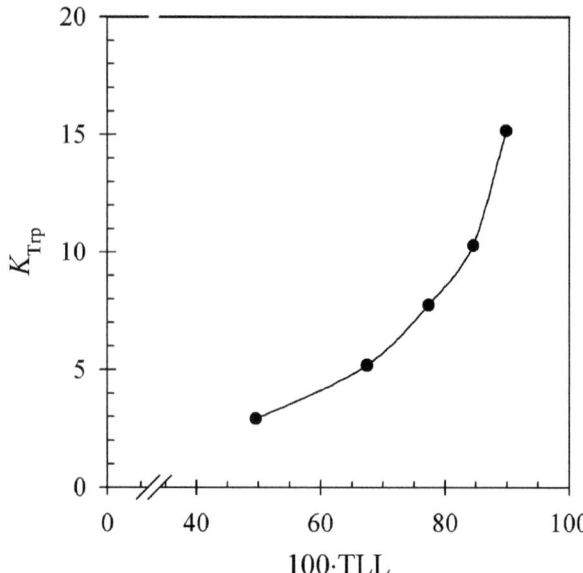

Fig. 5.5 Partition coefficients of L-tryptophan (K_{Trp}) within the {[C_4C_1im]Br + potassium citrate} ABS at $T = 298.15$ K and pH $= 7.00$ versus the TLL [15]

also strongly affects the partitioning of amino acids through its salting-out effects and its influence on the electrostatic character of amino acids by affecting the pH of aqueous media. Compared with potassium citrate, the higher efficacy of potassium phosphate in promoting the migration of the amino acid L-tryptophan into the IL-rich phase is related with its higher salting-out ability.

Our studies regarding the effect of the water-structuring salt concentration on the partitioning of L-tryptophan in {[C_4C_1im]Br + potassium citrate} ABS [15] indicate that with increasing the phase-forming salt concentration, the affinity of L-tryptophan for the IL-rich phase increases. Figure 5.5 depicts the obtained partition coefficients at $T = 298.15$ K and pH $= 7.00$ versus TLL. By increasing the TLL, as substantiated by Rogers and coworkers [19], the IL-rich phase becomes increasingly more hydrophobic, thus enhancing the partitioning of L-tryptophan for the IL-rich phase. On the other hand, as the divergence between the two phases increases, the bottom phase becomes increasingly more concentrated in the salt, thus causing a greater degree of bulky L-tryptophan salted out to the upper phase.

5.2.3 Influence of the Structural/Physicochemical Properties of Amino Acids

To gain an insight into the influence of structural/physicochemical properties of amino acids on their partitioning in IL-based ABS, we studied the extraction aptitude of ABS composed of [C_4C_1im]Br and potassium citrate for five model amino acids: L-tryptophan (Trp), L-phenylalanine (Phe), L-tyrosine (Tyr), L-leucine

(Leu), and L-valine (Val) at $T = 298.15$ K [15]. To examine how closely the distribution properties of amino acids may be related to their structural/physicochemical characteristics, other factors governing the partitioning of the amino acids, namely, the phase compositions and the electrostatic interactions, should be excluded. For this purpose, we studied the partition behavior of amino acids at pH = 6.00 [15]. The isoelectric point values, pI, for Trp, Phe, Tyr, Leu, and Val are, respectively, 5.94, 5.67, 5.66, 5.98, and 5.96 [26]. Thus, electrostatic interactions between amino acids and phase-forming components within the IL-based ABS are diminished at pH = 6.00 (i.e., the pH close to their isoelectric points) due to the development of a net charge of zero on the amino acid molecules. The partition behavior of amino acids at their isoelectric point provides an interesting approach to reveal influences of their structural/physicochemical properties on the extraction efficiency. The results obtained for the partition coefficients of the amino acids Trp, Phe, Tyr, Leu, and Val in the studied ABS at a common initial composition are presented in Table 5.3.

As can be seen, the partition coefficients of the amino acids at pH = 6.00 decrease in the following order: $K_{Trp} > K_{Tyr} > K_{Phe} > K_{Leu} > K_{Val}$. These results agree with the chemical structure of the amino acids. Trp, Tyr, and Phe have an aromatic π system that makes it possible for them to contribute in interactions with other aromatic rings. The imidazolium cation of the IL has also an aromatic π system. It appears thus that the electro-rich aromatic π system on the cationic moiety of the IL produces strong interactions with aromatic molecules capable of undergoing π⋯π interactions. Moreover, Trp has one more aromatic pyrrole ring than Phe and Tyr, fused to the benzene ring in its side chain, and thus displays stronger π⋯π interactions with the IL. As expected, π⋯π interactions between ILs and aromatic amino acids increase with increasing the accessible surface area of the side group of the amino acids in the order: ASA_{Trp}, 271.6 Å^2 > ASA_{Tyr}, 239.9 Å^2 > ASA_{Phe}, 228.6 Å^2 [27]. These results can be used to explain the observed partition coefficients of the aromatic model amino acids, which follow the order: $K_{Trp} > K_{Tyr} > K_{Phe}$. On the other hand, regarding the studied aliphatic amino acids, Val has a hydrophobic –CH2 group less than Leu. This actually makes the "small" molecule of Val relatively hydrophilic, originating a lower affinity for the hydrophobic IL-rich top phase ($K_{Val} < 1$). A similar behavior was observed for the partitioning of Val in the PEG-based ABS [4]. Studies on the amino acid partitioning within the PEG-based ABS by Zaslavsky et al. [28–30] demonstrated that the hydrophobic effect is the determining factor for separation. These studies further revealed that amino acids with relatively low hydrophobicity partition

Table 5.3 Partition coefficients of amino acids in the {[C$_4$C$_1$im]Br + potassium citrate} ABS at pH = 6.00 and T = 298.15 K

Amino acid	IL (wt.%)	Salt (wt.%)	K	Ref
Trp	35.14	21.92	3.664	[15]
Phe	34.76	21.83	1.967	[15]
Tyr	34.85	21.95	2.547	[15]
Leu	35.21	21.88	1.153	[15]
Val	34.77	21.77	0.823	[15]

favorably into the salt-rich phase in the PEG-salt ABS, whereas the amino acids with relatively high hydrophobicity predominantly partition into the PEG-rich phase [30]. Hence, among the various physicochemical descriptors applied in quantitative structure-activity relationships (QSAR) to describe the structure of a given substance, hydrophobicity is the most fundamental descriptor used frequently in studies of ABS separation mechanisms [3]. According to the definition, the hydrophobicity of a substance is a measure of the overall intensity of the total interactions of the substance with an aqueous medium (including hydrogen bonding, van der Waals, electrostatic interactions, etc.).

To examine the effect of hydrophobicity on the partitioning of amino acids within the studied IL-based ABS, we have plotted in Fig. 5.6 the logarithm of the partition coefficients as a function of the hydrophobicity scale of amino acids. A good correlation has been obtained. As can be seen, except for Phe and Tyr ($K_{Tyr} > K_{Phe}$; while, H_{Phe}, 5.12 (0.34) $\cong H_{Tyr}$, 5.10 (0.47)), the partition coefficients of amino acids, in general, increase with their increasing hydrophobicity, confirming the importance of the hydrophobic interactions as a driving force for their extraction in IL-based ABS. It is noticeable that the "relative hydrophobicity" values used in this discussion have been calculated from the partition behavior of amino acids in PEG-based ABS (the data in parenthesis are deviations) [3]. Therefore the differences between the properties of the coexisting phases in PEG- and IL-based ABS seem to be responsible for the discrepancy observed in the obtained trends for Tyr and Phe; the relative hydrophobicity of a solute is independent of the particular two-phase system used, provided that there are no specific interactions of the solute with the components present in the system. However, as mentioned above, Phe and Tyr display π⋯π interactions with imidazolium-based ILs.

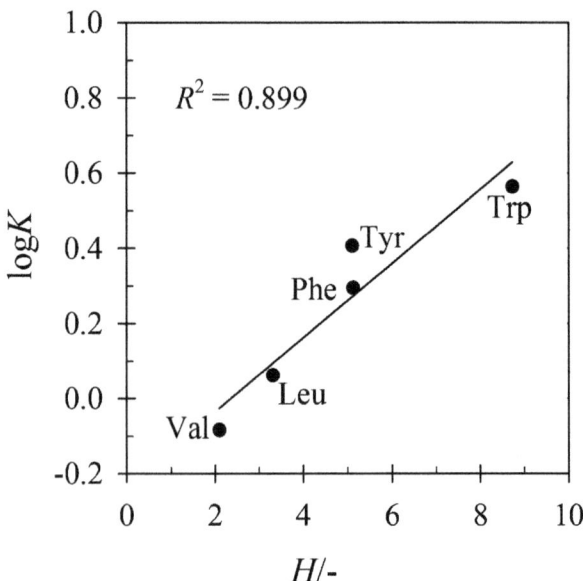

Fig. 5.6 Partition coefficients of the model amino acids within the {[C$_4$C$_1$im]Br + potassium citrate} ABS at $T = 298.15$ K and pH = 6.00 *versus* the hydrophobicity, H, of amino acids (Reproduced from Ref. [15] by permission of John Wiley & Sons Ltd)

5.2.4 Influence of the pH

To evaluate the effects of solution pH on the extraction capability of IL-based ABS for amino acids, we also studied the partitioning of five model amino acids, Trp, Phe, Tyr, Leu, and Val, within the {[C_4C_1im]Br + potassium citrate} ABS at different aqueous medium pH values (5.00, 6.00, and 7.00) and different phase compositions [15]. The partition coefficients and the corresponding mixture compositions are presented in Table 5.4.

Depending on the pH of the solution, the characteristic functional groups of amino acids (i.e., −COOH and −NH$_2$) can be ionized/protonated and lead to different equilibrium forms: a cationic form, a zwitterionic form, and an anionic form. The cationic form is predominant in the range of pH < pK_1, but the zwitterionic and the anionic forms are the main forms in the range of pK_1 < pH < pK_2 and pH > pK_2, respectively. The protonation constant values for the studied amino acids are (p$K_{a1,Trp}$ = 2.38 and p$K_{a2,Trp}$ = 9.34), (p$K_{a1,Phe}$ = 2.18 and p$K_{a2,Phe}$ = 9.09), (p$K_{a1,Tyr}$ = 2.24, p$K_{a2,Tyr}$ = 9.04 and p$K_{a3,Tyr}$ = 10.10), (p$K_{a1,Leu}$ = 2.32 and p$K_{a2,Leu}$ = 9.58), and (p$K_{a1,Val}$ = 2.27 and p$K_{a2,Val}$ = 9.52) [27]. Since the pH values of the mixtures used for the partitioning experiments are between pK_{a1} and pK_{a2}, all the studied amino acids are in their zwitterionic forms. Moreover, as discussed previously, according to the pI values, a net charge of zero on the amino acid molecules is promoted at pH = 6.00 (i.e., the pH close to their isoelectric points). Before the isoelectric point, with the decreasing of the pH of the aqueous phase to pH (=5.00), the percentage of the cationic form increases due to the protonation of the carboxylate group, whereas after the isoelectric point, pH increasing to pH (=7.00) causes the percentage of the anionic form to increase due to the dissociation of the ammonium group. Both carboxylate and ammonium groups provide a more hydrophilic character to the amino acids in contrast to their conjugated forms, i.e., carboxylic acid and amine groups, respectively [31]; it was also found that a carboxylate group is more hydrophilic than an amino group from the energy point of view [4]. In summary, the solution pH can affect the partitioning of amino acids within an ABS through its influences on the electrostatic and hydrophilicity/hydrophobicity characteristics of amino acids. On the other hand, the charged characteristics of the citrate anions (i.e., the salting-out agent) can greatly be affected by the pH of aqueous solutions. The citrate ions with higher valence that exist in the alkaline medium pH are more effective in salting out, compared to those with minor valence found in acidic medium [32].

Effects of the aqueous medium pH on the partition coefficients of amino acids in the studied IL-based ABS are shown in Fig. 5.7. As can be seen, the changes in the pH of aqueous solutions can produce significant changes in the partitioning of aromatic model amino acids, namely, Trp, Phe and Tyr; on the other hand, the same changes in the pH of aqueous solutions have no considerable effects on the partitioning of the aliphatic model amino acids, namely, Leu and Val.

With regard to the most hydrophobic aromatic amino acid, Trp, as it can be seen in Fig. 5.7, the partition coefficient increases in the order: K_{Trp} (pH = 7.00) > K_{Trp}

Table 5.4 pH effect on the partition coefficients of the amino acids, L-tryptophan, L-phenylalanine, L-tyrosine, L-leucine, and L-valine in the {[C_4C_1im]Br + potassium citrate} ABS at $T = 298.15$ K [15]

IL (wt.%)	Salt (wt.%)	100TLL	K
L-Tryptophan			
pH = 5.00			
35.02	23.93	44.24	1.849
35.02	25.70	60.52	2.436
35.11	27.06	71.16	2.737
35.03	29.06	78.82	3.155
35.01	30.92	85.06	3.549
pH = 6.00			
34.90	19.66	43.25	2.428
35.14	21.92	58.55	3.664
34.96	23.65	67.60	4.801
35.00	25.62	72.62	6.123
34.93	26.91	76.38	6.245
pH = 7.00			
34.88	18.22	49.51	2.909
34.90	21.91	67.40	5.183
34.78	25.45	77.29	7.748
34.80	27.66	84.53	10.287
35.56	29.46	89.82	15.146
L-Phenylalanine			
pH = 5.00			
34.70	23.79	35.97	1.397
34.79	25.64	53.70	1.416
35.44	27.01	66.26	1.449
34.55	28.36	69.06	1.497
33.91	28.76	70.15	1.470
pH = 6.00			
34.68	19.64	38.04	1.754
34.76	21.83	55.39	1.967
35.42	23.74	67.04	2.150
34.70	25.65	71.60	2.172
35.07	27.01	76.35	2.248
pH = 7.00			
34.77	18.12	44.18	1.586
34.58	21.88	63.73	1.838
34.85	25.37	75.26	2.374
34.26	29.74	85.05	2.877
33.74	32.52	87.53	3.040
33.56	32.64	89.86	3.182
L-Tyrosine			
pH = 5.00			
34.95	24.09	48.87	1.350

(continued)

Table 5.4 (continued)

IL (wt.%)	Salt (wt.%)	100TLL	K
34.93	25.86	60.92	1.467
34.97	27.14	70.17	1.618
35.24	29.01	80.10	1.691
34.96	30.93	84.18	1.859
pH = 6.00			
34.71	19.79	39.48	1.492
34.85	21.95	58.43	2.547
34.91	23.82	62.38	2.676
34.88	25.71	70.43	3.044
34.97	26.89	76.55	3.155
pH = 7.00			
35.09	18.16	50.12	1.845
35.90	21.97	70.45	2.143
34.76	25.38	76.88	2.273
34.84	27.70	82.50	2.543
34.86	29.78	87.38	2.684
L-Leucine			
pH = 5.00			
35.18	23.85	45.24	0.986
34.98	25.68	58.06	1.140
34.98	27.01	67.51	1.197
35.08	29.01	74.54	1.119
35.04	30.91	79.25	1.207
pH = 6.00			
35.03	19.80	50.91	1.050
35.21	21.88	62.89	1.153
34.80	23.69	68.98	1.261
34.92	25.83	74.81	1.227
34.91	27.05	76.33	1.299
pH = 7.00			
34.82	18.14	43.46	1.182
34.70	21.88	63.24	1.155
34.93	25.42	76.42	1.188
34.32	29.78	85.37	1.436
33.33	32.29	89.22	1.399
L-Valine			
pH = 5.00			
35.11	23.86	47.26	0.852
35.49	25.44	59.81	0.849
35.34	27.32	69.12	0.811
35.17	29.23	76.47	0.712
35.37	31.00	81.58	0.776

(continued)

Table 5.4 (continued)

IL (wt.%)	Salt (wt.%)	100TLL	K
pH = 6.00			
34.81	19.61	45.49	0.832
34.77	21.77	60.98	0.823
34.73	23.72	68.56	0.748
35.03	25.74	76.73	0.744
34.83	26.94	78.97	0.852
pH = 7.00			
34.71	18.11	43.89	0.828
34.59	21.91	64.64	0.704
35.21	25.42	78.70	0.687
34.34	29.72	87.64	0.715
33.24	32.26	91.07	0.647

(pH = 6.00) > K_{Trp} (pH = 5.00). This trend is the result of several factors: (1) decreasing the pH leads to an increase in the percentage of the citrate ions with minor valence, diminishing their salting-out effect, and hence the IL-rich top phase becomes less hydrophobic; (2) at lower pH values, the percentage of the cationic form of Trp increases dramatically, and hence a lower tendency toward extraction into the IL-rich phase is expected; (3) with the increase in the cationic form of Trp, the $\pi \cdots \pi$ interactions between the imidazolium cation, and the aromatic residue of Trp, which have already been mentioned before as possible driving forces for the enhanced extraction of aromatic amino acids, are weakened resulting in a decrease in the partitioning of Trp into the IL-rich top phase. Although at pH = 7.00 a higher tendency toward enrichment into the IL-rich phase is expected due to an increase in its hydrophobicity, it seems that, especially for less hydrophobic aromatic amino acids, such as Phe and Tyr, the hydrophobic/hydrophilic and the IL-amino acid-specific interactions are compensated by unfavorable electrostatic interactions.

In general, the results obtained in this study demonstrate that the partition of amino acids in ABS is driven by a combination of hydrophobic/hydrophilic, electrostatic, and IL-amino acid-specific interactions.

5.2.5 *Analysis of the Forces Governing the Partitioning of Amino Acids in Ionic Liquid-Based ABS*

The rational design of IL-based ABS to achieve the optimal partition conditions is clearly influenced by the ability to understand or predict the partitioning of a target molecule between the phases. If the partitioning of the target molecule could be

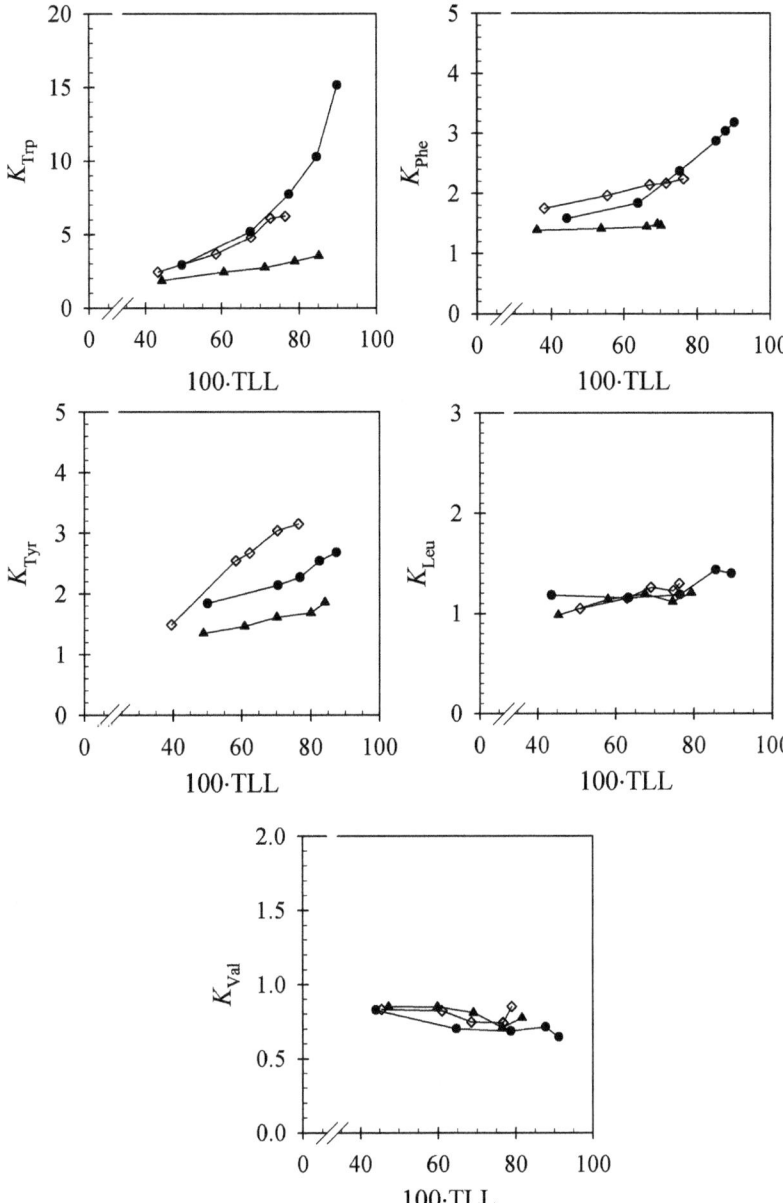

Fig. 5.7 Effect of pH on the partition coefficients of amino acids in the {[C_4C_1im]Br + potassium citrate} ABS at $T = 298.15$ K: ▲, pH = 5.00; ◇, pH = 6.00; ●, pH = 7.00 (Reproduced from Ref. [15] by permission of John Wiley & Sons Ltd)

reliably predicted, extraction in ABS could be optimized by calculations only. However, the quantitative modeling of partitioning in the complex IL-based ABS poses an extremely complex problem because of its dependence on a broad array of factors. Hence, it would be desirable to develop models and correlations which allow the prediction of the distribution properties of amino acids within IL-based ABS using the QSAR descriptors of amino acids and ABS characteristics.

To explore the contributions of the amino acid and ABS characteristics (i.e., the forces governing partitioning) to the observed partition coefficients, we applied the regression analysis of the experimental partition coefficient data. For this purpose, we first plotted in Figs. 5.6 and 5.8 the logarithm of the partition coefficients at $T = 298.15$ K and pH $= 6.00$, as a function of some amino acids physicochemical/structural descriptors (*see* Table 5.5) such as the relative hydrophobicity, H, normalized van der Waals volume, v, accessible surface area, ASA, molecular

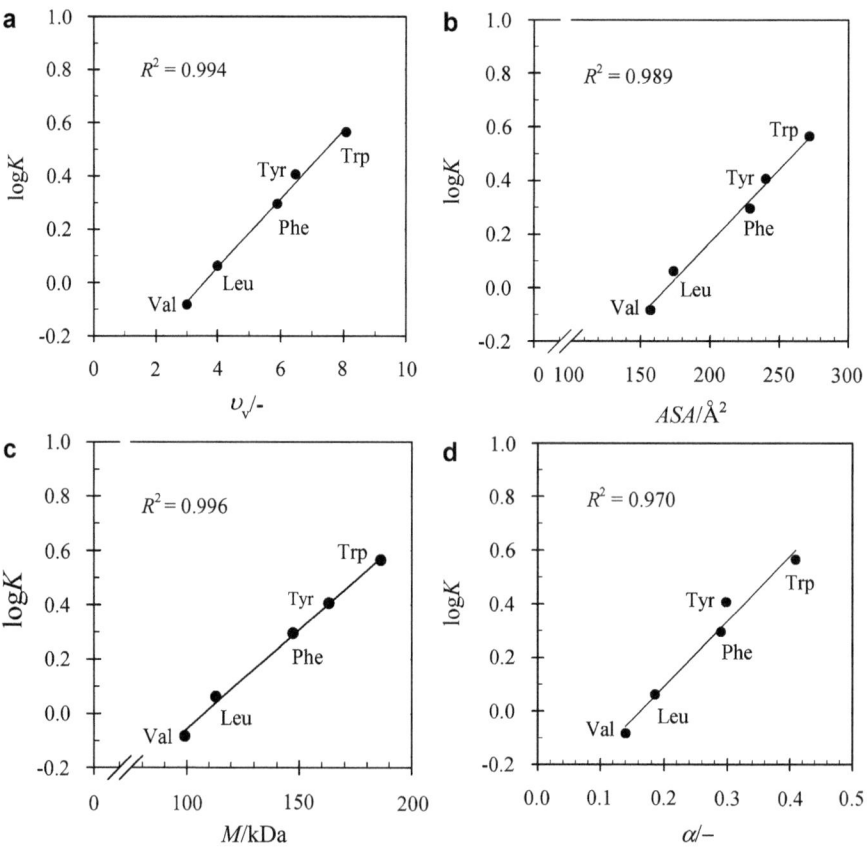

Fig. 5.8 Partition coefficients of the model amino acids within the $\{[C_4C_1im]Br + potassium\ citrate\}$ ABS at $T = 298.15$ K and pH $= 6.00$ *versus* the properties of amino acids. (**a**) Normalized van der Waals volume, v_v; (**b**) accessible surface area, ASA; (**c**) molecular weight, M; and (**d**) polarizability, α (Reproduced from Ref. [15] by permission of John Wiley & Sons Ltd)

Table 5.5 Properties of amino acids used for analysis of amino acid partitioning in IL-based ABS [15]

Amino acid	α^a	v_v^b	$ASA^c/Å^2$	H^d	M^e/(Dalton)
Trp	0.409	8.08	271.6	8.73 (0.39)	186.22
Phe	0.290	5.89	228.6	5.12 (0.34)	147.18
Tyr	0.298	6.47	239.9	5.10 (0.47)	163.18
Leu	0.186	4.00	173.7	3.31 (0.47)	113.16
Val	0.140	3.00	157.2	2.09 (0.22)	99.14

[a]Polarizability [27]
[b]Volume (normalized van der Waals volume) [33]
[c]Accessible surface area (ASA) [27]
[d]Average relative hydrophobicity of side chains of amino acid residues; data in parenthesis are deviations [3]
[e]Residue mass of amino acids. These values were calculated by subtracting the value of M (water) = 18.015 Da from the corresponding molecular weight [26] values

weight, M, and polarizability, α. As can be seen in Figs. 5.6 and 5.8, good correlations have been obtained. Thus, it was found that "log K" as a dependent variable could satisfactorily be explained as a function of the independent variables: the amino acids' structural/physicochemical properties, including x_H, x_v, x_{ASA}, x_M, and x_α, the charge of amino acids or phase components (represented by x_{pH} of the moiety), and the phase composition characteristic (x_{TLL}) as follows:

$$\log K = (b_0 + b_1 x_v + b_2 x_{ASA} + b_3 x_M + b_4 x_H + b_5 x_\alpha)\exp\left[b_6(x_{TLL})(x_{pH})\right] \quad (5.2)$$

where b_i (i from 0 to 6) represent adjustable parameters. It is noticeable that there is no simple linear correlation between TLL, pH values, and the partitioning of amino acids in IL-based ABS.

We started by evaluating the performance of the predictive model through its application to the correlation of the distribution behavior of three model amino acids (Trp, Phe, and Val) within the studied IL-based ABS. Values of the adjustable parameters obtained from nonlinear regression of the experimental "log K" of Trp, Phe, and Val to the model are collected in Table 5.6 for each working pH value. It was found that the proposed model can accurately be used to reproduce log K of three model amino acids. The corresponding correlation coefficients of regression, R^2, along with the absolute average relative deviations ($AARD$) for the calculated partition coefficients are also given in Table 5.6. To illustrate the reliability of the proposed model, the correlation between the calculated and experimentally determined K values for three model amino acids in the investigated IL-based ABS is shown in Fig. 5.9.

To verify the proposed predictive model, by the use of the parameters given in Table 5.6, the performance of Eq. (5.2) in predicting the partition behavior of two other amino acids, i.e., Tyr and Leu, within the investigated IL-based ABS has been examined. Since the corresponding experimental partition coefficients for Tyr and Leu have no contribution to obtaining the fitting parameters of Eq. (5.2), they can be

Table 5.6 Parameters of Eq. (5.2) for the correlation of the partition coefficient (K) in the {[C_4C_1im]Br + potassium citrate} ABS at $T = 298.15$ K and different total compositions [15]

	pH = 5.00	pH = 6.00	pH = 7.00
$b_0 \times 10^5$	214.757	−8946.0	1.25078
$b_1 \times 10^3$	69.92	59.12	49.18
$b_2 \times 10^7$	−8842.68	−6936.39	953.136
$b_3 \times 10^6$	−1151.41	−64.6214	−2300.36
$b_4 \times 10^6$	3547.50	378.134	18,870
$b_5 \times 10^8$	1670.50	−1427.33	−8276.21
$b_6 \times 10^3$	315.19	312.96	315.99
R^2	0.997	0.990	0.997
$AARD^a$	0.04	0.06	0.04
$AARD_{Trp}$	0.02	0.05	0.03
$AARD_{Phe}$	0.04	0.07	0.02
$AARD_{Val}$	0.03	0.05	0.05
$AARD_{Tyr}^b$	0.04	0.10	0.07
$AARD_{Leu}^b$	0.08	0.04	0.06
R^{2c}	0.991	0.987	0.997

$^aAARD = (1/N)\sum_{i=1}^{N} \left|K_i^{calcd} - K_i^{exptl}\right|/K_i^{exptl}$ where N represents the number of data

^{b}Corresponded to predicted values

^{c}The predicted values are also taken into account

used to verify the reliability of the predicted log K value. The corresponding $AARD$ values are also given in Table 5.6. On the basis of the obtained $AARD$ values given in Table 5.6, we conclude that Eq. (5.2) can be satisfactorily used to predict the distribution behavior of amino acids within the investigated ABS. To show the reliability of the model, a comparison between the experimental and predicted partition coefficients of Tyr and Leu are also shown in Fig. 5.9.

5.3 Extraction of Amino Acids with ABS Composed of Ionic Liquids and Carbohydrates

Freire et al. [14] have used a broad range of carbohydrates including monosaccharides, disaccharides, and polyols with aqueous solutions of 1-butyl-3-methylimidazolium trifluoromethanesulfonate ([C_4C_1im][CF_3SO_3]) as a water-stable IL to form ABS. Carbohydrates are noncharged, biodegradable, nontoxic, and a renewable feedstock. The mono and disaccharides are polyhydroxy aldehydes or ketones with a high affinity for water and salting-out aptitude (several –OH groups with dual donor/acceptor character that are involved in hydrogen bonding). The reduction of such aldoses and ketoses to an alcohol functionalized group leads to polyols with enhanced affinity for water and salting-out ability. Therefore, the

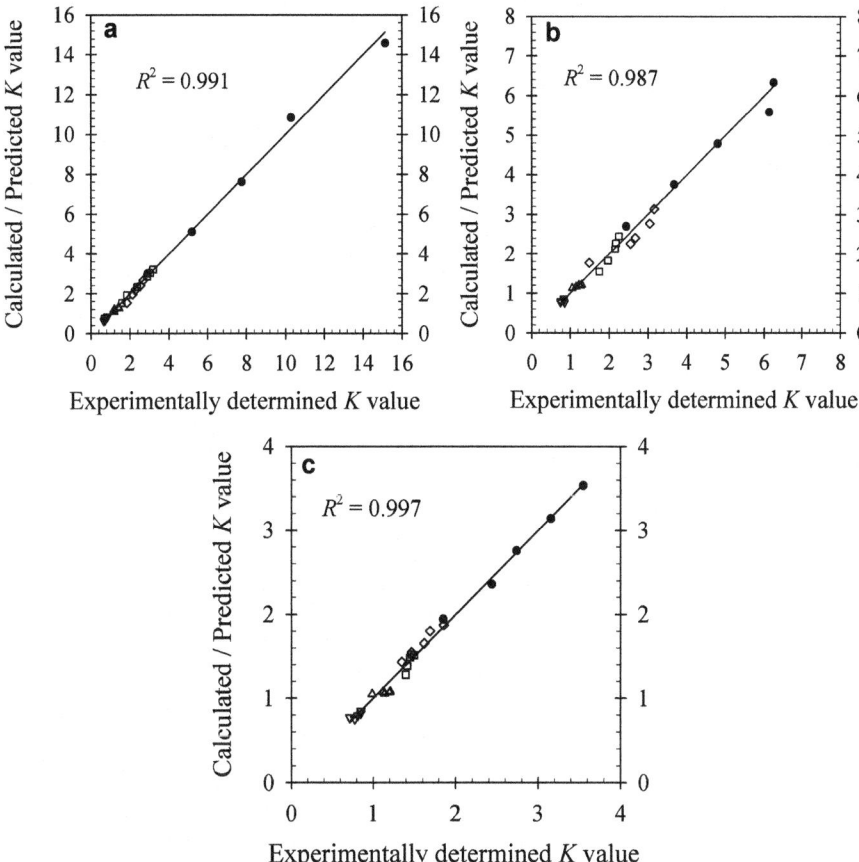

Fig. 5.9 Correlation between the calculated/predicted and experimentally determined K values for the partitioning of amino acids in the $\{[C_4C_1im]Br + \text{potassium citrate}\}$ ABS at $T = 298.15$ K and different phase compositions. ●, Calcd. Trp; □, Calcd. Phe; ◇, Pred. Tyr; △, Pred. Leu; ▽, Calcd. Val: (**a**) pH = 7.00; (**b**) pH = 6.00; (**c**) pH = 5.00 (Reproduced from Ref. [15] by permission of John Wiley & Sons Ltd)

use of these common carbohydrates in IL-based ABS instead of high charge density salts might lead to improved biotechnological routes [14].

Freire and coworkers [14] have also evaluated the extractive potential of these $\{[C_4C_1im][CF_3SO_3] + \text{carbohydrate}\}$ ABS for amino acids through the partitioning of the model amino acid L-tryptophan between the two aqueous phases. The partition coefficients, respective standard deviations, and exact weight fraction compositions of each component, are collected in Table 5.7.

Partition coefficients of L-tryptophan (K_{Trp}) in IL-carbohydrate ABS were found to be somewhat higher than those observed in PEG-polysaccharide ABS [23]. This fact leads to the conclusion that $[C_4C_1im][CF_3SO_3]$ presents an improved extraction ability over PEGs since weaker salting-out species are used (monosaccharides,

Table 5.7 Partition coefficients of L-tryptophan in ABS composed of [C$_4$C$_1$im][CF$_3$SO$_3$] and carbohydrates at $T = 298$ K [14]

Carbohydrate	[C$_4$C$_1$im][CF$_3$SO$_3$] (wt.%)	Carbohydrate (wt.%)	K_{Trp}
Sucrose	39.93	25.04	1.13 ± 0.02
D-(+)-glucose	40.11	24.94	1.06 ± 0.04
D-(−)-fructose	39.98	24.97	1.08 ± 0.02
D-(+)-mannose	39.84	24.91	1.08 ± 0.04
D-(+)-xylose	40.08	24.93	1.19 ± 0.04
D-maltitol	40.00	24.99	1.06 ± 0.01
Xylitol	39.82	24.89	1.06 ± 0.05
D-sorbitol	39.98	24.97	1.13 ± 0.06

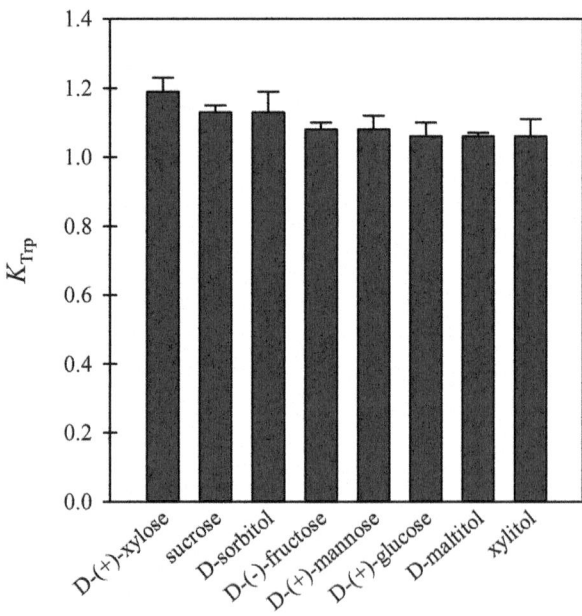

Fig. 5.10 Partition coefficients of L-tryptophan (K_{Trp}) in ABS composed of [C$_4$C$_1$im][CF$_3$SO$_3$] and carbohydrates at $T = 298$ K [14]

disaccharides, and polyols instead of polysaccharides). The enhanced ability for the extraction of L-tryptophan by hydrophilic imidazolium-based ILs seems to result from $\pi\cdots\pi$ and H-bonding interactions occurring between the IL cation and L-tryptophan [14]. The authors [14] also found that the partition coefficients for L-tryptophan are almost independent of the carbohydrate used (Fig. 5.10), meaning that the possible hydrophobic interactions between carbohydrates and tryptophan are not significant and have no impact on the extraction ability of these molecules.

The overall picture of this study reveals that IL-carbohydrate-based ABS extraction abilities depend on the hydrophobicity of the solute, the IL nature, and the IL-solute-specific interactions [14].

5.4 Extraction of Amino Acids with ABS Composed of Ionic Liquids and Polymers

After the pioneer work of Rebelo and coworkers [34] regarding the salting-in and salting-out effects of ILs on polymers dissolved in aqueous media, the work of Zafarani-Moattar et al. [17] is the only one concerning the evaluation of the extraction capability for amino acids of ABS composed of ILs and polymers. This kind of ABS can present interesting characteristics shared by polymers and ILs, such as low volatility, rapid phase disengagement, gentle biocompatible environment, good solvation ability, tunable physical properties, and high design capacity for achieving task-specific phase components to enhance the partitioning of target species.

Zafarani-moattar et al. [17] have investigated the partitioning behavior of two essential amino acids, L-tryptophan (Trp) and L-tyrosine (Tyr), within $\{[C_2C_1im]Br + PPG\ 400\}$ ABS at $T = 298.15$ K. The obtained results for the partition coefficients of Trp and Tyr together with the ABS total compositions at which the partition coefficients values were determined are collected in Table 5.8. It should be noted that the partition coefficients are in this case defined as the ratio between the concentrations of each amino acid in the PPG-rich phase to that in the IL-rich phase.

In the studied ABS, Trp displays K_{Trp} less than 1 for all compositions, denoting the Trp preferential affinity for the $[C_2C_1im]Br$-rich bottom phase; while, depending on the TLL, Tyr preferentially migrates either for the $[C_2C_1im]Br$-rich bottom phase ($K_{Tyr} < 1$ for the first tie-line) or PPG-rich top phase ($K_{Tyr} > 1$ for the other two tie-lines). Both of the amino acids, Trp and Tyr, have an aromatic π system that makes it possible for them to interact with the imidazolium cation of the $[C_2C_1im]Br$. Furthermore, Trp has one more aromatic pyrrole ring than Tyr in its side chain, and thus displays stronger $\pi\cdots\pi$ interactions with $[C_2C_1im]Br$. Consequently, a higher tendency toward enrichment into the $[C_2C_1im]Br$-rich bottom phase is expected for Trp. Regarding Tyr, the partitioning seems also to be governed by the hydrogen-bonding interactions occurring between the –OH group in its side chain and PPG. In a diluted ABS with a shorter TLL, the PPG-water hydrogen-bonding interactions are stronger, replacing the role of hydrogen-bonding interactions between PPG and Tyr leading to a lower affinity for the PPG-rich top phase ($K_{Tyr} < 1$). However, in a concentrated ABS with a longer TLL, the less available water molecules performing hydrogen-bonding with PPG causes a higher affinity of Tyr for the PPG-rich phase ($K_{Tyr} > 1$) [17]. The plot of the amino acid partition

Table 5.8 Partition coefficients of amino acids L-tryptophan and L-tyrosine in ABS composed of $[C_2C_1im]Br$ and PPG 400 at $T = 298$ K [17]

$[C_2C_1im]Br$ (wt.%)	PPG 400 (wt.%)	K_{Trp}	K_{Tyr}	α^a
38.01	38.00	0.145 ± 0.006	0.500 ± 0.005	3.4
43.23	43.22	0.075 ± 0.006	1.44 ± 0.007	19.2
47.50	47.00	0.054 ± 0.007	2.05 ± 0.015	38.0

[a]Separation factor which is defined by $\alpha = K_{Tyr}/K_{Trp}$

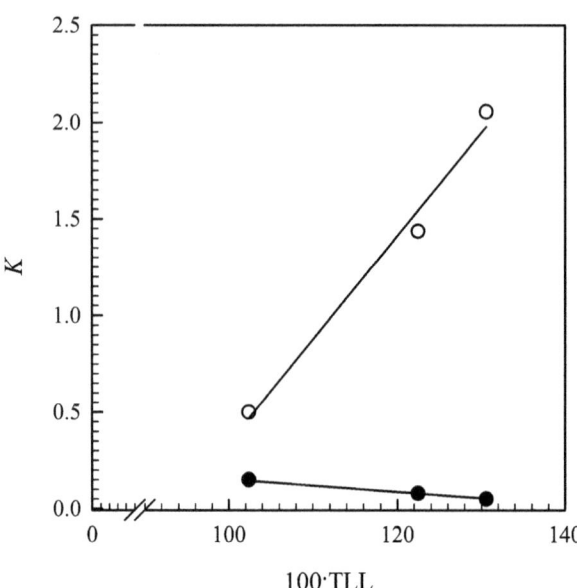

Fig. 5.11 Plots of partition coefficients of amino acids *versus* tie-line length (TLL) in the {PPG 400 + [C$_2$C$_1$im]Br} ABS at $T = 298.15$ K. ●, L-tryptophan; ○, L-tyrosine; —, best fitted linear curves with coefficients of determination, $R^2_{Trp} = 0.997$ and $R^2_{Tyr} = 0.985$ (Reproduced from Ref. [17] by permission of John Wiley & Sons Ltd)

coefficients *versus* the TLL is also shown in Fig. 5.11, demonstrating the existence of a linear correlation between them.

Tyr is one of the main impurity sources of amino acids in Trp fermentation broth. In order to illustrate the efficiency of separation, the separation factor, α, that is, the ratio between the two partition coefficients, was calculated at different compositions and collected in the last column of Table 5.8. The separation factors obtained indicate that an efficient separation of two amino acids, Trp and Tyr, can be obtained in a multiple step process using {[C$_2$C$_1$im]Br + PPG 400} ABS. These findings demonstrate that this kind of ABS may be used as viable alternatives for the purification of L-tryptophan from its fermentation broth.

5.5 Extraction of Amino Acids with ABS Composed of Ionic Liquids, Salts and Polymers

Apart from the application of ILs as main constituents of ABS, ILs can also be used as adjuvants for the extraction of biomolecules in typical polymer-salt ABS. Pereira and coworkers [35] were the first to study the effect of adding various imidazolium-based ILs to conventional PEG-salt ABS on the phase behavior and extraction capability for L-tryptophan. The goal of this work was to use ILs as additives to control the polarities of the phases of common polymer-salt ABS for enhanced yield of product recovery. Diverse ILs at 5 wt.% were added to the {PEG 600 + Na$_2$SO$_4$} ABS, and their extraction aptitude for L-tryptophan was evaluated. In this work, several structural features of ILs were evaluated, namely, the effect of the

Table 5.9 Partition coefficients of L-tryptophan (K_{Trp}) in the quaternary {PEG 600 + Na$_2$SO$_4$ + 5 wt.% IL} ABS at $T = 298.15$ K [35]

IL	PEG 600 (wt.%)	Na$_2$SO$_4$ (wt.%)	IL (wt.%)	K_{Trp}
No IL	40.00	5.04	—	14.93
[im]Cl	40.04	5.02	5.04	4.82
[C$_1$im]Cl	40.01	5.01	5.04	4.69
[C$_2$C$_1$im]Cl	39.97	5.10	5.01	7.63
[C$_4$C$_1$im]Cl	40.15	5.01	4.98	21.84
[C$_4$C$_1$C$_1$im]Cl	39.93	5.02	5.14	28.62
[OHC$_2$C$_1$im]Cl	40.01	5.07	5.00	4.72
[aC$_1$im]Cl	39.93	5.01	5.08	11.88
[C$_7$H$_7$C$_1$im]Cl	39.94	5.03	5.00	23.46
[C$_4$C$_1$im][C$_1$CO$_2$]	39.97	5.01	4.96	19.24
[C$_4$C$_1$im][C$_1$SO$_4$]	39.98	5.01	5.03	20.76
[C$_4$C$_1$im][HSO$_4$]	39.99	5.06	5.02	17.08

IL cation side alkyl chain length, the number of alkyl substitutions at the cation, the addition of functionalized groups at the aliphatic chain of the cation, and the anion nature. The partition coefficients obtained in this study for L-tryptophan (K_{Trp}) at a common initial composition are reported in Table 5.9. The K_{Trp} dependence on the IL cation and anion is displayed in Fig. 5.12. The partition coefficient corresponds to the ratio between the concentration of Trp at the PEG-rich phase and the Na$_2$SO$_4$-rich phase.

The obtained results demonstrate that K_{Trp} is larger than 1.0 for all cases, denoting the amino acid preferential partitioning for the PEG-rich phase (hydrophobic phase). Moreover, it was found that the K_{Trp} observed for the systems using ILs with salting-out inducing characteristics were smaller than those obtained for the ABS reference system without IL, but it showed a strong increase with the increase of the alkyl chain size of the cation. This trend follows the IL salting-in-/out inducing ability. While salting-out inducing ILs, such as [im]Cl, [C$_1$im]Cl, [C$_2$C$_1$im]Cl, [OHC$_2$C$_1$im]Cl, and [aC$_1$im]Cl, reduce K_{Trp}, the remaining ILs (salting-in inducing ILs) increase the partitioning of the amino acid. In addition, removing the most acidic hydrogen at the C$_2$ position of the ILs by increasing the number of alkyl chain substitutions, as in [C$_4$C$_1$C$_1$im]Cl, leads to a higher K_{Trp} when compared to [C$_4$C$_1$im]Cl. It was further found that it is mainly the IL partition to the PEG-rich phase that improves the partitioning of L-tryptophan (*see* Fig. 5.12). The presence of benzyl groups at the IL cation has also been observed to enhance the ability of the PEG-rich phase for extracting the amino acid. In this particular case, it appears that the additional contributions of π···π stacking interactions of a second aromatic ring are responsible for the increase in the K_{Trp} values. Furthermore, the observed effect of the IL anion follows the anion hydrophobicity [11], but it is minor when compared with the influence of the IL cation. This effect is identical to what was observed for IL-based ABS previously studied [10, 11].

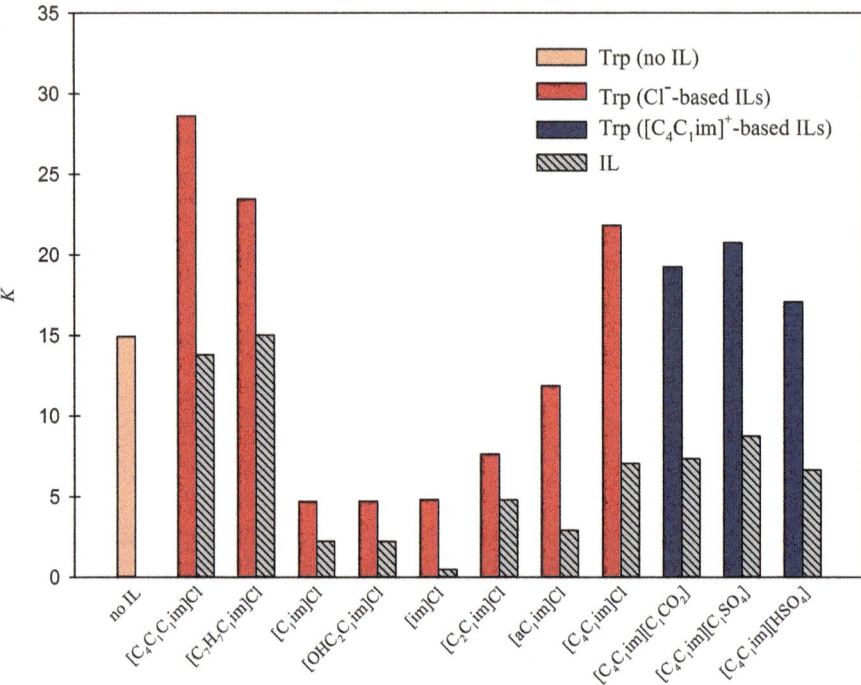

Fig. 5.12 Influence of the IL cation/anion on the partition coefficients of L-tryptophan (K_{Trp}) within ABS composed of PEG 600 and Na_2SO_4 at $T = 298.15$ K [35]

Hamzehzadeh and Vasiresh [36] studied the influence of [C_4C_1im]Br as an additive on the extraction aptitude for L-tryptophan of ABS composed of PEG 400 and $K_3C_6H_5O_7$ (as a common biodegradable salt) at $T = 298.15$ K and various TLLs. The partition coefficients obtained for L-tryptophan are summarized in Table 5.10. The graphical representation of the L-tryptophan partition coefficients *versus* TLL is presented in Fig. 5.13. The obtained results for the L-tryptophan partition coefficients reveal that the presence of [C_4C_1im]Br enhances up to twice the extraction ability by the polymer-rich phase. Moreover, it was found that it is the presence of the [C_4mim]Br on the PEG-rich phase that enhances the partitioning of L-tryptophan. As expected, the affinity of the amino acid to the preferred phase is governed by an increase on the TLL due to the higher phase divergences. In addition, comparing with our previous results using [C_4C_1im]Br as the main phase-forming component in the {[C_4C_1im]Br + $K_3C_6H_5O_7$} IL-based ABS, the partition of L-tryptophan into the extractive phase is substantially higher when using [C_4C_1im]Br as an adjuvant in the {PEG 400 + $K_3C_6H_5O_7$} polymer-based ABS. With the {[C_4C_1im]Br + $K_3C_6H_5O_7$} IL-based ABS, the highest K_{Trp} is 15.15 [15].

Hamzehzadeh and Abbasi [37] also studied the influence of [C_4C_1im]Br as an adjuvant on the extraction capacity for L-tyrosine of ABS composed of PEG600 and $K_3C_6H_5O_7$ at $T = 298.15$ K and various TLLs. The obtained results are collected in

Table 5.10 Partition coefficients of L-tryptophan (K_{Trp}) in the quaternary {PEG 400 + $K_3C_6H_5O_7$ + 5 wt.% [C_4C_1im]Br} ABS at $T = 298.15$ K [36]

PEG 400 (wt.%)	$K_3C_6H_5O_7$ (wt.%)	[C_4C_1im]Br (wt.%)	K_{Trp}
No IL			
25.11	25.03		6.51
26.99	25.99		8.01
29.79	26.86		19.00
32.92	28.98		49.47
With 5 wt.% [C_4C_1im]Br			
25.20	25.23	5.13	11.98
27.14	26.09	5.13	16.53
30.19	27.16	5.07	41.58
33.16	29.44	5.05	93.55

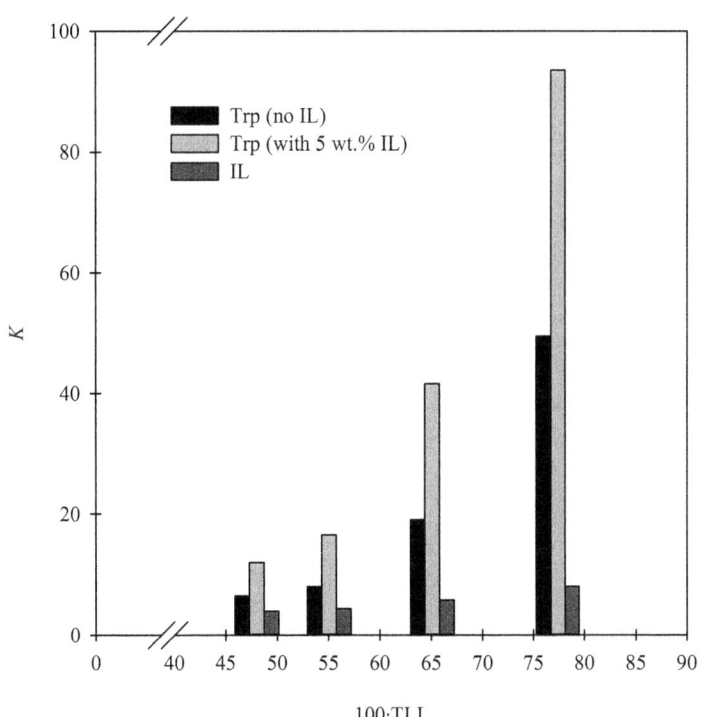

Fig. 5.13 Partition coefficients of [C_4C_1im]Br (dark gray) and L-tryptophan (gray) in ABS composed of PEG 400, $K_3C_6H_5O_7$, and 5 wt% [C_4C_1im]Br at $T = 298.15$ K and different compositions [36]

Table 5.11. It was found that [C_4C_1im]Br has a high affinity for the hydrophobic PEG-rich phase leading to higher extraction efficiencies by the polymer-rich phase for L-tyrosine (*see* Fig. 5.14). In addition, when [C_4C_1im]Br is used as adjuvant in

Table 5.11 Partition coefficients of L-tyrosine (K_{Tyr}) in the quaternary {PEG 600 + $K_3C_6H_5O_7$ + 5 wt.% [C_4C_1im]Br} ABS at $T = 298.15$ K [37]

PEG 600 (wt.%)	$K_3C_6H_5O_7$ (wt.%)	[C_4C_1im]Br (wt.%)	K_{Tyr}
No IL			
25.05	19.97		1.51
26.96	22.92		2.60
30.00	25.98		3.56
33.02	28.99		6.04
35.92	31.97		7.23
With 5 wt.% [C_4C_1im]Br			
25.01	20.01	4.97	1.84
27.03	23.04	4.96	3.12
30.06	26.00	4.96	4.52
33.01	29.00	4.97	6.95
35.32	31.39	4.93	8.16

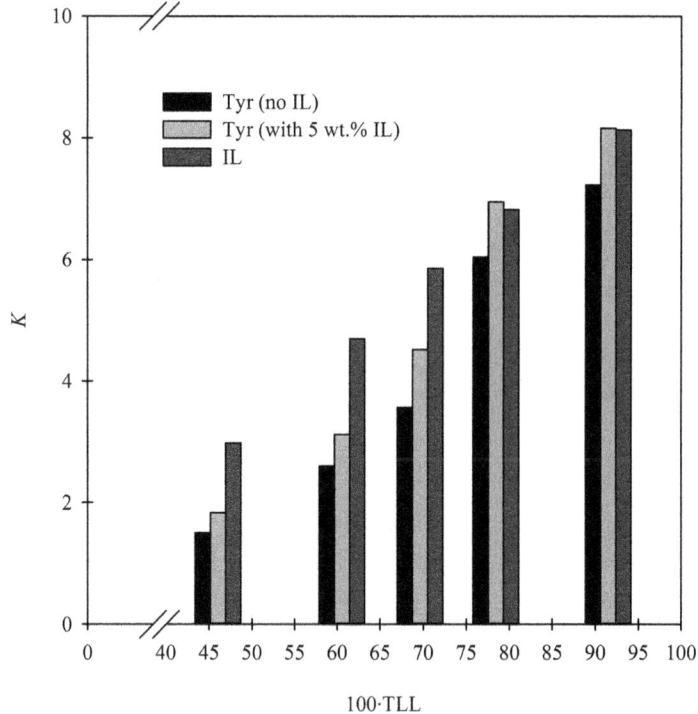

Fig. 5.14 Partition coefficients of [C_4C_1im]Br (dark gray) and L-tyrosine (gray) in ABS composed of PEG 600, $K_3C_6H_5O_7$, and 5 wt% [C_4C_1im]Br at $T = 298.15$ K and different compositions [37]

the {PEG 600 + $K_3C_6H_5O_7$} polymer-based ABS, the partition of L-tyrosine into the extractive phase is higher than that obtained with the {$[C_4C_1im]Br + K_3C_6H_5O_7$} IL-based ABS. With the {$[C_4C_1im]Br + K_3C_6H_5O_7$} IL-based ABS, the highest K_{Tyr} obtained was 3.14 [15].

In general, it can be concluded that the addition of small amounts of ILs to classical PEG-based ABS could largely control the extraction efficiency of amino acids, which further indicates that the application of ILs as adjuvants to modify the characteristics of the polymer-rich phase could be an interesting alternative to the common approach of PEG functionalization.

5.6 Conclusions

It is clearly patent that IL-based ABS are enhanced techniques for the extraction of amino acids when compared with conventional ABS. The partition coefficients obtained for amino acids using IL-based ABS are considerably higher than those observed in conventional polymer-polysaccharide and polymer-inorganic salts, especially for aromatic amino acids. Moreover, ILs can be fine-tuned in order to increase the extraction efficiencies and selectivities. The viscosity of IL-based ABS is also generally lower than that of traditional PEG systems, which can be an important advantage when envisaging the large-scale application of these systems.

The successful industrial application of IL-based ABS in downstream processing is still hampered by the ability to understand factors governing the partitioning of a target molecule between the phases. Although the mechanisms governing the partitioning of amino acids in IL-based ABS are complex and yet not fully understood, based on the partition coefficients reported in the literature, it can be concluded that hydrophobic interactions, hydrogen-bonding, electrostatic interactions, salting-out effects, and IL-amino acid-specific interactions, such as $\pi \cdots \pi$ interactions, play important roles and cannot be neglected.

References

1. Albertsson P-Å (1986) Partition of cell particles and macromolecules: separation and purification of biomolecules, cell organelles, membranes, and cells in aqueous polymer two-phase systems and their use in biochemical analysis and biotechnology. Wiley, New York
2. Walter H, Brooks DE, Fisher D (1986) Partitioning in aqueous two–phase system: theory, methods, uses, and applications to biotechnology. Elsevier, Orlando
3. Zaslavsky BY (1994) Aqueous two-phase partitioning: physical chemistry and bioanalytical applications. CRC Press, New York
4. Hatti-Kaul R (2000) Aqueous two-phase systems: methods and protocols, vol 11. Springer, New York
5. Hatti-Kaul R (2001) Aqueous two-phase systems. Mol Biotechnol 19(3):269–277

6. Albertsson P-Å (1958) Partition of proteins in liquid polymer-polymer two-phase systems. Nature 182:709–711
7. Adachi M, Harada M, Shioi A, Sato Y (1991) Extraction of amino acids to microemulsion. J Phys Chem 95(20):7925–7931
8. Itoh H, Thien M, Hatton T, Wang D (1990) A liquid emulsion membrane process for the separation of amino acids. Biotechnol Bioeng 35(9):853–860
9. Freire MG, Cláudio AFM, Araújo JM, Coutinho JAP, Marrucho IM, Lopes JNC, Rebelo LPN (2012) Aqueous biphasic systems: a boost brought about by using ionic liquids. Chem Soc Rev 41(14):4966–4995
10. Neves CMSS, Ventura SPM, Freire MG, Marrucho IM, Coutinho JAP (2009) Evaluation of cation influence on the formation and extraction capability of ionic-liquid-based aqueous biphasic systems. J Phys Chem B 113(15):5194–5199
11. Ventura SPM, Neves CMSS, Freire MG, Marrucho IM, Oliveira J, Coutinho JAP (2009) Evaluation of anion influence on the formation and extraction capacity of ionic-liquid-based aqueous biphasic systems. J Phys Chem B 113(27):9304–9310
12. Li Z, Pei Y, Liu L, Wang J (2010) (Liquid + liquid) equilibria for (acetate-based ionic liquids + inorganic salts) aqueous two-phase systems. J Chem Thermodyn 42(7):932–937
13. Louros CL, Cláudio AFM, Neves CMSS, Freire MG, Marrucho IM, Pauly J, Coutinho JAP (2010) Extraction of biomolecules using phosphonium-based ionic liquids + K3PO4 aqueous biphasic systems. Int J Mol Sci 11(4):1777–1791
14. Freire MG, Louros CL, Rebelo LPN, Coutinho JAP (2011) Aqueous biphasic systems composed of a water-stable ionic liquid + carbohydrates and their applications. Green Chem 13(6):1536–1545
15. Zafarani-Moattar MT, Hamzehzadeh S (2011) Partitioning of amino acids in the aqueous biphasic system containing the water-miscible ionic liquid 1-butyl-3-methylimidazolium bromide and the water-structuring salt potassium citrate. Biotechnol Prog 27(4):986–997
16. Passos H, Ferreira AR, Cláudio AFM, Coutinho JAP, Freire MG (2012) Characterization of aqueous biphasic systems composed of ionic liquids and a citrate-based biodegradable salt. Biochem Eng J 67:68–76
17. Zafarani-Moattar MT, Hamzehzadeh S, Nasiri S (2012) A new aqueous biphasic system containing polypropylene glycol and a water-miscible ionic liquid. Biotechnol Prog 28(1):146–156
18. Dupont J, de Souza RF, Suarez PA (2002) Ionic liquid (molten salt) phase organometallic catalysis. Chem Rev 102(10):3667–3692
19. Gutowski KE, Broker GA, Willauer HD, Huddleston JG, Swatloski RP, Holbrey JD, Rogers RD (2003) Controlling the aqueous miscibility of ionic liquids: aqueous biphasic systems of water-miscible ionic liquids and water-structuring salts for recycle, metathesis, and separations. J Am Chem Soc 125(22):6632–6633
20. Akama Y, Sali A (2002) Extraction mechanism of Cr (VI) on the aqueous two-phase system of tetrabutylammonium bromide and $(NH_4)_2SO_4$ mixture. Talanta 57(4):681–686
21. Neves CMSS, Carvalho PJ, Freire MG, Coutinho JAP (2011) Thermophysical properties of pure and water-saturated tetradecyltrihexylphosphonium-based ionic liquids. J Chem Thermodyn 43(6):948–957
22. Atefi F, Garcia MT, Singer RD, Scammells PJ (2009) Phosphonium ionic liquids: design, synthesis and evaluation of biodegradability. Green Chem 11(10):1595–1604
23. Lu M, Tjerneld F (1997) Interaction between tryptophan residues and hydrophobically modified dextran: effect on partitioning of peptides and proteins in aqueous two-phase systems. J Chromatogr A 766(1):99–108
24. Salabat A, Abnosi MH, Motahari A (2008) Investigation of amino acid partitioning in aqueous two-phase systems containing polyethylene glycol and inorganic salts. J Chem Eng Data 53(9):2018–2021
25. Wang J, Pei Y, Zhao Y, Hu Z (2005) Recovery of amino acids by imidazolium based ionic liquids from aqueous media. Green Chem 7(4):196–202. doi:10.1039/b415842c

26. Fauchère JL, Charton M, Kier LB, Verloop A, Pliska V (1988) Amino acid side chain parameters for correlation studies in biology and pharmacology. Int J Pept Protein Res 32(4):269–278
27. Lide D (2007) Handbook of chemistry and physics. Internet Version 2007. CRC Press 200:2006–2007
28. Zaslavsky BY, Mestechkina NM, Miheeva LM, Rogozhin SV (1982) Measurement of relative hydrophobicity of amino acid side-chains by partition in an aqueous two-phase polymeric system: Hydrophobicity scale for non-polar and ionogenic side-chains. J Chromatogr A 240(1):21–28. doi:http://dx.doi.org/10.1016/S0021-9673(01)84003-6
29. Zaslavsky BY, Miheeva LM, Gasanova GZ, Mahmudov AU (1987) Effect of polymer composition on the relative hydrophobicity of the phases of the biphasic system aqueous dextran-poly(ethylene glycol). J Chromatogr A 403(0):123–130. doi:http://dx.doi.org/10.1016/S0021-9673(00)96346-5
30. Zaslavsky BY, Gulaeva ND, Djafarov S, Masimov EA, Miheeva LM (1990) Phase separation in aqueous poly(ethylene glycol)-(NH4)2SO4 systems and some physicochemical properties of the phases. J Colloid Interface Sci 137(1):147–156. doi:http://dx.doi.org/10.1016/0021-9797(90)90051-O
31. Gulyaeva N, Zaslavsky A, Lechner P, Chait A, Zaslavsky B (2003) pH dependence of the relative hydrophobicity and lipophilicity of amino acids and peptides measured by aqueous two-phase and octanol–buffer partitioning. J Pept Res 61(2):71–79
32. Zafarani-Moattar MT, Hamzehzadeh S (2011) Effect of pH on the phase separation in the ternary aqueous system containing the hydrophilic ionic liquid 1-butyl-3-methylimidazolium bromide and the kosmotropic salt potassium citrate at $T = 298.15$ K. Fluid Phase Equilib 304(1):110–120
33. Charton M, Charton BI (1982) The structural dependence of amino acid hydrophobicity parameters. J Theor Biol 99(4):629–644
34. Visak ZP, Lopes JNC, Rebelo LPN (2007) Ionic liquids in polyethylene glycol aqueous solutions: salting-in and salting-out effects. Monatsh Chem 138(11):1153–1157
35. Pereira JFB, Lima ÁS, Freire MG, Coutinho JAP (2010) Ionic liquids as adjuvants for the tailored extraction of biomolecules in aqueous biphasic systems. Green Chem 12(9):1661–1669
36. Hamzehzadeh S, Vasiresh M (2014) Ionic liquid 1-butyl-3-methylimidazolium bromide as a promoter for the formation and extraction capability of poly (ethylene glycol)-potassium citrate aqueous biphasic system at $T = 298.15$ K. Fluid Phase Equilib 382:80–88
37. Hamzehzadeh S, Abbasi M (2015) The influence of 1-butyl-3-methyl-imidazolium bromide on the partitioning of l-tyrosine within the {polyethylene glycol 600+ potassium citrate} aqueous biphasic system at $T = 298.15$ K. J Chem Thermodyn 80:102–111

Chapter 6
Extraction of Proteins with ABS

Rupali K. Desai, Mathieu Streefland, Rene H. Wijffels, and Michel H.M. Eppink

Abstract Over the past years, there has been an increasing trend in research on the extraction and purification of proteins using aqueous biphasic systems (ABS) formed by polymers, e.g., polyethylene glycol (PEG). In general, when dealing with protein purification processes, it is essential to maintain their native structure and functional stability. In this context, ABS, liquid-liquid systems where both phases are water-rich, provide a biocompatible medium for such attempts. More recently, it was shown that the versatility offered by ABS is further enhanced by the introduction of ionic liquids (ILs) as alternative phase-forming components. This chapter describes and highlights the current progress on the field of protein extraction and purification using IL-based ABS. The general approach for protein extraction using IL-based ABS and factors influencing the partitioning are discussed. In addition, the challenges to overcome the use of IL-based ABS for protein extraction are also presented.

Keywords Proteins • Ionic liquids • Aqueous biphasic system • Extraction • Purification • Stability

6.1 Introduction

Proteins are an integral part of all living systems and have various applications in food and feed (both relatively low value) and pharmaceuticals (high value). Purification of proteins involves various unit operations, using low to high resolution techniques, to obtain proteins with desired purity and quality. Proteins, being fragile molecules, can be easily denatured by acid/base solutions, salts, and high temperature. Therefore, mild operation conditions for their recovery and purification are required, in order to maintain their nativity and functionality. With the current advances in biotechnology, a large increase in the titers of protein production was already observed; yet, the development of cost-effective purification methods is

---

R.K. Desai (✉) • M. Streefland • R.H. Wijffels • M.H.M. Eppink
Bioprocess Engineering Department, Wageningen University, 16, 6700 AA, Wageningen, The Netherlands
e-mail: rupali.desai@wur.nl; rene.wijffels@wur.nl; michel.eppink@wur.nl

still required. The high cost of protein purification continues to remain a bottleneck in downstream processing of proteins and mainly for protein value-added biopharmaceuticals. On the other hand, in the field of food and feed, proteins are obtained from, e.g., soya, and also there has been a growing interest in third-generation biofuels from microalgae. For instance, in fuel production processes, large amounts of proteins are generated which could be used for feed and food [1–3]. In fact, to make these processes economically feasible, it is necessary to refine other components from biomass. Proteins are a major fraction of algae biomass and are normally denatured by the solvents used for lipids extraction. The main challenge therefore lies in separating the proteins in their native form without affecting their functionality. Thus, depending on the biomass or initial medium, protein purification protocols vary and drive the development of more specific, robust, and cost-effective methods [4].

Aqueous biphasic systems (ABS) based on polymers were first proposed by Albertsson [5], who studied their applicability in protein extraction and purification. ABS allow the integration of concentration and purification processing steps and serve as an alternative approach to the traditional processes. Typical ABS are formed by mixing polymer-polymer and polymer-salt combinations above given concentrations to form two distinct aqueous phases, each one enriched in one of the phase-forming components. Both phases are water-rich (~80–90 % w/w), and thus ABS can provide a mild and gentle environment for protein separation without affecting their native structure and stability [6, 7]. In addition to the largely investigated polymer-based ABS, in the last decade, ionic liquids (ILs) were proposed as alternative phase-forming components of ABS [8]. And because of the inherent properties of ILs, this possibility allowed the use of ABS in a new range of applications.

The interest on ILs as extractive solvents increased primarily because of their nonvolatile nature, which is the major advantage over traditional organic volatile solvents. In addition to their nonvolatility, ILs, being composed of cations and anions, can be more easily tuned to achieve specific properties, such as a tunable polarity, viscosity, and solvent miscibility. Their tunable polarity enabled them to be used in biotransformations to increase substrate solubility, to dissolve enzymes, and to tailor the reaction rate [9]. Moreover, due to their tailoring ability, ILs are also able to form ABS not only with inorganic salts but also with polymers [10], carbohydrates [11], and amino acids [12]. The main advantage of IL-based ABS over the conventional systems comprises their ability to tune and tailor the properties of the coexisting phases by permutation and combination of different cations and anions, thereby improving the selectivity of these systems for a wide variety of solutes [13].

Based on the advantages and large recent interest on IL-based ABS for separation purposes, this chapter describes the general approaches of protein purification described in the literature using IL-based ABS and factors that influence the partitioning of proteins in these systems. The challenges in developing a successful ABS for extraction of proteins are also discussed. Finally, this chapter aims a better understanding on the mechanisms ruling protein extraction using IL-based ABS.

Fig. 6.1 Approach required for protein extraction using IL-based ABS

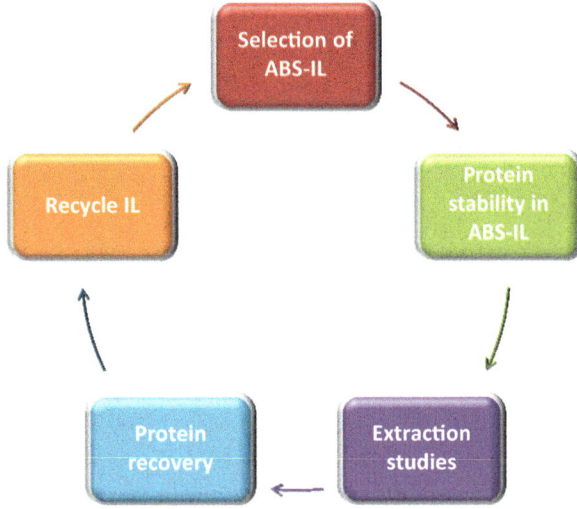

Figure 6.1. depicts a scheme on the approach required to use IL-based ABS for the extraction of proteins.

6.2 Extraction of Proteins/Enzymes Using IL-Based ABS

The extraction of proteins using IL-based ABS has been studied by different research groups and for which a summary is given in Table 6.1. This table was adapted from [14] and updated to include more recent studies.

As a first point, only water-miscible ILs are able to form ABS since water-immiscible ILs do not form two aqueous-rich phases (see Table 6.2). IL-based ABS are formed by mixing water-miscible ILs with salts, carbohydrates, amino acids, and polymers [8, 10, 12, 15]. The ability to form ABS with solutes other than salts has indeed been studied [8, 10, 12, 15], but their efficiency in extracting proteins is however scarcely studied. Although more promising than inorganic salts in what concerns the use of more biocompatible systems, these alternative systems suffer the drawback of only being able to form ABS with a limited number of ILs due to their low salting-out ability (carbohydrates, amino acids, and polymers versus salts).

In general, most of the studies reported in the literature deal with imidazolium-based ILs composed of halogens or $[BF_4]^-$ anions. Recently, ABS based on guanidinium-based ILs have been studied for protein extraction [16, 17] and where it was shown that model proteins, such as BSA, could be extracted with high efficiency for the IL-rich phase without losing its native structure and maintain its stability. IL-based ABS could thus serve as potential platforms for protein extraction if the stability of proteins at the IL-rich phase is maintained. While

Table 6.1 Investigated IL-based ABS for the extraction of proteins

Protein	IL-based ABS	References
Bovine serum albumin (BSA)	[C_4mim]Cl/K_2HPO_4, Ammoeng 110™/K_2HPO_4-KH_2PO_4, [$C_n$mim]Br($n=4,6,8$)/K_2HPO_4, [C_4mim][N(CN)$_2$]/K_2HPO_4, ILs^a/K_2HPO_4, guanidinium-based ILs/K_2HPO_4	[16, 17, 18, 19, 36, 37, 40, 41]
Ovalbumin	[C_4mim]Cl/K_2HPO_4, ILs^a/K_2HPO_4, guanidinium-based ILs/K_2HPO_4	[16, 36, 37, 43]
Lysozyme	Ammoeng 110™/K_2HPO_4-KH_2PO_4, guanidinium-based ILs/K_2HPO_4	[16, 18]
γ-globulin	[$C_n$mim]Br($n=4,6,8$)/K_2HPO_4	[19]
Myoglobin	[C_4mim]Cl/K_2HPO_4, Ammoeng 110™/K_2HPO_4-KH_2PO_4,	[18, 43]
Hemoglobin	[C_4mim]Cl/K_2HPO_4, ILs^a/K_2HPO_4, hydroxyl ammonium ionic liquids^a/K_2HPO_4	[36, 37, 43]
Cytochrome c	[C_4mim]Cl/K_2HPO_4, [$C_n$mim]Br ($n=4,6,8$)/K_2HPO_4, amino-based ILs^a/K_3PO_4, glycine-based ILs^a/K_2HPO_4	[19, 43, 44, 45]
Fungal proteins	[C_4mim]Cl/K_3PO_4	[46]
Trypsin	Ammoeng 110™/K_2HPO_4-KH_2PO_4, [$C_n$mim]Br ($n=4,6,8$)/K_2HPO_4, ILs^a/K_2HPO_4, guanidinium-based ILs/K_2HPO_4	[16, 18, 19, 36]
Lipase CaL-A	[C_2mim][C_4SO_4]/(NH_3)$_2SO_4$	[47]
Lipase CaL-B	Imidazolium-based ILs^a/K_2HPO_4-KH_2PO_4	[48]
Thermomyces lanuginosus lipase (TlL)	[C_2mim][C_2SO_4]/K_2CO_3	[49]
Alcohol dehydrogenases	(Ammoeng 100™/Ammoeng 101™/Ammoeng 110™)/K_2HPO_4-KH_2PO_4	[13]
Horseradish peroxidase	[C_4mim]Cl/K_2HPO_4	[39]
RuBisCo	Iolilyte 221PG/KH_2PO_4-Na_2HPO_4	[25]
Wheat esterase	[C_4mim][BF_4]/NaH_2PO_4	[50]

Updated from [14]
^{a}Details of ILs used can be found in the corresponding literature

Table 6.2 Commonly used cation/anions combination of water-miscible versus water-immiscible ILs

Water-miscible ILs	Cations	Imidazolium, pyridinium, ammonium, phosphonium
	Anions	Chloride, bromide, fluoride, alkylsulfate, tosylate, tetrafluoroborate, dicyanamide
Water-immiscible ILs^a	Cations	Imidazolium, pyridinium, ammonium, phosphonium
	Anions	Bistriflimide, hexafluorophosphate

^{a}Water-immiscible ILs also contain water (~2–6 %)

most of the studies focused on the extraction efficiency of model proteins/enzymes, the studies carried out by Dreyer [18] and Pei [19] made an attempt to understand the mechanisms responsible for the high extractions attained.

6.2.1 Stability of Proteins in IL-Rich Phases

Proteins are complex macromolecules and require a gentle environment to maintain their structural and functional integrity. Changes in this environment, such as solvent concentration, pH, ionic strength, and temperature, could result in denaturation of proteins. Thus, the primary criterion for any protein purification process is the ability to maintain the proteins' structural integrity and functionality. In this context, when using IL-based ABS for the purification of proteins, it is necessary to understand their stability in aqueous solutions of ILs. There are some studies carried out to infer on protein-IL interactions [20, 21] and where model proteins have been used, namely, BSA, lysozyme, and cytochrome c. However, there are other proteins with higher commercial value, such as monoclonal antibodies, RuBisCo (ribulose-1,5-bisphosphate carboxylase/oxygenase), etc., that should be studied in what concerns their stability in aqueous solutions of ILs and their feasibility to be extracted by ABS. Although some hydrophobic ILs are able to stabilize enzymes [22, 23], they are not discussed in this chapter since these do not form ABS.

In the studies regarding the stability of proteins in aqueous solutions of ILs, the techniques employed to monitor the proteins' structural and thermal stability include UV spectroscopy, fluorescence, circular dichroism (CD), small-angle neutron scattering (SANS), differential scanning calorimetry (DSC), dynamic light scattering (DLS), and size exclusion chromatography (SEC). The stability studies were designed to address the factors that influence the formation of ABS and the stability of proteins, such as (*i*) type of IL; (*ii*) concentration of IL; (*iii*) other process conditions, such as pH, ionic strength, and temperature; and (*iv*) protein properties, such as size, charge, and surface hydrophobicity.

ABS consist of two aqueous-rich phases: an IL-rich phase and a phase rich in salt, polymer, amino acid, or carbohydrate. The concentration of IL in the IL-rich phase of ABS can vary from 1.5 to 3.0 mol/kg [24], and thus it is prudent to study IL-protein interactions in aqueous solutions. Moreover, it was already shown that the concentration of IL has a strong influence on the protein stability [21, 25]. In our recent study, we have shown that the protein's stability in aqueous solutions of ILs is influenced by the concentration of IL and by the protein properties, such as size and complexity of the molecule [25]. In this study, the stability of BSA, IgG, and RuBisCo was studied in aqueous solutions of two ILs, Iolilyte 221 PG and Cyphos 108, at different concentrations [25]. It was found that as the concentration of the IL increases (0–50 %, v/v), the proteins start forming aggregates. RuBisCo (~540 kDa), being a large complex protein/enzyme that consists of eight large and small subunits, begins to aggregate at lower IL concentrations (~30 %, v/v), while BSA (~67 kDa), a smaller protein, forms no aggregates or only negligible aggregates at 50 % v/v of IL(Iolilyte 221 PG). IgG (~150 kDa), with an intermediate size, forms aggregates at 50 % (v/v) of Iolilyte 221 PG. In this study [25], the aggregate formation was monitored using SEC and DLS studies. In an additional study, SANS results showed that human serum albumin and cytochrome c form aggregates at high concentrations (50 %, v/v) of [C_4mim]Cl and retain their high-order structure at lower IL concentrations (25 %, v/v) [26]. Lysozyme and interleukin-2 (IL-2) showed increased thermal stability in aqueous solutions up to 40 % (w/w) of IL, although it is dependent on the pH [27], indicating thus that the charge of the protein also influences its stability in IL aqueous solutions. Different ILs with varying "kosmotropicity" were also investigated for their effect on protein structure and long-term stability [28]. In this study, cytochrome c showed no significant changes in its structure when dissolved in hydrated choline dihydrogen phosphate (containing 20 % w/w of water). Cytochrome c additionally showed a higher thermal and long-term stability, leading the authors to conclude that the "kosmotropicity" of ILs has strong implications on the proteins' stability [28].

The influence of inorganic salts and ion-specific-induced precipitation of proteins is well described by the Hofmeister series [29]. Since ILs are also composed of ions, their influence on protein stability in aqueous solutions can also be explained, to some extent, by the Hofmeister series [20, 28, 30, 31]. Ions can be classified as kosmotropes (water structure makers) which stabilize proteins and chaotropes (water structure breakers) which destabilize proteins. The rank of kosmotrope-chaotrope ions according to the Hofmeister series is shown in Fig. 6.2. The most suitable combination to enhance protein stability comprises a kosmotropic anion and a chaotropic cation [32–34]. According to the example described above, the cytochrome c stability in hydrated choline dihydrogen phosphate is a result of this type of ion combination. Though the stability of proteins in ILs can be explained by the Hofmeister series, some deviations were also found while following a reverse trend [31]. In fact, a large array of factors is responsible for the protein stability in aqueous solutions of ILs, such as the ability to establish hydrogen bond, electrostatic, and dispersive interactions and hydrophobicity.

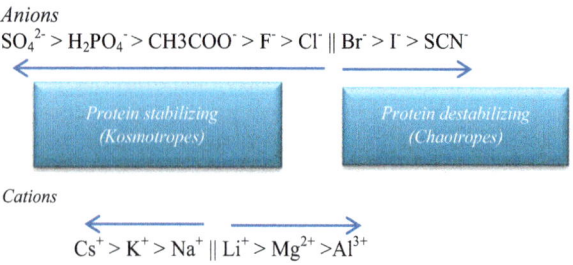

Fig. 6.2 Hofmeister series and protein stability

6.2.2 Partitioning Behavior of Proteins in IL-Based ABS

It has been shown that the partitioning of solutes in typical PEG-based ABS is primarily governed by the system properties, such as type and concentration of phase-forming components, pH, and temperature and the solute properties, such as hydrophobicity, charge, molecular weight, etc. [35]. Thus, partition coefficients and selectivity can be tuned by modifying these parameters. The extraction of proteins using IL-based ABS has been studied by several authors (see Table 6.1), revealing that protein partition preferentially to the IL-rich phase. Most of these studies are however empirical, and to be able to use IL-based ABS as a separation tool on a preparative scale, it is mandatory to understand the mechanisms and factors influencing the partitioning of proteins in these systems.

Different authors have studied the influence of the phase-forming components, concentration, pH, and temperature on the partitioning of proteins in IL-based ABS [25, 36–38]. Protein distribution in ABS depends on their ability to interact with the phase-forming components and extraction conditions, so that the separation could be protein specific. Cao et.al. [39] studied the extraction of horseradish peroxidase in four alkylimidazolium-based ABS. The enzyme partitioned to the IL-rich phase, but its activity decreases with the increase in the alkyl side chain length of the IL. In the same study, increasing the IL concentration favors the maintenance of the enzyme activity. Dreyer et al. [13] studied the feasibility of ABS formation with ammonium-based ILs and showed that Ammoeng 110 forms ABS more easily than Ammoeng 100 and Ammoeng 101. Ammoeng ILs contain an oligo-ethylene side chain in the cation which was expected to have a stabilizing effect on the enzyme (alcohol dehydrogenase) extracted. A low temperature for ABS formation together with ILs with oligo-ethylene side chains demonstrated to provide a gentle environment for protein extractions [13]. Desai et al. [25] showed that the partition coefficient of RuBisCo increases as the IL and salt concentration increases; however, a decrease in the enzyme activity was observed with higher concentrations (>20 %, w/w) of IL. In summary, all these results indicate that the chaotropicity of the IL and its concentration influence the stability of the protein to be extracted.

The system parameters (pH and temperature) also influence the partitioning of proteins to the IL-rich phase through the modification on the protein charge and surface properties. Protein properties contributing to their partitioning in ABS can be summarized as follows [5]:

$$\log K = \log K_0 + \log K_{el} + \log K_{hphob} + \log K_{size} + \log K_{biosp} + \log K_{conf} \quad (6.1)$$

where K is the partition coefficient and $_{el}$, $_{hphob}$, $_{biosp}$, and $_{conf}$ are, respectively, electrostatic, hydrophobicity, biospecificity, and configuration which contribute to the partition coefficient value, while K_0 represents additional factors.

Partitioning of proteins is governed, in a large extent, by the pH of the system. Depending on their isoelectric point, proteins carry a net positive or net negative charge at a given pH. The extraction of BSA, myoglobin, lysozyme, and trypsin using IL-based ABS at different pH values showed that proteins are preferentially transferred to the IL-rich phase as the pH increases [18]. On the other hand, the molecular weight of the protein also influences its partitioning in the biphasic system. Dreyer et al. [18] showed that larger proteins, such as BSA, are better extracted in the IL-rich phase, while smaller proteins, like myoglobin, remain in the salt-buffer-rich phase. In a separate study, RuBisCo, which is large protein (540 kDa), is also extracted into the IL-rich phase [25].

Du et al. [40] studied the extraction of BSA from biological fluids using imidazolium-based ABS and observed that the electrostatic interactions and salting-out effect are the driving forces in protein partitioning. In summary, both research groups [18, 40] have shown that there is a strong correlation between the protein charge and its partitioning in IL-based ABS. Thus, indicating electrostatic interaction between the amino acids on the protein surface and IL cations to be the main driving force. On the other hand, Pei et al. [41] have shown that hydrophobic interactions are the main driving force for protein extraction in IL-based ABS. In the same study, the influence of temperature on the extraction of BSA was evaluated demonstrating that higher temperatures favor the partitioning of proteins to the IL-rich phase.

It could be summarized that the partitioning of proteins in IL-based ABS can be tuned by changing the phase components and the composition, pH, and temperature of the system. Nevertheless, partitioning in IL-based ABS is a quite complex phenomenon not influenced by a single factor, yet it is a result of a combined effect of these factors.

6.3 Recovery of Proteins from the IL-Rich Phase

Like conventional ABS, protein extraction using IL-based ABS involves two main steps: (i) forward extraction, i.e., extraction of the protein from the initial source/matrix into one of the phases (here, IL-rich phase), and (ii) recovery of the (purified) protein from the IL-rich phase.

In conventional ABS, proteins can be recovered by modification of system parameters, such as pH, change in salt concentration, or addition of other salts. The main goal is to achieve a high recovery of a protein with a high purity level without affecting the functionality of the protein. This is indeed one of the major lacunas in the literature since there are almost no attempts on the literature to this end. An isolated work was recently published by Pereira et al. [42] where the protein (BSA) was recovered by dialysis from the IL-rich phase, allowing the further use of the ABS in a new extraction step. The authors [42] demonstrated the recovery of the protein and the IL reusability in three-step consecutive extractions, concluding that IL-based ABS can be adequately reused without losses on their extraction performance.

ILs being salts and proteins being macromolecules, their separation can be achieved by ultrafiltration and/or nanofiltration, induced precipitation, and chromatographic techniques, such as size exclusion chromatography and by the use of affinity tags (HisTags) able to help in recovering the protein from the IL-rich phase by Immobilized Metal Affinity Chromatography (IMAC). Protein recovery studies are thus one of the major lacunas in the IL-based ABS field and must be investigated in the near future.

6.4 Conclusions and Future Perspectives

IL-based ABS are a promising platform for the extraction and purification of proteins. However, there are still some issues which need to be addressed to be able to use IL-based ABS on a commercial scale, namely:

1. Currently, studies on proteins of commercial importance are scarce; only few studies were performed, for instance, for RuBisCo and alcohol dehydrogenases. Most studies in the literature address model proteins (BSA, lysozyme, etc.).
2. With a plethora of ILs available and the complex and variable nature of proteins, it is difficult to generalize or to predict the behavior of proteins in IL-based ABS. However, the setup of a well-defined guideline with respect to some protein classes would be useful. A mechanistic modeling approach still seems to be far off.
3. Stability of proteins in ILs is the prime requirement to guarantee the viability of IL-based ABS for protein separation. Most studies on this line are focused on model proteins, such as BSA and lysozyme. A pragmatic approach would be to create a public and free available (online) database with respect to the functional stability of commercial proteins in IL-based systems.
4. More sophisticated analytical methods to quantify proteins in the IL-rich phase should be attempted to avoid interferences from the IL. Also for preparative chromatography, the stability and functionality of currently available resins need to be determined.
5. The high costs of ILs are one of the major drawbacks when envisaging the large-scale application of IL-based ABS. The reuse of ILs in large-scale applications is essential to guarantee the economic viability.
6. The ILs used for ABS formation are water-soluble and hence can enter into the ecosystem. Thus, toxicity and biodegradation of ILs pose another concern and must be considered while designing protein extraction and separation processes.

All these points require not only extra efforts to study different IL-based ABS but more focused studies on the use of biodegradable and biocompatible ILs and efficient IL recycling processes. Since polymers, such as PEG, are able to maintain and even increase the stability of some proteins, IL-PEG ABS seem as an interesting option for protein extraction. Progress in IL-based ABS would open up new applications on their use, especially in biorefinery of third-generation biomass feedstocks (e.g., microalgae), where proteins could be separated from more hydrophobic components. IL-based ABS are novel systems and their use for protein extraction is still in an early stage. Thus, there is ample scope for improvement in protein extractions using IL-based ABS and a strong requirement for further in-depth investigations.

References

1. Kiron V, Phromkunthong W, Huntley M, Archibald I, De Scheemaker G (2012) Marine microalgae from biorefinery as a potential feed protein source for Atlantic salmon, common carp and whiteleg shrimp. Aquac Nutr 18(5):521–531. doi:10.1111/j.1365-2095.2011.00923.x
2. Schwenzfeier A, Wierenga PA, Gruppen H (2011) Isolation and characterization of soluble protein from the green microalgae Tetraselmis sp. Bioresour Technol 102(19):9121–9127. doi: http://dx.doi.org/10.1016/j.biortech.2011.07.046
3. Wijffels RH, Barbosa MJ (2010) An outlook on microalgal biofuels. Science 329 (5993):796–799. doi:10.1126/science.1189003
4. Labrou NE (2014) Protein downstream processing: design, development and application of high and low-resolution methods. Humana Press, Totowa
5. Albertsson P-A (1958) Partition of proteins in liquid polymer-polymer two-phase systems. Nature 182(4637):709–711
6. Rosa PAJ, Azevedo AM, Sommerfeld S, Bäcker W, Aires-Barros MR (2011) Aqueous two-phase extraction as a platform in the biomanufacturing industry: economical and environmental sustainability. Biotechnol Adv 29(6):559–567. doi:http://dx.doi.org/10.1016/j.biotechadv.2011.03.006
7. Rosa PAJ, Ferreira IF, Azevedo AM, Aires-Barros MR (2010) Aqueous two-phase systems: a viable platform in the manufacturing of biopharmaceuticals (extraction techniques). J Chromatogr A 1217(16):2296–2305. doi:http://dx.doi.org/10.1016/j.chroma.2009.11.034
8. Gutowski KE, Broker GA, Willauer HD, Huddleston JG, Swatloski RP, Holbrey JD, Rogers RD (2003) Controlling the aqueous miscibility of ionic liquids: aqueous biphasic systems of water-miscible ionic liquids and water-structuring salts for recycle, metathesis, and separations. J Am Chem Soc 125(22):6632–6633. doi:10.1021/ja0351802
9. Yang Z (2012) Ionic liquids and proteins: academic and some practical interactions. In: Ionic liquids in biotransformations and organocatalysis. Wiley, pp 15–71. doi:10.1002/9781118158753.ch2
10. Freire MG, Pereira JFB, Francisco M, Rodríguez H, Rebelo LPN, Rogers RD, Coutinho JAP (2012) Insight into the interactions that control the phase behaviour of new aqueous biphasic systems composed of polyethylene glycol polymers and ionic liquids. Chem Eur J 18 (6):1831–1839. doi:10.1002/chem.201101780
11. Wu B, Zhang YM, Wang HP (2008) Aqueous biphasic systems of hydrophilic ionic liquids + sucrose for separation. J Chem Eng Data 53(4):983–985. doi:10.1021/je700729p
12. Zhang J, Zhang Y, Chen Y, Zhang S (2007) Mutual coexistence curve measurement of aqueous biphasic systems composed of [bmim][BF4] and glycine, l-serine, and l-proline, respectively. J Chem Eng Data 52(6):2488–2490. doi:10.1021/je0601053
13. Dreyer S, Kragl U (2008) Ionic liquids for aqueous two-phase extraction and stabilization of enzymes. Biotechnol Bioeng 99(6):1416–1424. doi:10.1002/bit.21720
14. Freire MG, Cláudio AFM, Araújo JMM, Coutinho JAP, Marrucho IM, Lopes JNC, Rebelo LPN (2012) Aqueous biphasic systems: a boost brought about by using ionic liquids. Chem Soc Rev 41(14):4966–4995
15. Zhang Y, Zhang S, Chen Y, Zhang J (2007) Aqueous biphasic systems composed of ionic liquid and fructose (4th MTMS 4th international symposium on molecular thermodynamics and molecular simulation). Fluid Phase Equilib 257(2):173–176. doi:http://dx.doi.org/10.1016/j.fluid.2007.01.027

16. Ding X, Wang Y, Zeng Q, Chen J, Huang Y, Xu K (2014) Design of functional guanidinium ionic liquid aqueous two-phase systems for the efficient purification of protein. Analytica Chimica Acta 815(0):22–32. doi:http://dx.doi.org/10.1016/j.aca.2014.01.030
17. Zeng Q, Wang Y, Li N, Huang X, Ding X, Lin X, Huang S, Liu X (2013) Extraction of proteins with ionic liquid aqueous two-phase system based on guanidine ionic liquid. Talanta 116 (0):409–416. doi:http://dx.doi.org/10.1016/j.talanta.2013.06.011
18. Dreyer S, Salim P, Kragl U (2009) Driving forces of protein partitioning in an ionic liquid-based aqueous two-phase system. Biochem Eng J 46(2):176–185
19. Pei Y, Wang J, Wu K, Xuan X, Lu X (2009) Ionic liquid-based aqueous two-phase extraction of selected proteins. Sep Purif Technol 64(3):288–295
20. Constantinescu D, Weingärtner H, Herrmann C (2007) Protein denaturation by ionic liquids and the Hofmeister series: a case study of aqueous solutions of ribonuclease A. Angew Chem Int Ed 46(46):8887–8889. doi:10.1002/anie.200702295
21. Takekiyo T, Yamazaki K, Yamaguchi E, Abe H, Yoshimura Y (2012) High ionic liquid concentration-induced structural change of protein in aqueous solution: a case study of lysozyme. J Phys Chem B 116(36):11092–11097. doi:10.1021/jp3057064
22. Zhang W-G, Wei D-Z, Yang X-P, Song Q-X (2006) Penicillin acylase catalysis in the presence of ionic liquids. Bioprocess Biosyst Eng 29(5–6):379–383. doi:10.1007/s00449-006-0085-9
23. Nara SJ, Harjani JR, Salunkhe MM (2002) Lipase-catalysed transesterification in ionic liquids and organic solvents: a comparative study. Tetrahedron Letters 43(16):2979–2982. doi:http://dx.doi.org/10.1016/S0040-4039(02)00420-3
24. Li Z, Pei Y, Wang H, Fan J, Wang J (2010) Ionic liquid-based aqueous two-phase systems and their applications in green separation processes (bioCop – monitoring chemical contaminants in foods). TrAC Trends Anal Chem 29(11):1336–1346. doi:http://dx.doi.org/10.1016/j.trac.2010.07.014
25. Desai R, Streefland M, Wijffels RH, Eppink M (2014) Extraction and stability of selected proteins in ionic liquid based aqueous two phase systems. Green Chem 16:2670. doi:10.1039/c3gc42631a
26. Baker GA, Heller WT (2009) Small-angle neutron scattering studies of model protein denaturation in aqueous solutions of the ionic liquid 1-butyl-3-methylimidazolium chloride (application of ionic liquids in chemical and environmental engineering). Chem Eng J 147(1):6–12. doi:http://dx.doi.org/10.1016/j.cej.2008.11.033
27. Weaver KD, Vrikkis RM, Van Vorst MP, Trullinger J, Vijayaraghavan R, Foureau DM, McKillop IH, MacFarlane DR, Krueger JK, Elliott GD (2012) Structure and function of proteins in hydrated choline dihydrogen phosphate ionic liquid. Phys Chem Chem Phys 14 (2):790–801
28. Fujita K, MacFarlane DR, Forsyth M, Yoshizawa-Fujita M, Murata K, Nakamura N, Ohno H (2007) Solubility and stability of cytochrome c in hydrated ionic liquids: effect of oxo acid residues and kosmotropicity. Biomacromolecules 8(7):2080–2086. doi:10.1021/bm070041o
29. Fukaya Y, Hayashi K, Wada M, Ohno H (2008) Cellulose dissolution with polar ionic liquids under mild conditions: required factors for anions. Green Chem 10(1):44–46. doi:10.1039/b713289a
30. Kilpeläinen I, Xie H, King A, Granstrom M, Heikkinen S, Argyropoulos DS (2007) Dissolution of wood in ionic liquids. J Agric Food Chem 55(22):9142–9148. doi:10.1021/jf071692e
31. Debeljuh N, Barrow CJ, Byrne N (2011) The impact of ionic liquids on amyloid fibrilization of A[small beta]16-22: tuning the rate of fibrilization using a reverse Hofmeister strategy. Phys Chem Chem Phys 13(37):16534–16536. doi:10.1039/c1cp22256b
32. Zhao H (2005) Effect of ions and other compatible solutes on enzyme activity, and its implication for biocatalysis using ionic liquids. J Mol Catal B Enzym 37(1–6):16–25. doi:10.1016/j.molcatb.2005.08.007
33. Baldwin RL (1996) How Hofmeister ion interactions affect protein stability. Biophys J 71 (4):2056–2063. doi:S0006-3495(96)79404-3 [pii] 10.1016/S0006-3495(96)79404-3

34. Collins KD, Washabaugh MW (1985) The Hofmeister effect and the behaviour of water at interfaces. Q Rev Biophys 18(4):323–422
35. Albertsson PÅ (1986) Partition of cell particles and macromolecules: separation and purification of biomolecules, cell organelles, membranes, and cells in aqueous polymer two-phase systems and their use in biochemical analysis and biotechnology. Wiley, New York [etc.]
36. Lin X, Wang Y, Zeng Q, Ding X, Chen J (2013) Extraction and separation of proteins by ionic liquid aqueous two-phase system. Analyst 138(21):6445–6453. doi:10.1039/c3an01301d
37. Chen J, Wang Y, Zeng Q, Ding X, Huang Y (2014) Partition of proteins with extraction in aqueous two-phase system by hydroxyl ammonium-based ionic liquid. Anal Methods 6(12):4067–4076. doi:10.1039/c4ay00233d
38. Dreyer SE (2008) Aqueous two phase extraction of proteins and enzymes using tetraalkylammonium-based ionic liquids. PhD thesis, University of Rostock
39. Cao Q, Quan L, He C, Li N, Li K, Liu F (2008) Partition of horseradish peroxidase with maintained activity in aqueous biphasic system based on ionic liquid. Talanta 77(1):160–165. doi:http://dx.doi.org/10.1016/j.talanta.2008.05.055
40. Du Z, Yu Y-L, Wang J-H (2007) Extraction of proteins from biological fluids by use of an ionic liquid/aqueous two-phase system. Chem Eur J 13(7):2130–2137. doi:10.1002/chem. 200601234
41. Pei YC, Liu L, Li ZY, Wang JJ, Wang HY (2010) Selective separation of protein and saccharides by ionic liquids aqueous two-phase systems. SCIENCE CHINA Chem 53(7):1554–1560. doi:10.1007/s11426-010-4025-9
42. Pereira MM, Pedro SN, Quental MV, Lima AS, Coutinho JAP, Freire MG (2015) Enhanced extraction of bovine serum albumin with aqueous biphasic systems of phosphonium- and ammonium-based ionic liquids. J Biotechnol 206:17–25
43. Ruiz-Angel MJ, Pino V, Carda-Broch S, Berthod A (2007) Solvent systems for countercurrent chromatography: an aqueous two phase liquid system based on a room temperature ionic liquid (4th international conference on countercurrent chromatography). J Chromatogr A 1151 (1–2):65–73
44. Wu C, Wang J, Wang H, Pei Y, Li Z (2011) Effect of anionic structure on the phase formation and hydrophobicity of amino acid ionic liquids aqueous two-phase systems. J Chromatogr A 1218(48):8587–8593. doi:http://dx.doi.org/10.1016/j.chroma.2011.10.003
45. Wu C, Wang J, Li Z, Jing J, Wang H (2013) Relative hydrophobicity between the phases and partition of cytochrome-c in glycine ionic liquids aqueous two-phase systems. J Chromatogr A 1305(0):1–6. doi:http://dx.doi.org/10.1016/j.chroma.2013.06.066
46. Yan J-K, Ma H-L, Pei J-J, Wang Z-B, Wu J-Y (2014) Facile and effective separation of polysaccharides and proteins from Cordyceps sinensis mycelia by ionic liquid aqueous two-phase system. Sep Purif Technol. doi:http://dx.doi.org/10.1016/j.seppur.2014.03.020
47. Deive FJ, Rodríguez A, Rebelo LPN, Marrucho IM (2012) Extraction of Candida antarctica lipase A from aqueous solutions using imidazolium-based ionic liquids (ILSEPT2011 special issue). Sep Purif Technol 97(0):205–210. doi:http://dx.doi.org/10.1016/j.seppur.2011.12.013
48. Ventura SPM, Sousa SG, Freire MG, Serafim LS, Lima ÁS, Coutinho JAP (2011) Design of ionic liquids for lipase purification. J Chromatogr B 879(26):2679–2687. doi:http://dx.doi.org/10.1016/j.jchromb.2011.07.022
49. Deive FJ, Rodriguez A, Pereiro AB, Araujo JMM, Longo MA, Coelho MAZ, Lopes JNC, Esperanca JMSS, Rebelo LPN, Marrucho IM (2011) Ionic liquid-based aqueous biphasic system for lipase extraction. Green Chem 13(2):390–396. doi:10.1039/c0gc00075b
50. Jiang B, Feng Z, Liu C, Xu Y, Li D, Ji G (2014) Extraction and purification of wheat-esterase using aqueous two-phase systems of ionic liquid and salt. J Food Sci Technol 1–8. doi:10. 1007/s13197-014-1319-5

Chapter 7
Extraction of Alcohols, Phenols, and Aromatic Compounds with ABS

María J. Trujillo-Rodríguez, Verónica Pino, and Juan H. Ayala

Abstract Ionic-liquid-based aqueous biphasic systems (IL-based ABS) combine the advantages of both ionic liquids (ILs) and aqueous biphasic systems (ABS) in extraction approaches. Most ILs are liquid solvents exclusively composed of ions, characterized for low to negligible vapor pressures at ambient conditions, extraordinary chemical and thermal stabilities, and impressive solvation abilities for polar to nonpolar compounds, among many other properties. ABS permit the development of extraction methods without the need of using toxic and volatile organic solvents, while both immiscible phases involved in the extraction procedure are aqueous rich. Thus, the combination of hydrophilic ILs with ABS has implied numerous effective applications for different target compounds. In this chapter, attention is focused on applications of IL-based ABS for the extraction of alcohols, phenols, aromatic dyes, and other aromatic compounds. The discussion is centered on the nature of the IL-based ABS, their characterization, and the performance of the extraction procedure, among other features.

Keywords Ionic liquids • Aqueous biphasic systems • Alcohols • Phenols • Aromatic compounds • Extraction methods

List of Abbreviations

ABS	Aqueous biphasic systems
CCC	Countercurrent chromatography
EE_{CA}	Extraction efficiency of chloranilic acid
EU	European Union
HPLC	High-performance liquid chromatography
IL	Ionic liquid
K_{CA}	Partition coefficient of chloranilic acid
K_{OW}	Octanol–water partition coefficient

M.J. Trujillo-Rodríguez • V. Pino (✉) • J.H. Ayala
Departamento de Química, Área de Química Analítica, Universidad de La Laguna (ULL), La Laguna (Tenerife) 38206, Spain
e-mail: veropino@ull.edu.es

© Springer-Verlag Berlin Heidelberg 2016
M.G. Freire (ed.), *Ionic-Liquid-Based Aqueous Biphasic Systems*, Green Chemistry and Sustainable Technology, DOI 10.1007/978-3-662-52875-4_7

PEG	Polyethylene glycol
RID	Refractive index detection
RTIL	Room temperature ionic liquid
TLL	Tie-line length
US EPA	Environmental Protection Agency of United States
UVD	Ultraviolet detection
VOC	Volatile organic compound
β-CD	β-Cyclodextrin

7.1 Introduction

7.1.1 Ionic Liquids and Applications

Ionic liquids (ILs) undoubtedly constitute an exciting group of nonmolecular solvents with melting points below 100 °C [1–3]. Several ILs are able to retain their liquid nature at room temperature and even down to -96 °C, being termed, in this case, as room temperature ILs (RTILs) [1–3].

ILs are mainly composed of large and asymmetric organic cations containing nitrogen or phosphorous atoms, associated with inorganic/organic anions, as summarized in Fig. 7.1. A high number of ILs can be formed by different combinations of cations/anions, which can also be properly functionalized. The enormous interest surrounding ILs is totally linked to their properties: negligible vapor pressure at ambient conditions, high chemical and electrochemical stabilities, nonflammability, and impressive solvation ability for polar to nonpolar compounds, among others. In addition to this, it is possible to observe dramatic modifications in such properties (for instance, from water soluble to water insoluble) by simply carrying out minimum modifications in their chemical structures. Therefore, ILs can be tailored for specific applications in a quite simple manner, which in turn results enormously advantageous [4–6].

ILs have been pointed out in many cases as "green solvents" because they do not generate volatile organic compounds (VOCs) and so they constitute a nice alternative to replace toxic chlorinated organic solvents [7]. Nevertheless, recent studies have pointed out the toxicity of certain ILs, particularly those containing imidazolium cations, certain anions (hexafluorophosphate as an example), or long substituents in their structures [8–12]. The synthetic preparation of ILs has improved from a green chemistry point of view with the application of new raw materials and improved reaction conditions (e.g., solvent-free synthesis and reduced energy input) [12]. Therefore, trends are now focused on their chemical modification to obtain biodegradable ILs [9]. In any case, hydrophilic ILs (those that are completely water soluble at room temperature) normally have lower toxicity than hydrophobic ones, and there are also a higher number of available hydrophilic ILs.

ILs have been utilized in applications belonging to diverse areas, such as energy [13, 14], materials science [14, 15], organic chemical synthesis [16],

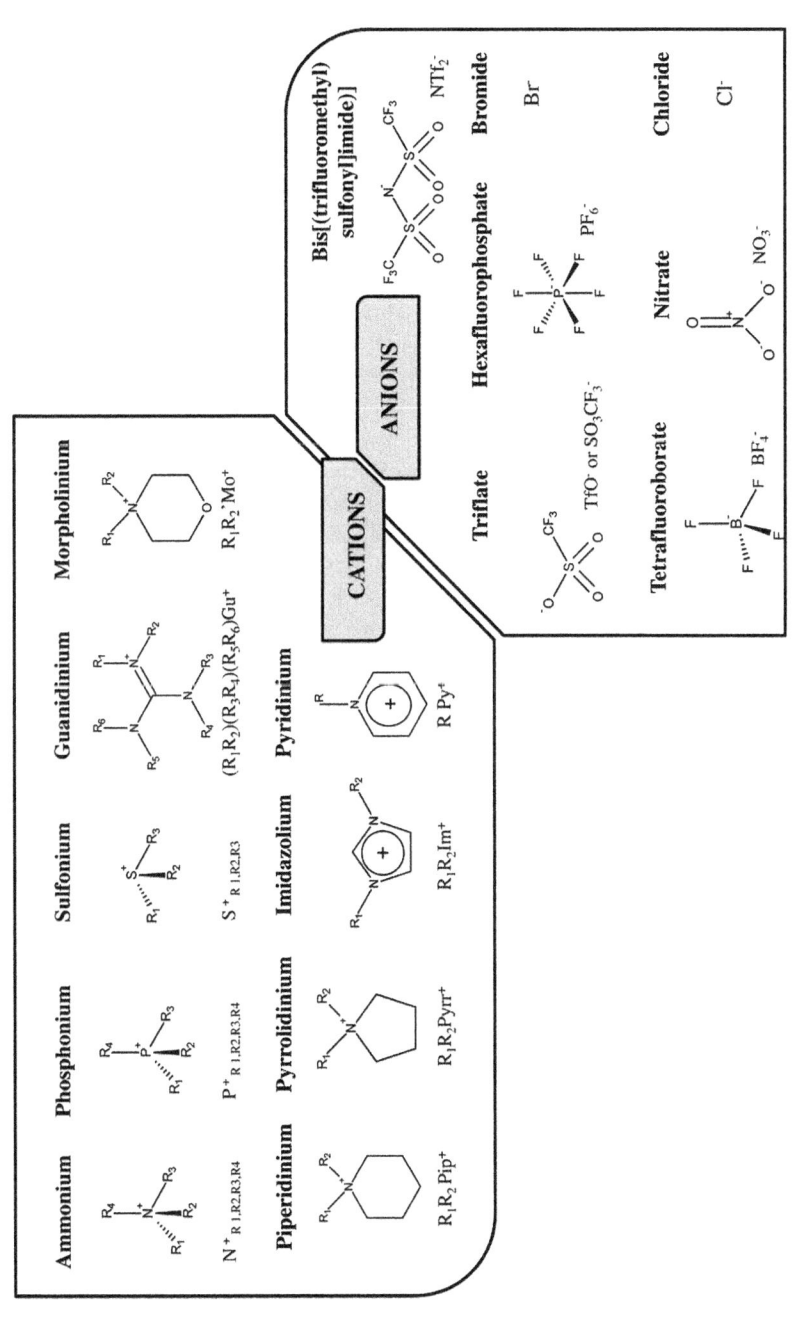

Fig. 7.1 Structures and abbreviations of main cations/anions of ILs

electrochemistry [17, 18], chemical engineering [19], analytical chemistry [4, 20, 21], and even in the pharmaceutical and medical field [14, 22]. Furthermore, enormous interest has been devoted to their applications in extraction and pre-concentration schemes for compounds of interest, mainly related to hydrophobic ILs [23–26].

7.1.2 Ionic Liquids in Aqueous Biphasic Systems

The definition of aqueous biphasic systems (ABS) is quite implicit, two phases mainly composed of water but being actually immiscible [27]. The aqueous phases can be composed of certain combinations of polymers, salts, or even polymers with salts. The separation of both phases takes place at a specific concentration. Analytes can experience partitioning to one of the phases in higher extent with respect to the other, which represents the basis of an enrichment/extraction procedure. Given the fact that the overall extraction procedure is organic volatile solvent-free, it results in a quite attractive "green" procedure. Polymer-based ABS have been particularly successful in applications with biomolecules [28]. A problem of polymer–polymer-based ABS is the fact that the two immiscible phases are quite hydrophobic and the minimum differences in polarities are related to their different water contents. If polymer–salt-based ABS are used, there is one more hydrophobic phase (the polymer) and one highly hydrophilic phase (the salt).

Given the outstanding properties of ILs, it is not a surprise that they have been also quite successful when involved in ABS [27, 29, 30], by forming IL-based ABS, in which the IL plays an important role in changing the characteristics of at least one of the phases while providing a better solvation performance. In this context, these systems have been used not only for extraction purposes of certain analytes but also indirectly to recover ILs from aqueous solutions. In any case, it must be taken into account that these systems necessarily require a hydrophilic IL, which is advantageous from an environmental point of view. In addition to this, the viscosity of ILs dramatically decreases with the addition of water. Thus, IL-based ABS are also characterized by low viscosities, which in turn favor the transfer phenomenon of the analyte from one phase to another.

Among applications of IL-based ABS, this book chapter is focused on alcohols, phenols, aromatic dyes, and other aromatic compounds. Attention is paid to the nature of IL-based ABS, their characterization, and the performance of the extraction procedure, among other features. In this chapter, the most common abbreviations for ILs will be utilized. For example, alkyl substituents of the IL cation will first be written showing their length (M for methyl; C_n for the rest of alkyl chains, being n the number of carbon atoms; Al for the allyl group; and MOC_2 for methoxyethyl); followed by the terms Im for imidazolium, Pyrr for pyrrolidinium, Py for pyridinium, Pip for piperidinium, and Mo for morpholinium; and, finally, by the anion (Cl^-, chloride; Br^-, bromide; $CF_3SO_3^-$, trifluoromethanesulfonate; $N(CN)_2^-$, dicyanamide; SCN^-, thiocyanate; BF_4^-, tetrafluoroborate; MSO_4^-, methylsulfate;

and Tos⁻, tosylate). In the case of Im-based ILs, the substituent located in position 1 will be written first, followed by the substituent in position 3. Quaternary ammonium- and phosphonium-based ILs are written by N or P, respectively, followed by their substituents in subscripts using mainly numbers (1, methyl; 4, butyl; 2OH, 2-hydroxyethyl; and i4, isobutyl), and finally by the anion. Table 7.1 [31–42] includes a summary of the main cations/anions of ILs reported in ABS applications involving alcohols, phenols, aromatic dyes, and other aromatic compounds.

Figure 7.2 shows the main variants of IL-based ABS schemes when utilized in the extraction of alcohols, phenols, aromatic dyes, and other aromatic compounds. The specific applications will be described as a function of the nature of the group of compounds studied.

7.2 Extraction of Aromatic Dyes Using IL-Based ABS

Dyes are widely utilized in textile, leather, food, cosmetic, and printing industries, among others [43–45]. In most cases, they have a synthetic origin, being formed by complex aromatic structures responsible for their color and their high stability. Thus, aromatic dyes are designed to be resistant to light, to high temperatures, and to oxidizing agents [46].

Aromatic dyes can be classified in three groups: anionic, cationic, and nonionic dyes [44]. In the case of anionic and nonionic dyes, the chromophore is mostly an azo- or an anthraquinone-type group. These structures are quite resistant to degradation even in wastewaters [46, 47]. Besides, if azo-based dyes suffer reduction in the environment, they generate hazardous substances, such as aromatic amines, which are characterized for their high toxicity and potential carcinogenicity [46]. Thereby, dye removal has received target attention due to their toxicity and nonbiodegradability issues.

IL-based ABS already demonstrated to be efficient alternatives for the removal of aromatic dyes from environmental waters. Recent works reported favorable partition coefficients for the extraction of chloranilic acid [31, 32, 34], indigo blue [31, 34], indigo carmine [31], rhodamine 6G [32], Sudan III [34], reactive red 120 [42], 4-(2-pyridylazo)resorcinol [42], methyl orange [42], and thymol blue [38] utilizing IL-based ABS. These works are summarized in Table 7.2 [31, 32, 34, 38, 42]. Furthermore, the structures and octanol–water partition coefficients of these aromatic dyes are shown in Fig. 7.3.

The first report of an IL-based ABS for the separation of an aromatic dye was published by Hiramaya et al. in 2012 [38]. An atypical ABS was created by a quaternary mixture of water–tetrahydrofuran–C_4MIm-Cl/NaCl for the extraction of thymol blue in its red neutral form, as shown in Fig. 7.2a. The mixture of salts C_4MIm-Cl and NaCl acted as a salting-out agent causing the phase separation of the initially homogeneous water–tetrahydrofuran mixture. The distribution ratio and the extractability of the dye into the tetrahydrofuran phase were studied as a function of the molar fraction of the IL under a fixed NaCl amount. Results showed

Table 7.1 Structures of the cations and anions of the ILs most commonly used in IL-based ABS applications for the extraction of aromatic dyes, alcohols, phenols, and other aromatic compounds

Cation name/abbreviation	Structure of the cation	Anions[a]
1-Ethyl-3-methylimidazolium, C_2MIm^+		Cl^- [31–33]
		$CF_3SO_3^-$ [34, 35]
		MSO_4^- [36]
1-Propyl-3-methylimidazolium, C_3MIm^+		BF_4^- [37]
1-Butyl-3-methylimidazolium, C_4MIm^+		Cl^- [31–33, 38–41]
		Br^- [42]
		$CF_3SO_3^-$ [34, 35]
		MSO_4^- [35, 36]
		$N(CN)_2^-$ [34, 35]
		SCN^- [35]
		Tos^- [34]
		BF_4^- [37]
1-Pentyl-3-methylimidazolium, C_5MIm^+		BF_4^- [37]
1-Hexyl-3-methylimidazolium, C_6MIm^+		Cl^- [32, 33]
		BF_4^- [37]
1-Heptyl-3-methylimidazolium, C_7MIm^+		BF_4^- [37]
1-Octyl-3-methylimidazolium, C_8MIm^+		Cl^- [32]
		BF_4^- [37]
1-Methoxyethyl-3-methylimidazolium, $(MOC_2)MIm^+$		$CF_3SO_3^-$ [35]
N-Methylpyridinium, MPy^+		MSO_4^- [36]
N-Butyl-N-methylpyrrolidinium, C_4MPyrr^+		Cl^- [31, 33]
		$CF_3SO_3^-$ [35]

(continued)

Table 7.1 (continued)

Cation name/abbreviation	Structure of the cation	Anions[a]
N-Butyl-N-methylpiperidinium, C_4MPip^+	$CH_3(CH_2)_3$—N$^+$—CH_3 (piperidinium ring)	Cl$^-$ [31]
N-Butyl-N-methylmorpholinium, C_4MMo^+	$CH_3(CH_2)_3$—N$^+$—CH_3 (morpholinium ring)	$CF_3SO_3^-$ [35]
Tetrabutylammonium, $N^+_{4,4,4,4}$	$CH_3(CH_2)_3$—N$^+$(($CH_2)_3CH_3$)$_3$	Cl$^-$ [33]
(2-Hydroxyethyl)trimethylammonium or choline, $N^+_{1,1,1,2OH}$	$HO(CH_2)_2$—N$^+$(CH_3)$_3$	Cl$^-$ [33]
Tetrabutylphosphonium, $P^+_{4,4,4,4}$	$CH_3(CH_2)_3$—P$^+$(($CH_2)_3CH_3$)$_3$	Cl$^-$ [33, 34]; Br$^-$ [34]
Tributylmethylphosphonium, $P^+_{4,4,4,1}$	CH_3—P$^+$(($CH_2)_3CH_3$)$_3$	MSO_4^- [34, 36]
Tri(isobutyl)methylphosphonium, $P^+_{i4,i4,i4,1}$	CH_3—P$^+$(i-Bu)$_3$	Tos$^-$ [34]

[a]Being Cl^- chloride, Br^- bromide, $CF_3SO_3^-$ trifluoromethanesulfonate, $N(CN)_2^-$ dicyanamide, SCN^- thiocyanate, BF_4^- tetrafluoroborate, MSO_4^- methylsulfate, and Tos^- tosylate

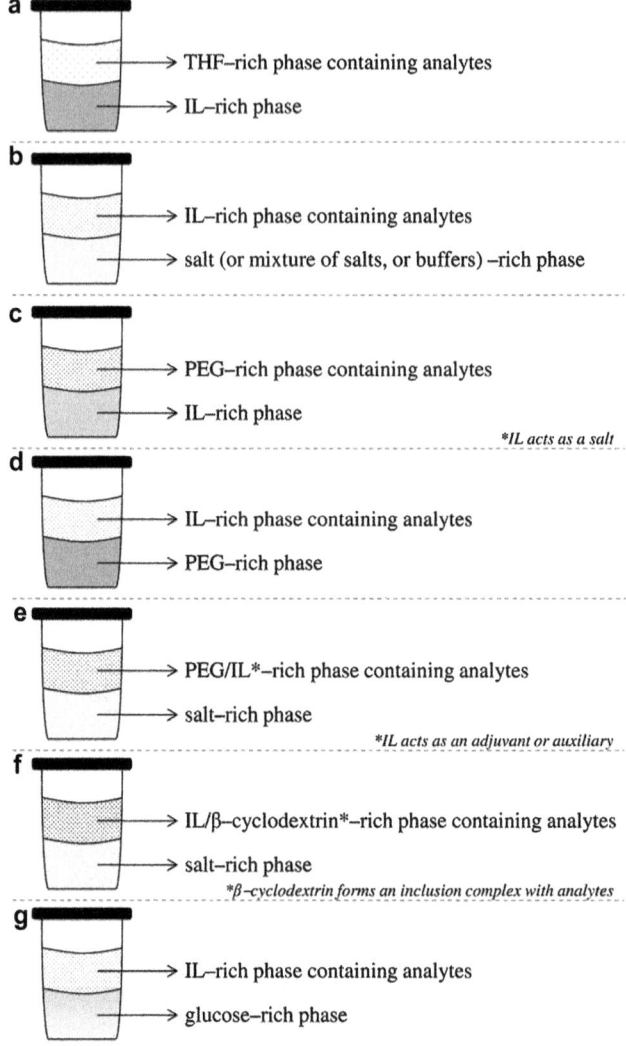

Fig. 7.2 Main variants of IL-based ABS schemes when utilized for the extraction of alcohols, phenols, aromatic dyes, and other aromatic compounds. The real position of each phase (at the *top* or at the *bottom* of the tube) depends on the relative density of each one

a decrease in both parameters (distribution ratio and extractability) when increasing the IL content. The addition of the IL contributed to lower the polarity differences of the two phases in the ABS thus suggesting that the C_4MIm-Cl remaining in the water-rich phase was acting not only as a salting-out agent but also as a component of a mixed solvent [38].

Since this first report on the subject [38], other works studied the behavior of aromatic dyes in IL-based ABS [31, 32, 34, 42]. The main purposes of these works

Table 7.2 Applications of IL-based ABS for the extraction of aromatic dyes

Aromatic dye	IL phase (content in the ABS)	2nd phase (content in the ABS)	Extraction conditions/operational temp. (K)	Analytical technique[a]	Log K_{max}[b] (analyte or comment)/extra comments	Ref.
Thymol blue	*Mixture of salts* C$_4$MIm-Cl NaCl (−)	Tetrahydrofuran (−)	Agitation (10 min), centrifugation and phase separation/298.15	UV–Vis spectroscopy	~1.30 (−) Initial aqueous solution contains HNO$_3$ (2 mol·L^{-1}) Studies of the ABS formed only with the IL or with NaCl	[38]
Reactive red 120 4-(2-Pyridylazo)-resorcinol Methyl orange	C$_4$MIm-Br (−)	*Salts* Potassium phosphate buffers (40 %, w/w, for both pH 9.7 and 11.6, and 35.4 %, w/w, for pH 7.02)	Agitation (1 min), centrifugation (1 min), and phase separation Addition of C$_4$MIm-PF$_6$ to the IL phase: the analytes remain in the aqueous phase and the IL phase can be separated/298.15	UV–Vis spectroscopy	4.08 (methyl orange)/ −	[42]
Chloranilic acid Indigo blue Sudan III	C$_2$MIm-CF$_3$SO$_3$ C$_4$MIm-CF$_3$SO$_3$ C$_4$MIm-Tos C$_4$MIm-N(CN)$_2$ P$_{4,4,4,4}$-Br P$_{4,4,4,4}$-Cl P$_{4,4,4,1}$-MSO$_4$[c] P$_{i4,i4,i4,1}$-Tos (45 %, w/w)	*Salts* Al$_2$(SO$_4$)$_3$ Potassium citrate (10 %, w/w)	Agitation, equilibration step (12 h) and phase separation Heating and filtration steps of the IL phase for recovering the dyes 298.15	UV–Vis spectroscopy (332 nm for chloranilic acid, 335 nm for indigo blue, and 348 nm for Sudan III)	∞ (Complete extraction of the dye) Determination of real recovery of the dye	[34]

(continued)

Table 7.2 (continued)

Aromatic dye	IL phase (content in the ABS)	2nd phase (content in the ABS)	Extraction conditions/operational temp. (K)	Analytical technique[a]	Log K_{max}[b] (analyte or comment)/extra comments	Ref.
Chloranilic acid Indigo carmine Indigo blue	C_2MIm-Cl C_4MIm-Cl C_4MPip-Cl C_4MPyrr-Cl (~52 %, w/w)	*Polymer* PEG 1500 (~38 %, w/w)	Equilibrium step (3 h) and phase separation 323 K	UV–Vis spectroscopy	~1.8 (indigo carmine) Comparison with other PEG-based ABS systems (PEG 1500–Na_2SO_4 and PEG 4000–dextran)	[31]
Chloranilic acid Rhodamine 6G	*Mixture of a-Polymer* PEG 1500 PEG 4000 PEG 6000 PEG 8000 (40 %, w/w) *An IL as adjuvant* C_2MIm-Cl C_4MIm-Cl C_6MIm-Cl C_8MIm-Cl (5 %, w/w)	*Mixture of Salts* K_3PO_4 K_2HPO_4 K_2HPO_4/KH_2PO_4 buffer (pH 7) (25 %, w/w) *An IL as adjuvant* C_2MIm-Cl C_4MIm-Cl C_6MIm-Cl C_8MIm-Cl (5 %, w/w)	Equilibrium step (12 h) and phase separation 298 (±1)	UV–Vis spectroscopy (332 nm for chloranilic acid and 527 nm for rhodamine 6G)	~2.28 (rhodamine 6G) The IL acts as an adjuvant	[32]

[a] Analytical technique selected for the determination of the dye(s)
[b] Maximum partition coefficient (expressed as logarithm) achieved, under the best conditions
[c] Different compositions were utilized for the $P_{4,4,4,1}$-MSO_4 and citrate-based ABS system: 35 % (w/w) of IL, 19 % (w/w) of salt, and 46 % (w/w) of water

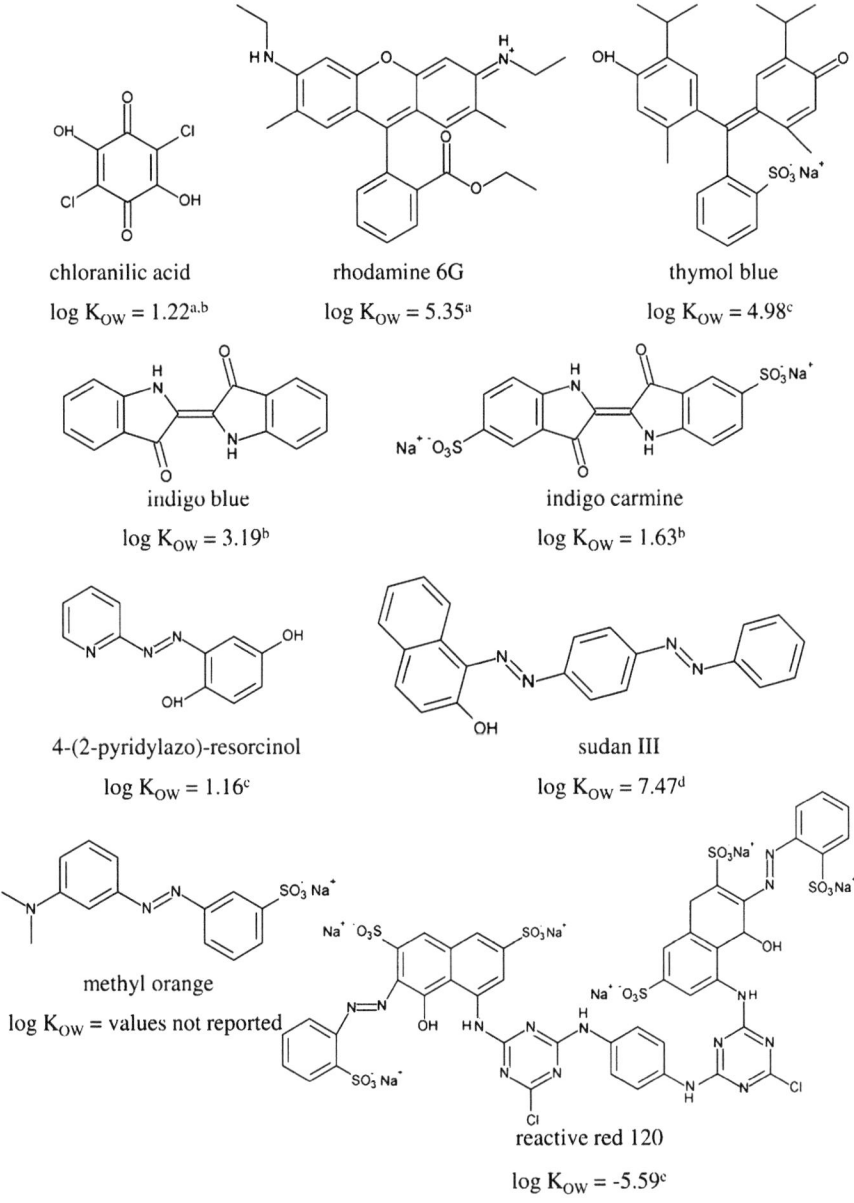

Fig. 7.3 Structures of aromatic dyes extracted with IL-based ABS and their corresponding octanol–water partition coefficients (expressed as log K_{OW}). Log K_{OW} values were obtained from [32][a], [31][b], SciFinder© Scholar database 2014[c], [34][d], and ChemSpider© database 2014[e]

were the following: (1) to comprehensively study the ternary IL-based ABS by constructing the binodal curves and (2) to determine the partition coefficients of the aromatic dyes between the two phases and their variations with the concentration of the IL (or variations with any the other component of the ABS, pH, etc.) to find the driving forces behind a successful extraction.

The utilization of aromatic dyes as probes in IL-based ABS has an important advantage since the dye-rich phase is colored. Thus, a simple analytical technique such as UV–Vis spectroscopy can be employed for the quantification of the dye in all the reported works.

Attending to the preparation of IL-based ABS for the extraction of aromatic dyes, different strategies have been followed. Several works studied IL–salt combinations [34, 42], whereas others focused on IL–polymer combinations [31, 32]. Sheikhian et al. [42] studied ABS composed of the IL C_4MIm-Br and potassium phosphate buffers (pH 7.2, 9.7, and 11.6). The study was devoted to the extraction of three anionic dyes, namely, reactive red 120 ($pK_a = 2.1$), 4-(2-pyridylazo)-resorcinol ($pK_{a,1} = 3.1$, $pK_{a,2} = 5.6$, and $pK_{a,3} = 11.9$), and methyl orange ($pK_a = 3.5$) [42]. An enrichment of the dyes in the IL-rich phase (top phase) was observed, as it is shown in Fig. 7.2b. Results showed that high pH buffers were more favorable for the extraction of the dyes. Besides, the work reported the recycling of the C_4MIm-Br from the IL-rich phase (containing C_4MIm-Br and dyes) by the addition of the hydrophobic IL C_4MIm-PF_6. After 2 h of extraction, almost 96.8 % of the C_4MIm-Br is recycled into the C_4MIm-PF_6 phase, while dyes remain in the aqueous phase. Finally, the C_4MIm-PF_6 phase, which is now rich in C_4MIm-Br, was vigorously mixed with deionized water to perform a back extraction of the C_4MIm-Br hydrophilic IL [42].

Ferreira et al. [34] reported the utilization of other IL–salt ABS, based on phosphonium and imidazolium ILs and an inorganic (aluminum sulfate) or organic (potassium citrate) salt. The extracted dyes were chloranilic acid (in its anionic form) and indigo blue and Sudan III (in their nonionic forms). The IL-rich phase containing dyes was also located at the top of the tube (Fig. 7.2b). Partition coefficients and extraction efficiencies were evaluated for all ABS tested. These results proved that the phosphonium-based ILs were more effective for the complete extraction of the dyes. The four aliphatic chains present in the IL cation (phosphonium based) confer a higher hydrophobicity to the IL, thus favoring the extraction ability for the hydrophobic dyes studied [34]. In this work, the dye Sudan III was completely extracted in both phosphonium- and imidazolium-based ABS, but the dye indigo blue was only extracted in the phosphonium-based ABS. Moreover, the imidazolium-based ABS caused the precipitation of the dye indigo blue. With regard to the dye chloranilic acid, it displayed a partitioning trend more dependent on the IL nature due to its anionic nature at the pH values studied (from 1 to 8). Figure 7.4 shows the obtained partition coefficients (as K_{CA}) and extraction efficiencies (as EE_{CA} in %) for chloranilic acid with different ILs and salts [34]. It is worth mentioning that the removal procedure of the dyes from the IL-rich phase was conducted in a single step by their precipitation using changes of temperature. This also permitted the recycling of the IL [34].

7 Extraction of Alcohols, Phenols, and Aromatic Compounds with ABS

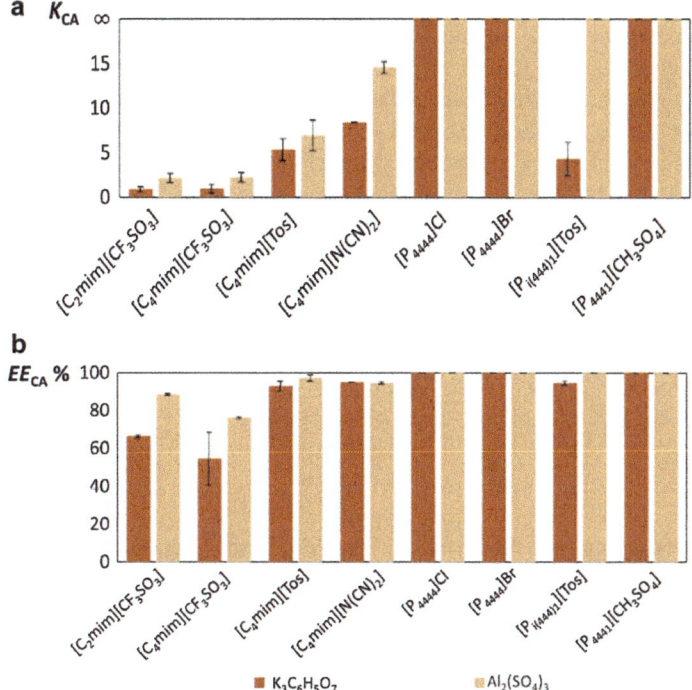

Fig. 7.4 Representation of partition coefficients (K_{CA}) (**a**) and extraction efficiencies (EE_{CA}, in %) (**b**) obtained for the extraction of chloranilic acid using IL-based ABS composed of phosphonium or imidazolium IL cations and aluminum sulfate or potassium citrate salts at 298.15 K. The rest of conditions are described in the text. K_{CA} values equal to ∞ refer to the complete extraction of the dye (or its non-detection in the salt-rich phase) (Figure reproduced from Ferreira et al. [34] with permission of Elsevier)

Pereira et al. [31] described the utilization of ABS composed of the polyethylene glycol (PEG) polymer, with an average molecular weight of 1500 g · mol^{-1} (PEG 1500) and different ILs containing chloride anions (C$_4$MPip-Cl, C$_4$MPyrr-Cl, C$_2$MIm-Cl, and C$_4$MIm-Cl) [31]. The partition coefficients of three dyes (chloranilic acid, indigo carmine, and indigo blue, in their nonionic forms) were determined in these PEG–IL-based ABS, as well as in PEG 1500–Na$_2$SO$_4$ and PEG 4000–dextran for comparison purposes. In general, results showed that the utilization of the most hydrophobic ILs caused an inversion on the migration of the dyes. Thus, dyes prefer the PEG-rich phase in the case of the ABS composed of PEG 1500–Na$_2$SO$_4$ and PEG 4000–dextran and even in IL-based ABS formed by the most hydrophilic ILs (C$_2$MIm-Cl and C$_4$MIm-Cl) (see Fig. 7.2c). However, an enrichment of the dyes in the IL-rich phase was obtained when employing C$_4$MPip-Cl and C$_4$MPyrr-Cl as ILs in the PEG–IL-based ABS (see Fig. 7.2d). The authors [31] concluded that it is possible to tune the PEG-rich phase in terms of being more or less hydrophilic simply by modifying the IL cation in PEG–IL-based ABS.

ILs have also been utilized as adjuvants in polymer–salt ABS for the extraction of chloranilic acid (in its anionic form) and rhodamine 6G (in its nonionic form) by de Souza et al. [32]. In this case, different ABS composed of PEG (with different molecular weights) and different potassium salts were studied in the presence or absence of different chloride-based ILs. An enrichment of the dyes in the PEG-rich phase was always achieved (see Fig. 7.2e). Results showed that all ILs interacted with the PEG-rich phase by decreasing its hydrophobicity. The migration of the IL depended on the molecular weight of the PEG, potassium salt content, and pH, as well as on the combination of the selected cation/anion of the IL. Furthermore, the partition coefficients obtained for the dyes in all tested ABS suggested that their extraction was controlled by two main forces: (1) hydrophobic interactions, due to the preference of the dyes for the PEG-rich phase, and (2) electrostatic interactions, because opposite effects in the partition of the two dyes were observed when increasing the size of the alkyl chain of the ILs tested [32].

7.3 Extraction of Alcohols, Phenols, and Other Aromatic Compounds Using IL-Based ABS

Alcohols, phenols, and other related aromatic compounds constitute a diverse group of analytes with important applications in the chemical industry [48]. They are widely used in perfumes, cosmetics, and other personal care products [49], acting as fragrances or disinfectants [50]. They have also been utilized as solvents in industries, in pharmaceuticals, as plasticizers, or in the development of pesticides [49, 51]. In some cases, they can have a natural origin [52]. Due to their extensive use and presence, these compounds will eventually find their way to wastewaters, lakes, aquifers, etc. [53]. Many of them present toxic, mutagenic, and carcinogenic effects [54], and so they are regarded as contaminants in waters. The removal of these compounds from waters and wastewater constitutes an important issue, in order to protect not only the environment but also public health [55, 56].

Most phenolic compounds are listed as priority pollutants by the Environmental Protection Agency of United States (US EPA) [55, 57], which limits the phenol content in wastewaters to less than $1 \text{ mg} \cdot \text{L}^{-1}$ and the pentachlorophenol in other waters to $0.1 \text{ µg} \cdot \text{L}^{-1}$ [57, 58], as representative examples. The European Union (EU) has also set limits such as $0.1 \text{ µg} \cdot \text{L}^{-1}$ for individual phenols in drinking waters, around $5 \text{ µg} \cdot \text{L}^{-1}$ in bathing waters, and for pentachlorophenol of $1 \text{ mg} \cdot \text{L}^{-1}$ in industrial effluents and around $2 \text{ µg} \cdot \text{L}^{-1}$ in superficial waters [58].

IL-based ABS have been employed for the extraction of alcohols, phenols, and other related aromatic compounds from water media. Table 7.3 [33, 35–37, 39–41] summarizes these applications. We would like to highlight that literature works involving phenols or phenol derivatives with antioxidants properties have been excluded in this chapter because they are detailed in Chap. 8. The compounds covered in this chapter include phenol [37, 39], 4-nitrophenol [39],

7 Extraction of Alcohols, Phenols, and Aromatic Compounds with ABS

Table 7.3 Applications of IL-based ABS for the extraction of alcohols, phenols, and other related aromatic compounds

Aromatic compound or alcohol (spiked level)	IL phase (content in the ABS)	2nd phase (content in the ABS)	Extraction conditions/operational temp. (K)	Analytical technique[a]	Log K_{max}[b] (analyte, or comment)/extra comments	Ref.
4-nitrophenol Phenol Nitrobenzene Aniline (1000 mg·L^{-1})	C$_4$MIm-Cl (70 %, w/w)	*Salts* K$_2$HPO$_4$ (2.0–3.5 mol·L^{-1}) K$_3$PO$_4$ (2.5–4.5 mol·L^{-1}) K$_2$CO$_3$ (2.5–4.5 mol·L^{-1})	Vortex (10 min), equilibrium step (30 min), phase separation, and filtration Room temperature	HPLC–UV (4-nitrophenol at 316 nm; phenol at 269 nm; nitrobenzene at 267 nm)	~2.6 (nitrobenzene) Comparison with a hydrophobic enrichment method with C$_8$MIm-PF$_6$	[39]
Pentachlorophenol (<15 mg·L^{-1})	C$_2$MIm-MSO$_4$ C$_4$MIm-MSO$_4$ P$_{4,4,4,1}$-MSO$_4$[b] MPy-MSO$_4$ (–)	*Salts* K$_2$CO$_3$ K$_2$HPO$_4$[b] (–)	Agitation, equilibrium step (24 h), phase separation, and filtration 298.15	HPLC-UV (280 nm)	3.06 (–) Determination of the extraction performance	[36]
Bisphenol A (4.3·10^{-4} mol·L^{-1})	C$_2$MIm-Cl[c] C$_4$MIm-Cl C$_6$MIm-Cl AlMIm-Cl C$_4$MPyr-Cl P$_{4,4,4,4}$-Cl N$_{4,4,4,4}$-Cl N$_{1,1,1,2OH}$-Cl[c] (different contents)	*Salts* K$_3$PO$_4$ (different contents)	Agitation, equilibrium step (12 h), phase separation, and filtration 298	UV–Vis spectroscopy	Determination of the extraction performance Application for the extraction of bisphenol A from water and artificial human urine	[33]
Methylparaben Ethylparaben Propylparaben Benzylparaben (–)	C$_4$MIm-Cl (5 %, w/w)	*Salt* K$_3$PO$_4$ (pH 9^d) (Different contents)	Adjusting pH, agitation, equilibrium step (2 min), and phase separation Room temperature	HPLC-UV (254 nm)	~4.8 (benzylparaben) Effect of the addition of β-cyclodextrin (1.0 mL, 10 mg·L^{-1}) to the ABS, which formed an inclusion complex with parabens Development and	[40]

(continued)

Table 7.3 (continued)

Aromatic compound or alcohol (spiked level)	IL phase (content in the ABS)	2nd phase (content in the ABS)	Extraction conditions/ operational temp. (K)	Analytical technique[a]	Log K_{max}[b] (analyte, or comment)/extra comments	Ref.
					validation of the analytical methods Application for the extraction of parabens from river water, seawater, and wastewater Determination of the extraction recoveries	
Phenol ($1.0 \cdot 10^{-2}$ mol·L^{-1})	C_3MIm-BF$_4$ C_4MIm-BF$_4$ C_5MIm-BF$_4$ C_6MIm-BF$_4$ C_7MIm-BF$_4$ C_8MIm-BF$_4$ (2 g)	*Glucose* 6-(Hydroxymethyl)oxane-2,3,4,5-tetrol (2 g)	Agitation (3 h), equilibrium step (24 h), phase separation, and dilution with water 298.15	UV–Vis spectroscopy (270 nm)	1.89 (–) –	[37]
Methanol Ethanol Propanol Butanol Pentanol Hexanol (–)	C_4MIm-Cl (–)	*Salt* K$_2$HPO$_4$ (–)	The ABS constitute the mobile phase in the CCC Room temperature	CCC-UV	2.04 (hexanol)/ Comparison with PEG-based ABS systems Determination of the extraction performance	[41]

1,3-Propanediol (–)	C_4MIm-CF_3SO_3 C_4MIm-$N(CN)_2$ C_4MIm-SCN C_4MIm-MSO_4[c] C_4MMo-CF_3SO_3 C_4MPy-CF_3SO_3 C_2MIm-CF_3SO_3 $(MOC_2)MImCF_3SO_3$ (–)	Mixture of salts K_2HPO_4 and KH_2PO_4 (pH 7) (–)	Magnetic stirring (16 h), equilibrium step (3 h) and, phase separation 310.15	HPLC–RI	~1.41 (–) –	[35]

[a]Analytical technique selected for the monitoring of the compound(s)
[b]Maximum partition coefficient (expressed as logarithms) achieved, under the best conditions
[c]The IL or the 2nd phase that presented a higher ability to phase separation
[d]pH was studied in the range of 2–12

pentachlorophenol [36], bisphenol A [33], nitrobenzene [39], parabens (methylparaben, ethylparaben, propylparaben, and benzylparaben) [40], n-alcohols (methanol, ethanol, propanol, butanol, pentanol, and hexanol) [41], and 1,3-propanediol [35]. The chemical structures of these compounds are shown in Fig. 7.5 as well as their octanol–water partition coefficients. The monitoring of the majority of these compounds has been carried out by UV–Vis spectroscopy [33, 36, 37]. In other cases, high-performance liquid chromatography (HPLC) in combination with UV–Vis spectroscopy has been employed [39, 40] or coupled with refractive index detection (RID) [35]. Besides, in one work, the IL-based ABS constitute the mobile phase of a countercurrent chromatographic (CCC) system, with a UV–Vis detector [41]. In the majority of applications, analytes are majorly enriched in the IL-rich phase, as shown in Fig. 7.2b.

IL-based ABS for the applications described above are normally made by a water–IL–salt combination [33, 35, 36, 39–41]. However, the water–IL–glucose combination [37] or even an artificial urine solution–IL–salt ternary system has also been reported [33].

The first report of the utilization of an IL-based ABS method for this group of compounds was due to Zhang et al. in 2010 [39]. Authors initially studied the extraction of 4-nitrophenol, phenol, and nitrobenzene from aqueous solutions using IL–salt ABS [39]. The hydrophilic C_4MIm-Cl was always tested in combination with three different salts: K_2HPO_4, K_3PO_4, and K_2CO_3. In all cases, the IL-rich phase containing the extracted aromatic compounds corresponds to the top phase. Authors [39] observed that the salting-out trend observed followed the Hofmeister series: $K_3PO_4 > K_2HPO > K_2CO_3$. The authors [39] also compared, in terms of partition coefficients, the IL-based ABS performance with an enrichment method using the hydrophobic IL C_4MIm-PF_6. Despite better efficiencies attained with the hydrophobic IL, both methods proved to be successful in the extraction of the group of aromatic compounds selected [39]. The advantages of the utilization of the IL-based ABS procedure include the hydrophilic nature of the IL (more versatile and less toxic), lower cost, and the possibility of recycling the IL.

IL–salt-based ABS have also been utilized for the extraction of pentachlorophenol [36]. ILs composed of the methylsulfate anion and different cations based on imidazolium, phosphonium, and pyridinium were tested to generate the ABS, in combination with different salts: K_2CO_3 and K_2HPO_4 (an alkaline medium). The binodal curves obtained in this study are shown in Fig. 7.6.

From Fig. 7.6, it is possible to observe that the salt K_2HPO_4 allows a greater immiscibility region. Thus, K_2HPO_4 was the salt selected as the salting-out agent for the extraction of pentachlorophenol [36]. Regarding the cation comprising the IL of the IL-based ABS, two effects were studied: the effect of cation nature (C_4MIm^+, $P_{4,4,4,1}^+$, and MPy^+, all with the same anion MSO_4^-) and the influence of the length of the alkyl chain of the cation for the ILs C_2MIm-MSO_4 and C_4MIm-MSO_4. Regarding the cationic nature, the ability to phase separation follows the order $P_{4,4,4,1}^+ > MPy^+ > C_4MIm^+$. Authors [36] claimed that the absence of aromaticity in phosphonium cations favors the concentration of the charge in the heteroatom, thus increasing the ability to undergo phase separation when inorganic salts

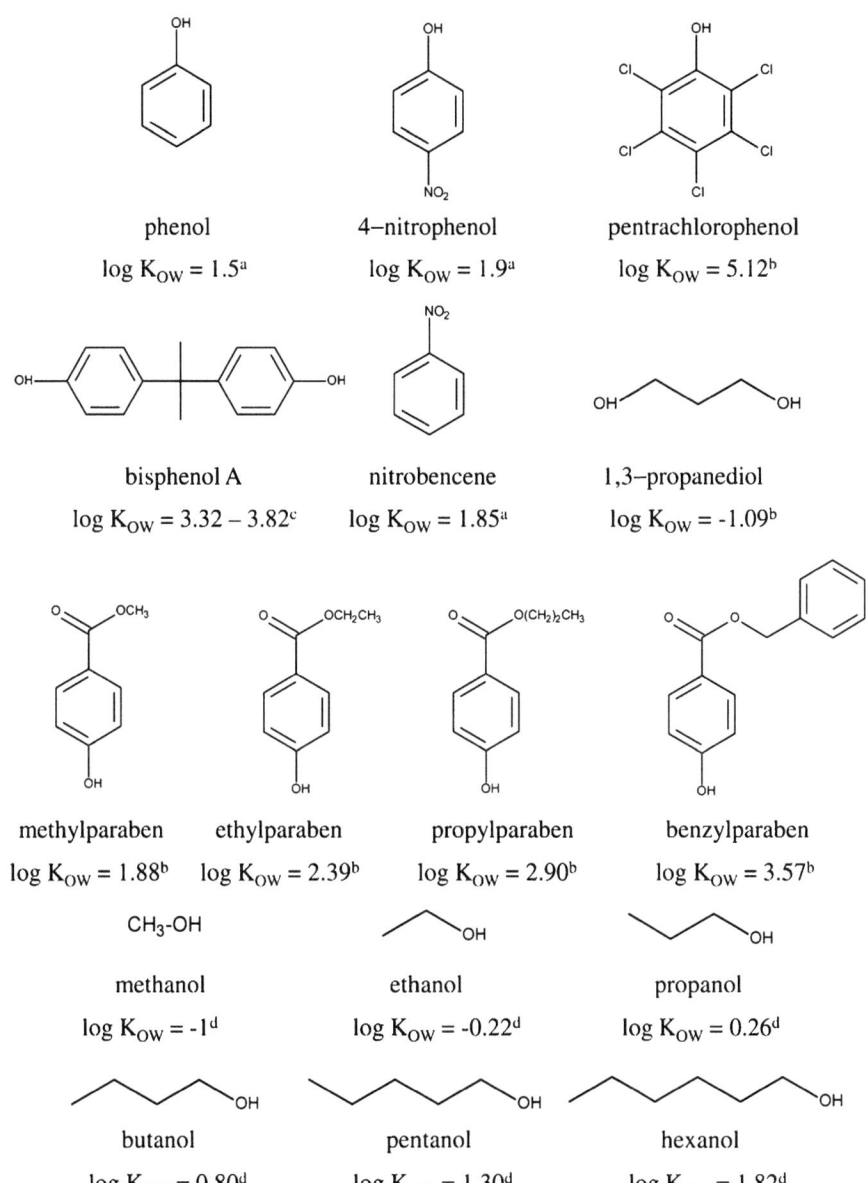

Fig. 7.5 Structures of alcohols, phenols, and other aromatic compounds extracted with IL-based ABS and their corresponding octanol–water partition coefficients (expressed as log K_{OW}) (Log K_{OW} values were obtained from [39][a], SciFinder© Scholar database 2014[b], [33][c], and [41][d])

are added. Regarding the alkyl chain size, a higher ability to phase separation for cations with larger side chains was observed. This is linked to a decrease in solubility of an IL when increasing the aliphatic moieties. The extraction

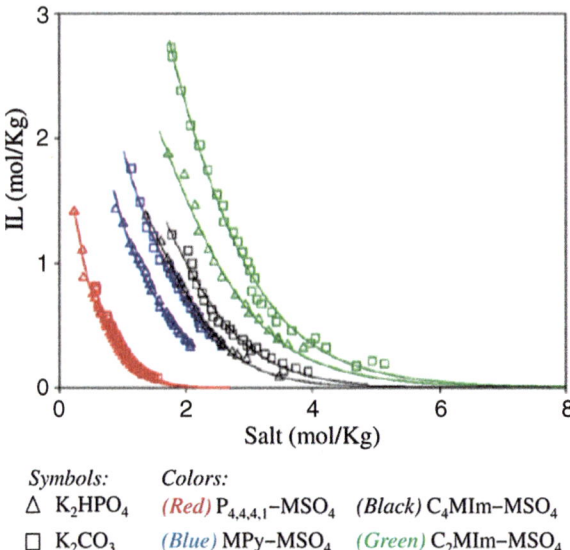

Fig. 7.6 Binodal curves for water–IL–salt ternary systems composed of the ILs $P_{4,4,4,1}$-MSO_4 (*red*), MPy-MSO_4 (*blue*), C_4MIm-MSO_4 (*black*), and C_2MIm-MSO_4 (*green*) and the salts K_2HPO_4 (▲) and K_2CO_3 (□) at 298.15 K. The rest of conditions are described in the text (Figure reproduced from Moscoso et al. [36], MDPI Open Access)

Symbols: *Colors:*
▲ K_2HPO_4 (*Red*) $P_{4,4,4,1}$–MSO_4 (*Black*) C_4MIm–MSO_4
□ K_2CO_3 (*Blue*) MPy–MSO_4 (*Green*) C_2MIm–MSO_4

efficiencies for pentachlorophenol with all IL-based ABS were higher than 92 %. Best results were achieved with the phosphonium IL, which was capable to remove more than 99 % of pentachlorophenol [36].

Passos et al. [33] used IL-based ABS for the extraction of bisphenol A from water and artificial human urine (in alkaline medium) [33]. An IL−salt combination was selected for the enrichment of the compound. Different chloride-based ILs were selected, composed of imidazolium, pyrrolidinium, phosphonium, ammonium, or choline cations, and the extraction efficiency was the selected parameter to test all ABS. In aqueous solutions, best results were obtained with ABS composed of the ILs $N_{1,1,1,2OH}$-Cl and C_2MIm-Cl [33]. In any case, extraction efficiencies were always higher than 98.5 %. Authors [33] also studied the possibility of reducing the volume of the IL-rich phase, in order to apply the method to artificial human urine. Results showed that the complete extraction of BPA from this complex matrix was possible, with the concentration of BPA up to 100-fold [33].

Noorashikin et al. [40] employed IL-based ABS for the extraction of four parabens from river waters, seawaters, and wastewaters. Two types of IL−salt-based ABS were studied: one formed by C_4MIm-Cl and K_2PO_4, as in Fig. 7.2b, and others composed of C_4MIm-Cl and K_2PO_4 in the presence of β-cyclodextrins (β-CDs) (Fig. 7.2f). β-CDs form inclusion complexes with parabens, thus favoring their extraction. Authors [40] obtained, for ABS, the partition coefficients, the extraction recoveries, and the pre-concentration factors, with parabens being enriched in the top IL-rich phase. The IL/β-cyclodextrin-based ABS method gave the best results for parabens [40]. Calibration curves for parabens were constructed with the two IL-based ABS. The methods were validated in terms of precision, linearity, and limit of detections. Linear ranges from 0.01 to 0.10 mg · L^{-1} were established with all parabens and both methods. Relative standard deviation values down to 1.1 % were obtained in all cases, showing the high precision of the procedure. Limits of detection between 0.17 and 0.51 mg·L^{-1} for the parabens, when using the IL-based ABS method, and between 0.022 and 0.075 mg·L^{-1}, when using the IL/β-cyclodextrin-based ABS method, were obtained [40]. To the best of our knowledge, this is the only work in literature which reports the analytical performance of an IL-based ABS method including detection limits and precision, for the whole number of works related to the group of compounds detailed in this chapter, namely, alcohols, phenols, aromatic dyes, and other related aromatic compounds.

The use of imidazolium-based ILs has also been reported, with the BF$_4^-$ anion and cations with alkyl chains of different lengths, enabling to form ABS in the presence of 6-(hydroxymethyl)oxane-2,3,4,5-tetrol [37]. Phenol can effectively be extracted in the IL-rich phase of these systems (see Fig. 7.2g), even from dilute solutions. The obtained partition coefficients showed an increase of the phase separation with the increase of the glucose content and with the length of the alkyl chain attached to the IL [37]. Partition coefficients values up to 79 were obtained [37]. These values are totally comparable to those obtained when extracting phenol with other methods, such as by its enrichment using hydrophobic ILs or other conventional procedures using amines, alcohols, or acids as extractants.

Regarding the extraction of alcohols using IL-based ABS, Ruiz–Ángel et al. [41] reported the partition coefficients of a series of n-alcohols (methanol, ethanol, propanol, butanol, pentanol, and hexanol) with ABS composed of the C$_4$MIm-Cl IL and the K$_2$HPO$_4$ salt. A comparison was also carried out with the more conventional system PEG 1000-K$_2$HPO$_4$ ABS [41]. Both ABS were utilized as mobile phases in CCC. Based on the obtained results, it was concluded that there is a significant polarity difference between the PEG- and the IL-based ABS. Thereby, the discrimination factor of the IL-based ABS was three times higher, and its relative hydrophobicity was ten times lower in respect to the PEG-based ABS [41]. A large discrimination factor means that the system is able to discriminate the polar solutes in the lower inorganic salt phase and the nonpolar ones in the upper phase. Extraction efficiencies of 87.5 %, 92.8 %, 96.8 %, 98.3 %, and 99.1 % for the extraction of ethanol to hexanol were obtained, respectively [41].

The extraction of 1,3-propanediol was also investigated using IL-based ABS by Müller and Górak [35]. This compound is utilized as monomer in the production of the polymer polytrimethylene terephthalate [59]. The principal route of production of 1,3-propanediol involves microbiological methods in which a final step of purification is necessary. Distillation or conventional extraction methods with organic solvents are not adequate routes due to the high boiling point and low solubility of the compound. As an alternative to these methods, authors [35] proposed the utilization of IL-based ABS composed of an IL and a mixture of two salts (pH 7). Authors [35] studied ILs composed of imidazolium, morpholinium, or pyrrolidinium cations (with different substituents) and dicyanamide, thiocyanate, and methylsulfate anions. Partition coefficients for 1,3-propanediol in all ABS were obtained. The obtained values ranged from

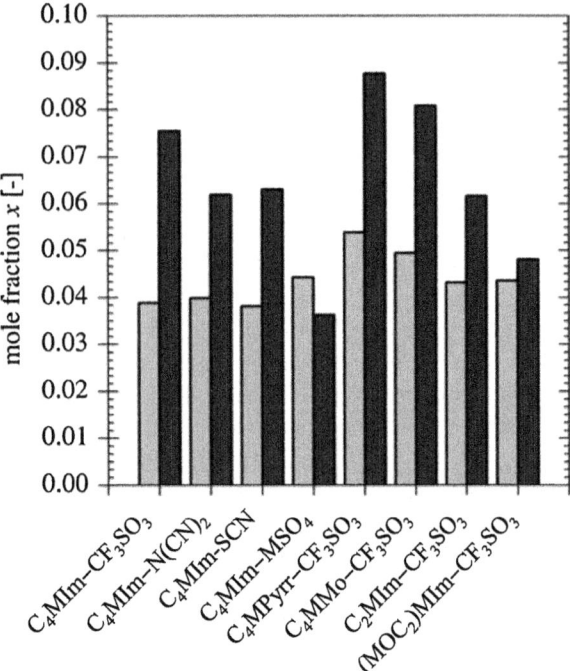

Fig. 7.7 Representation of the mole fraction of phosphate buffer (*gray*) and the mole fraction of IL (*black*) necessary to achieve a partition coefficient of 1,3-pentachlorophenol equal to 7 in the IL-based ABS composed of a phosphate buffer (pH 7) and different ILs: C_4MIm-CF_3SO_3, C_4MIm-$N(CN)_2$, C_4MIm-SCN, C_4MIm-MSO_4, C_4MMo-CF_3SO_3, C_4MPyrr-CF_3SO_3, C_2MIm-CF_3SO_3, and $(MOC_2)MIm$-CF_3SO_3 at 310.15 K. The rest of conditions are described in the text (Figure reproduced from Müller et al. [35] with permission of Elsevier)

1.5 to 27.7. The influence of the nature of the anion and the cation of the IL and the relative polarity of the cation were studied using different tie-line lengths (TLL) [35]. Figure 7.7 shows the mole fraction of phosphate buffer and IL necessary to achieve a partition coefficient equal to 7. These results showed that the amount of the IL required depends on its polarity: more polar ILs, like C_4MIm-CF_3SO_3, $(MOC_2)MIm$-CF_3SO_3, and C_4MIm-MSO_4, require lower mole fractions to achieve the extraction of 1,3-propanediol than less polar ILs [35].

7.4 Conclusions

IL-based ABS have been proved to be quite successful systems to carry out environmentally-friendly extractions of alcohols, phenols, aromatic dyes, and other aromatic compounds. A large number of IL-based ABS have been utilized in these applications, in which the IL plays an important role in one of the phases, or by modifying polymer-based ABS, providing in this latter case better solvation characteristics. This chapter has intended to give an overview of the existing applications, paying attention to the nature of the IL-based ABS and its performance. Different ILs have also been reported in these applications, and it is expected that ILs with greener features will be tested in the near future. We believe that the number of applications for alcohols, phenols, aromatic dyes, and other aromatic compounds involving IL-based ABS would continue to flourish.

References

1. Welton T (1999) Room–temperature ionic liquids. Solvents for synthesis and catalysis. Chem Rev 99:2071–2083
2. Rogers RD, Seddon KR (2003) Ionic liquids–solvents of the future? Science 302:792–793
3. Hallett JP, Welton T (2011) Room–temperature ionic liquids: solvents for synthesis and catalysis. Chem Rev 111:3508–3576
4. Ho TD, Zhang C, Hantao LW, Anderson JL (2014) Ionic liquids in analytical chemistry: fundamentals, advances, and perspectives. Anal Chem 86:262–285
5. Müller K, Albert J (2014) Contribution of the individual ions to the heat capacity of ionic liquids. Ind Eng Chem Res 53:10343–10346
6. Karunanithi AT, Mehrkesh A (2013) Computer–aided design of tailor–made ionic liquids. AICHE J 59:4627–4640
7. Davis JH (2004) Task–specific ionic liquids. Chem Lett 33:1072–1077
8. Swatloski RP, Holbrey JD, Rogers RD (2003) Ionic liquids are not always green: hydrolysis of 1–butyl–3–methylimidazolium hexafluorophosphate. Green Chem 5:361–363
9. Pham TPT, Cho CW, Yun YS (2010) Environmental fate and toxicity of ionic liquids: a review. Water Res 44:352–372
10. Cevasco G, Chiappe C (2014) Are ionic liquids a proper solution to current environmental challenges? Green Chem 16:2375–2385
11. Welton T (2011) Ionic liquids in green chemistry. Green Chem 13:225
12. Bubalo MC, Radošević K, Redovniković IR, Halambek J, Srček VG (2014) A brief overview of the potential environmental hazards of ionic liquids. Ecotoxicol Environ Saf 99:1–12
13. MacFarlane DR, Tachikawa N, Forsyth M, Pringle JM, Howlett PC, Elliott GD, Davis JH, Watanabe M, Simon P, Angell CA (2014) Energy applications of ionic liquids. Energy Environ Sci 7:232–250
14. Smiglak M, Pringle JM, Lu X, Han L, Zhang S, Gao H, MacFarlane DR, Rogers RD (2014) Ionic liquids for energy, materials, and medicine. Chem Commun 50:9228–9250
15. Armand M, Endres F, MacFarlane DR, Ohno H, Scrosati B (2009) Ionic–liquid materials for the electrochemical challenges of the future. Nat Mater 8:621–629
16. Nevagi RJ, Dighe SN, Dighe SN, Chaskar PK, Srinivasan KV, Jain KS (2014) Use of ionic liquids as neoteric solvents in the synthesis of fused heterocycles. Arch Pharm Chem Life Sci 347:540–551
17. Díaz M, Ortiz A, Ortiz I (2014) Progress in the use of ionic liquids as electrolyte membranes in fuel cells. J Membr Sci 469:379–396
18. Kar M, Simons TJ, Forsyth M, MacFarlane DR (2014) Ionic liquid electrolytes as a platform for rechargeable metal–air batteries: a perspective. Phys Chem Chem Phys 16:18658–18674
19. Werner S, Haumann M, Wasserscheid P (2010) Ionic liquids in chemical engineering. Annu Rev Chem Biomol Eng 1:203–230
20. Tan Z-Q, Liu J-F, Pang L (2012) Advances in analytical chemistry using the unique properties of ionic liquids. Trends Anal Chem 39:218–227
21. Joshi MD, Anderson JL (2012) Recent advances of ionic liquids in separation science and mass spectrometry. RSC Adv 2:5470–5484
22. Marrucho IM, Branco LC, Rebelo LPN (2014) Ionic liquids in pharmaceutical applications. Annu Rev Chem Biomol Eng 5:527–546
23. Fontanals N, Borrull F, Marcé RM (2012) Ionic liquids in solid–phase extraction. Trends Anal Chem 41:15–26

24. Trujillo–Rodríguez MJ, Rocío–Bautista P, Pino V, Afonso AM (2013) Ionic liquids in dispersive liquid–liquid microextraction. Trends Anal Chem 51:87–106
25. Liu R, Liu J-F, Yin Y-G, Hu X-L, Jiang G-B (2009) Ionic liquids in sample preparation. Anal Bioanal Chem 393:871–883
26. Poole CF, Poole SK (2010) Extraction of organic compounds with room temperature ionic liquids. J Chromatogr A 1217:2268–2286
27. Freire MG, Cláudio AFM, Araújo JMM, Coutinho JAP, Marrucho IM, Canongia Lopes JN, Rebelo LPN (2012) Aqueous biphasic systems: a boost brought about by using ionic liquids. Chem Soc Rev 41:4966–4995
28. Pereira JBF, Ventura SPM, e Silva FA, Shahriari S, Freire MG, Coutinho JAP (2013) Aqueous biphasic systems composed of ionic liquids and polymers: a platform for the purification of biomolecules. Sep Purif Technol 113:83–89
29. Gutowski KE, Broker GA, Willauer HD, Huddleston JG, Swatloski RP, Holbrey JD, Rogers RD (2003) Controlling the aqueous miscibility of ionic liquids: aqueous biphasic systems of water–miscible ionic liquids and water–structuring salts for recycle, metathesis, and separations. J Am Chem Soc 125:6632–6633
30. Li Z, Pei Y, Wang H, Fan J, Wang J (2010) Ionic liquid–based aqueous two-phase systems and their applications in green separation processes. Trends Anal Chem 29:1336–1346
31. Pereira JBF, Rebelo LPN, Rogers RD, Coutinho JAP, Freire MG (2013) Combining ionic liquids and polyethylene glycols to boost the hydrophobic–hydrophilic range of aqueous biphasic systems. Phys Chem Chem Phys 15:19580–19583
32. de Souza RL, Campos VC, Ventura SPM, Soares CMF, Coutinho JAP, Lima AS (2014) Effect of ionic liquids as adjuvants on PEG–based ABS formation and the extraction of two probe dyes. Fluid Phase Equilib 375:30–36
33. Passos H, Sousa ACA, Pastorinho MR, Nogueira AJA, Rebelo LPN, Coutinho JAP, Freire MG (2012) Ionic–liquid–based aqueous biphasic systems for improved detection of bisphenol A in human fluids. Anal Methods 4:2664–2667
34. Ferreira AM, Coutinho JAP, Fernandes AM, Freire MG (2014) Complete removal of textile dyes from aqueous media using ionic–liquid–based aqueous two–phase systems. Sep Purif Technol 128:58–66
35. Müller A, Górak A (2012) Extraction of 1,3–propanediol from aqueous solutions using different ionic liquid–based aqueous two-phase systems. Sep Purif Technol 97:130–136
36. Moscoso F, Deive FJ, Esperança JMSS, Rodríguez A (2013) Pesticide removal from aqueous solutions by adding salting out agents. Int J Mol Sci 14:20954–20965
37. Chen Y, Meng Y, Yang J, Li H, Liu X (2012) Phenol distribution behavior in aqueous biphasic systems composed of ionic liquids–carbohydrate–water. J Chem Eng Data 57:1910–1914
38. Hirayama N, Higo T, Imura H (2012) Evaluation of a hydrophilic ionic liquid as a salting–out phase separation agent to a water–tetrahydrofuran homogeneous system for aqueous biphasic extraction separation. Anal Sci 28:541–543
39. Zhang D, Deng Y, Chen J (2010) Enrichment of aromatic compounds using ionic liquid and ionic liquid–based aqueous biphasic systems. Sep Sci Technol 45:663–669
40. Noorashikin MS, Mohamad S, Abas MR (2014) Extraction and determination of parabens in water samples using an aqueous two–phase system of ionic liquid and salts with beta–cyclodextrin as the modifier coupled with high performance liquid chromatography. Anal Methods 6:419–425
41. Ruiz–Angel MJ, Pino V, Carda–Broch S, Berthod A (2007) Solvent systems for countercurrent chromatography: an aqueous two phase liquid system based on a room temperature ionic liquid. J Chromatogr A 1151:65–73
42. Sheikhian L, Akhond M, Absalan G (2014) Partitioning of reactive red–120, 4– (2–pyridylazo) –resorcinol, and methyl orange in ionic liquid–based aqueous biphasic systems. J Environ Chem Eng 2:137–142
43. Amini M, Arami M, Mahmoodi NM, Akbari A (2011) Dye removal from colored textile wastewater using acrylic grafted nanomembrane. Desalination 267:107–113

44. Harrelkas F, Azizi A, Yaacoubi A, Benhammou A, Pons MN (2009) Treatment of textile dye effluents using coagulation–flocculation coupled with membrane processes or adsorption on powdered activated carbon. Desalination 235:330–339
45. Charumathi D, Das N (2012) Packed bed column studies for the removal of synthetic dyes from textile wastewater using immobilised dead *C. tropicalis*. Desalination 285:22–30
46. Aksu Z, Tezer S (2005) Biosorption of reactive dyes on the green alga *Chlorella vulgaris*. Process Biochem 40:1347–1361
47. Kiran I, Akar T, Ozcan AS, Ozcan A, Tunali S (2006) Biosorption kinetics and isotherm studies of Acid Red 57 by dried *Cephalosporium aphidicola* cells from aqueous solutions. Biochem Eng J 31:197–203
48. Taylor P, Gagan M (eds) (2002) Alkenes and aromatics. In: The molecular word. The Open University, Cambridge
49. Ortiz de García S, Pinto Pinto G, García Encina P, Irusta Mata R (2013) Consumption and occurrence of pharmaceutical and personal care products in the aquatic environment in Spain. Sci Total Environ 444:451–465
50. Etschmann MMW, Bluemke W, Sell D, Schrader J (2002) Biotechnological production of 2-phenylethanol. Appl Microbiol Biotechnol 59:1–8
51. Flint S, Markle T, Thompson S, Wallace E (2012) Bisphenol a exposure, effects, and policy: a wildlife perspective. J Environ Manag 104:19–34
52. Feng Y, Su G, Zhao H, Cai Y, Cui C, Sun-Waterhouse D, Zhao M (2015) Characterisation of aroma profiles of commercial soy sauce by odour activity value and omission test. Food Chem 167:220–228
53. Pal A, He Y, Jekel M, Reinhard M, Gin KY-H (2014) Emerging contaminants of public health significance as water quality indicator compounds in the urban water cycle. Environ Int 71:46–62
54. Kovacic P, Somanathan R (2014) Nitroaromatic compounds: environmental toxicity, carcinogenicity, mutagenicity, therapy and mechanism. J Appl Toxicol 34:810–824
55. Banat FA, Al–Bailey B, Al-Asheh S, Hayajneh O (2000) Adsorption of phenol by bentonite. Environ Pollut 107:391–398
56. Dutta NN, Brothakur S, Baruaha R (1998) A novel process for recovery of phenol from alkaline wastewater: laboratory study and predesign cost estimate. Water Environ Res 70:4–9
57. United States Environmental Protection Agency EPA (2013) Alphabetical list of national primary and secondary drinking water standards. http://water.epa.gov/lawsregs/rulesregs/regulatingcontaminants/basicinformation.cfm. Accessed 6 Jul 2016
58. Roger MS (eds) (2001) Handbook of analytical separations, Vol 3. Environmental analysis. Elsevier. Amsterdam
59. Kraus G (2008) Synthetic methods for the preparation of 1,3–propanediol. Clean–Soil Air Water 36:648–651

Chapter 8
Extraction of Natural Phenolic Compounds with ABS

Milen G. Bogdanov and Ivan Svinyarov

Abstract Amongst natural compounds, phenolic compounds – the most abundant class of secondary plant metabolites – are commonly employed as additives in the manufacturing of various human commodities, such as dietary supplements, food and pharmaceutical and cosmetic products, and hence can be considered of high importance from an industrial standpoint. The extraction of these compounds proceeds according to well-established procedures, which are multistage, laborious and time and energy consuming and use organic solvents which are often flammable, volatile and toxic. Accordingly, there is the need to develop novel and sustainable procedures, so that the production of such value-added chemicals can be achieved in a more efficient and environmentally friendly manner. This chapter represents a comprehensive overview on the recent achievements in the extraction of natural compounds, and mainly phenolic compounds, by means of ionic-liquid-based aqueous biphasic systems (IL-based ABS). It considers factors that influence the extraction efficiency, such as inorganic salt type and concentration, IL type and concentration, pH value and temperature, and provides some clues towards the extraction yield improvement.

Keywords Ionic liquids • Liquid–liquid extraction • Separation • Innovative technologies • Recovery • Natural products • Antioxidants

8.1 Introduction

Phenols represent a large class of organic compounds comprising at least one hydroxyl group attached to a benzene ring (see Fig. 8.1a). The simplest representative is phenol, but more complex species such as stilbenoids, flavonoids, phenolic acids, lignans, tannins, etc., build up separate classes of phenolic compounds [1]. These species are widely distributed in the plant kingdom and are the most abundant secondary metabolites that are biosynthesized by plants and fruits de novo

---

M.G. Bogdanov (✉) • I. Svinyarov
Faculty of Chemistry and Pharmacy, University of Sofia "St. Kl. Ohridski", 1, James Bourchier Blvd, 1164 Sofia, Bulgaria
e-mail: mbogdanov@chem.uni-sofia.bg

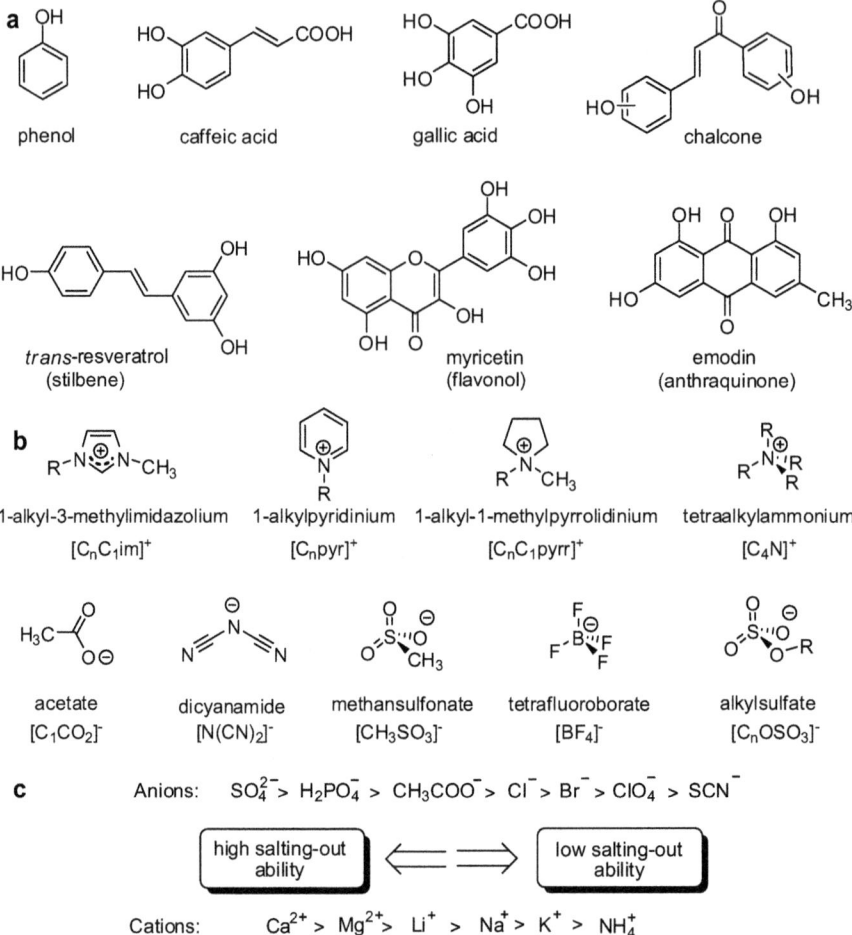

Fig. 8.1 (a) Chemical structures of phenol and selected natural polyphenolic compounds; (b) structure, name and abbreviation of commonly used cations and anions in ionic liquids; and (c) order of effectiveness in salting-out of selected inorganic ions

in response to stress (drought and cold), injury, viral infection or invasion by bacterial or fungal pathogens [2]. Being widespread constituents of foods of plant origin, phenolic compounds are an integral part of our daily diet, and it is nowadays truly believed that they are responsible for the prevention of various diseases associated with the oxidative stress [3–7]. Indeed, their preventive effects regarding cardiovascular, neurodegenerative diseases and cancer, as well as their modulating effect towards various enzymes and cell receptors, were proved by a number of in vitro and in vivo studies [8–15]. Therefore, phenolic compounds that are employed as additives in the production of dietary supplements, food and pharmaceutical and cosmetic products can be considered of high importance from an industrial standpoint.

The production of phenolic compounds from their native sources proceeds according to well-established methods [16, 17], which include initial solid–liquid extractions with molecular organic volatile solvents (VOCs), e.g. saturated hydrocarbons, alcohols, halogenoalkanes, etc., followed by the additional treatment of the obtained crude extracts in order to isolate the compounds of interest in a pure form. These procedures are multistage, laborious and time and energy consuming, and the organic solvents typically employed are flammable, volatile and toxic, the latter being in a contradiction with the nowadays universally accepted 12 principles of the green chemistry [18]. Thus, the need of extractants of improved characteristics from safety, ecological, toxicological and technological standpoint can be suggested [19, 20].

Ionic liquids (ILs) are promising candidates that could meet the abovementioned requirements. ILs, also named as designer' solvents, received a significant attention from the scientific community in the last two decades [21]. Consisting entirely of ions (usually nonsymmetrical charge-stabilized organic cations and inorganic or organic anions, see Fig. 8.1b), most of them can be liquid at room temperature and display a wide range of unique properties, such as negligible vapour pressure, non-flammability and high thermal stability, amongst others. These unique properties make of ILs potential solvents to replace the commonly employed VOCs [22]. Moreover, the versatility of possible ion combinations, each of them hypothetically resulting in a new IL, allows their physicochemical properties to be fine-"tuned" by careful selection of the ions [22–26], thereby allowing "IL tailoring" for a particular application.

To date, ILs have been successfully introduced instead of VOCs in numerous processes such as synthesis [27], catalysis [28], electrochemistry [29] and analytical chemistry [30]. Additionally, ILs have proved to be efficient solvents for the solid–liquid extraction of a wide variety of natural products [19, 20], in all cases providing enhanced extraction yields and significant reduction of the extraction time and solvent consumption. The latter was believed to be due to the stronger dissolving power of ILs, but it was recently shown [31] that their role in the solid–liquid extraction processes is not limited only to the enhanced interactions provided by the ions, but can be rather attributed to the pronounced solvent-matrix interactions leading to the plant matrix disruption and permeability modification [31]. However, in the majority of these cases, neither attempts for IL recycling, an important issue that addresses the economics of their use, have been done, nor possible ways for the isolation of target solutes in their pure form after the solid–liquid extraction step have been examined. Indeed, a short retrospection of the recent literature shows that these issues are studied to a very limited extent and that the methods employed depend on specific properties of both ILs and solutes to be extracted. Particularly, an anti-solvent-induced precipitation step proved successful for the isolation of neutral compounds [32] and hydrodistillation for the recovery of volatile compounds [33]. Additionally, back extraction with organic solvents [34–36], the use of ion exchange resins [37] or resins for selective trapping [38] and, in some cases, the solute partitioning in IL-based aqueous biphasic systems (IL-ABS) [39–41] had given satisfactory results.

It was recently shown, by Gutowski et al. [42], that aqueous solutions of hydrophilic ILs can be switched into IL-based ABS by introduction of an inorganic salt (Fig. 8.1c). The driving force of the separation can be attributed to the salting-out effect of the inorganic salt, resulting in the formation of two water-rich immiscible phases, where the upper phase (in majority of the cases) consists mainly of IL and the lower one of concentrated inorganic salt. This way, IL-ABS can be considered as an attractive alternative to the widely applied conventional liquid–liquid extraction with organic solvents, since they ensure simultaneous enrichment, separation, purification and isolation of a particular molecule. Indeed, IL-ABS have been utilized to extract a wide range of compounds, such as proteins [43], saccharides [44], enzymes [45], antibiotics [46] and alkaloids [47, 48], to name just a few. It is noteworthy that in some cases a controlled partition of the solutes of interest had been achieved by variation of the system composition or its pH value, thus broadening the scope of IL-ABS applications.

This chapter represents a comprehensive overview on the recent achievements in the extraction of natural phenolic compounds by means of IL-ABS. Table 8.1 provides a summary of the works existent in the literature [39, 40, 49–61]. Factors that influence the extraction efficiency, e.g. inorganic salt type and concentration, IL type and concentration, pH value and temperature, are presented. All these factors are discussed in a consequent manner, so that some clues towards the extraction yield improvement can be given.

8.2 Influence of Extraction Parameters

The successful extraction of a certain molecule with ABS simultaneously depends on the physicochemical properties of the compound to be extracted and on the phase-forming components of the system. Thus, several factors have to be studied in order to obtain a better distribution and selectivity. The most studied factors controlling the partition of molecules within ABS are IL and salt type and concentration, i.e. the ABS composition, temperature and pH value of the system. The performance of the system under study is commonly assessed by calculating the corresponding partition coefficients (K) according to the following relationship:

$$K = C_{IL}/C_{salt} \qquad (8.1)$$

where C_{IL} and C_{salt} represent the concentration of the compound of interest in the IL-rich and salt-rich phases, respectively.

It should be stressed, however, that an increase in K does not necessarily mean an increase in the extraction efficiency ($EE\%$), since the ratio between the two phases is not taken into account in the determination of K. Hence, $EE\%$ are calculated according to the following equation:

8 Extraction of Natural Phenolic Compounds with ABS

Table 8.1 Recent application of IL-based ABS in liquid–liquid extractions of phenolic compounds

Extracted compound(s)	ILs tested[a]	Salting-out agent	Best IL-based ABS[b]	References
Phenol 4-Nitrophenol	$[C_4C_1im]Cl$	K_2CO_3, K_2HPO_4, K_3PO_4	Forward extraction	[49]
			$[C_4C_1im]Cl/$ K_2HPO_4	
Phenol	$[C_nC_1im][BF_4]$, (n = 3, 4, 5, 6, 7, 8)	Glucose	Forward extraction	[50]
			15 wt% $[C_4C_1im]$ $[BF_4]$ + 35 wt% glucose	
			298 K	
Chlorophenols	$[C_4C_1im][BF_4]$	NaH_2PO_4	Forward extraction	[51]
			$[C_4C_1im]$ $[BF_4]/$ NaH_2PO_4	
			pH = 4.0	
Bisphenol A	$[C_nC_1im]Cl$, (n = 2, 4, 6);	K_3PO_4	Forward extraction	[52]
	$[(C_1=C_2)C_1im]Cl$, $[C_4C_1pyrr]Cl$, $[(C_4)_4P]Cl$, $[(C_4)_4N]Cl$, $[(HO)^2C_2(C_1)_3N]Cl$		2.5 wt% $[(HO)^2C_2(C_1)_3N]Cl$ + 45 wt% K_3PO_4	
			2.7 wt% $[C_nC_1im]Cl$ + 37 wt% K_3PO_4	
			298 K, pH = 13	
Vanillin	$[C_nC_1im]Cl$, (n = 2, 4, 6, 7, 10), $[BzC_1im]Cl$, $[(HO)^2C_2C_1im]Cl$, $[(C_1=C_2)C_1im]Cl$	K_3PO_4	Forward extraction	[53]
			25 wt% $[C_4C_1im]Cl$ + 15 wt% K_3PO_4	
	$[C_4C_1im]X$, {X = Cl, Br, $[N(CN)_2]$, $[C_1OSO_3]$, $[CF_3SO_3]$, $[C_1SO_3]$, $[C_1CO_2]$}		298 K, pH = 13	
Paracetamol	$[(C_2)_4N]X$, $[(C_3)_4N]X$, $[(C_4)_4N]X$, (X = Cl, Br)	Potassium citrate, K_2CO_3, $K_2HPO_4/$ KH_2PO_4	Forward extraction	[54]
			31 wt% $[(C_2)_4N]X$ + 27 wt%	

(continued)

Table 8.1 (continued)

Extracted compound(s)	ILs tested[a]	Salting-out agent	Best IL-based ABS[b]	References
			potassium citrate	
			298 K, pH = 7	
Parabens	[C$_4$C$_1$im]Cl	K$_3$PO$_4$	*Forward extraction*	[55]
			5 wt% [C$_4$C$_1$im]Cl + K$_3$PO$_4$, 298 K, pH = 9	
Eugenol Propyl gallate	[C$_n$C$_1$im]Cl, (n = 2, 4, 6, 8), [C$_4$C$_1$pyrr]Cl, [C$_4$C$_1$pip]Cl, [(C$_4$)$_4$N]Cl	Potassium citrate, K$_2$HPO$_4$/KH$_2$PO$_4$	*Forward extraction*	[56]
			40 wt% [C$_4$C$_1$im]Cl + 20 wt% citrate buffer	
			298 K, pH = 7	
Gallic acid	[C$_2$C$_1$im][CF$_3$SO$_3$], [C$_n$C$_1$im]Cl, (n = 7, 8) [C$_4$C$_1$im]X, {X = Br, [N(CN)$_2$], [C$_1$OSO$_3$], [C$_2$OSO$_3$], [C$_8$OSO$_3$], [CF$_3$SO$_3$]}	Na$_2$SO$_4$, K$_2$HPO$_4$/KH$_2$PO$_4$, K$_3$PO$_4$	*Forward extraction*	[57]
			25 wt% [C$_4$C$_1$im][CF$_3$SO$_3$] + 15 wt% Na$_2$SO$_4$	
			298 K, pH = 4	
			Back extraction	
			25 wt% [C$_4$C$_1$im][CF$_3$SO$_3$] + 15 wt% Na$_2$SO$_4$	
			298 K, pH = 13 (NaOH)	
Gallic acid Vanillic acid Syringic acid	[C$_4$C$_1$im]X, {X = Cl, Br, [N(CN)$_2$], [C$_1$OSO$_3$], [C$_2$OSO$_3$], [C$_1$CO$_2$], [C$_1$SO$_3$], [CF$_3$SO$_3$], [OTs], [(C$_1$O)$_2$PO$_2$], [SCN]}	Na$_2$SO$_4$, Na$_2$CO$_3$	*Forward extraction*	[40]
			25 wt% [C$_4$C$_1$im][CF$_3$SO$_3$] + 20 wt% Na$_2$SO$_4$	

(continued)

8 Extraction of Natural Phenolic Compounds with ABS

Table 8.1 (continued)

Extracted compound(s)	ILs tested[a]	Salting-out agent	Best IL-based ABS[b]	References
			298 K, pH = 3.3	
			Back extraction	
			20 wt% [C$_4$C$_1$im][CF$_3$SO$_3$] + 10 wt% Na$_2$CO$_3$	
			298 K, pH = 11.1	
Chlorogenic acid Hydroxycinnamic acid (derivatives)	[C$_2$C$_1$im][C$_1$CO$_2$]	K$_3$PO$_4$	*Forward extraction*	[58]
			30 wt% [C$_2$C$_1$im][C$_1$CO$_2$] + 15 wt% K$_3$PO$_4$	
			298 K, pH = 9	
Puerarin (isoflavone)	[C$_4$C$_1$im]X, (X = Br, OH), [(HO)^{2}C$_2$C$_1$im][BF$_4$], [(HO$_2$C)C$_1$C$_1$im][BF$_4$]	Na$_2$CO$_3$, K$_2$CO$_3$, (NH$_4$)$_2$SO$_4$, K$_2$HPO$_4$	*Forward extraction*	[59]
			30 wt% [C$_4$C$_1$im]X + 30 wt% K$_2$HPO$_4$	
			298 K	
4-(2-Pyridyazo)-resorcinol	[C$_4$C$_1$im]Br	Phosphate salts	*Forward extraction*	[60]
			20 wt% [C$_4$C$_1$im]Br + 25 wt% salt	
			298 K, pH = 9.7	
Natural phenols (not specified)	[(HO)^{2}C$_2$(C$_1$)$_3$N]Cl	K$_3$PO$_4$, Na$_2$CO$_3$	*Forward extraction*	[61]
			30 wt% [(HO)^{2}C$_2$(C$_1$)$_3$N]Cl + 20 wt% K$_3$PO$_4$	

(continued)

Table 8.1 (continued)

Extracted compound(s)	ILs tested[a]	Salting-out agent	Best IL-based ABS[b]	References
			298 K	
			Back extraction	
			$[(HO)^2C_2(C_1)_3N]Cl + [(HO)^2C_2(C_1)_3N][NTf_2]$	
			298 K	
Anthraquinones	$[C_nC_1im]Br$, ($n=2$, 4, 6), $[C_4C_1im][BF_4]$, $[C_4C_1im][N(CN)_2]$	Na_2SO_4, NaH_2PO_4, $(NH_4)_2SO_4$, $MgSO_4$	Forward extraction	[39]
			15 wt% $[C_4C_1im][BF_4]$ + 15 wt% Na_2SO_4	
			298 K, pH = 6.1	
			Back extraction	
			15 wt% $[C_4C_1im][BF_4]$ + 15 wt% Na_2SO_4	
			298 K, pH = 14	

[a]Cations: $[C_nC_1im]$ (1-alkyl-3-methylimidazolium), $[(C_1=C_2)C_1im]$ (1-allyl-3-methylimidazolium), $[BzC_1im]$ (1-benzyl-3-methylimidazolium), $[(HO)^2C_2C_1im]$ (1-(2-hydroxyethyl)-3-methylimidazolium), $[(HO_2C)C_1C_1im]$ (1-acetic-3-methylimidazolium), $[C_4C_1pyrr]$ (1-butyl-1-methylpyrrolidinium), $[(C_1)_4N]$ (N,N,N,N-tetramethylammonium), $[(HO)^2C_2(C_1)_3N]$ (cholinium). Anions: Cl (chloride), Br (bromide), I (iodide), $[BF_4]$ (tetrafluoroborate), $[PF_6]$ (hexafluorophosphate), $[OTs]$ (tosylate), $[C_nCO_2]$ (alkyl carboxylate), $[OTf]$ (trifluoromethanesulfonate), $[NTf_2]$ (bis(trifluoromethanesulfonyl)imide or triflimide), $[N(CN)_2]$ (dicyanamide), $[C_nSO_3]$ (alkyl sulfonate), $[C_nOSO_3]$ (alkyl sulphate), $[OH]$ (hydroxide), $[SCN]$ (thiocyanate). [b]IL-based ABS ensuring highest recoveries for the forward or back extractions, respectively

$$EE\% = [C_{IL} \times V_{IL}/(C_{IL} \times V_{IL} + C_{salt} \times V_{salt})] \times 100 \quad (8.2)$$

where V_{IL} and V_{salt} are the corresponding volumes of the IL-rich and salt-rich phases.

In the following sections, both factors are discussed, depending on their availability in the articles under consideration.

8.2.1 Inorganic Salt

Despite organic compounds being hydrophobic in nature, they are soluble in water to same extent, especially if functional groups which can be easily hydrated are present in their structure. Thus, in order to achieve the quantitative recovery by means of liquid–liquid extraction, the procedure has to be repeated several times. However, the distribution of such species into the hydrophobic phase, even being of high affinity to water, can be facilitated by the addition of inorganic salts to the system. This effect is well known as the salting-out effect and it is frequently used in the laboratory practice.

In this direction, Yu et al. [62] conducted a comparative study and showed that the introduction of salting-out agents, such as KH_2PO_4, $MgSO_4$ and $CuSO_4$, does not alter the extraction efficiency of hydrophobic IL/water systems, namely, $[C_4C_1im][PF_6]$/water and $[C_6C_1im][PF_6]$/water, towards caffeic acid and ferulic acid. In contrast to these results, Fan et al. [63] studied the influence of a series of inorganic salts, such as NaCl, $NaNO_3$, Na_2SO_4, NaI, $NaClO_4$, $ZnSO_4$ and $Al_2(SO_4)_3$, on the distribution ratio of phenol-based endocrine-disrupting agents in a system composed of $[C_6C_1im][PF_6]$/water. Although phenol-based endocrine-disrupting agents and some phenol compounds discussed below are not from a natural source, they are here discussed for comparison purposes within the phenol class of compounds. It was found that the partition of these phenols into the IL-rich phase increases significantly with the concentration of the salts, except in the presence of NaCl, $NaNO_3$ and NaI, the latter being attributed to the salting-in capacity of the salts under study. Considering the free Gibbs energy of hydration, it was concluded that the salts with higher hydration ability will coordinate water molecules better, which in turn will push the organic molecules to be preferably partitioned into the more hydrophobic IL-rich phase [63].

Similarly, Zhang et al. [49] performed a comparative analysis of both hydrophobic and hydrophilic enriched systems, with $[C_8C_1im][PF_6]$-water and $[C_4C_1im]$Cl-based ABS, to extract aromatic compounds, including phenol and 4-nitrophenol. The inorganic salts employed in this case were of stronger salting-out capacity, leading to the following order of phase-forming ability for the same salt cation: $K_3PO_4 > K_2HPO_4 > K_2CO_3$, being consistent with the Hofmeister series [64]. As in the above discussed case, the distribution coefficients for the phenolic compounds were found to increase with the inorganic salt concentration, and this was also attributed to the ability of the salt to interact with the water molecules. It is noteworthy that the extraction capacity of the hydrophobic $[C_8C_1im][PF_6]$ was found better than that demonstrated by all $[C_4C_1im]$Cl-based ABS; however, the authors [64] concluded that the potential of the hydrophilic system is worth to be further studied since it is less expensive. In the same line, Wang et al. [51] evaluated the influence of the concentration of NaH_2PO_4 on the partition of chlorinated phenols in $[C_4C_1im][BF_4]$-based ABS. An unusual behaviour was observed in this case, namely, that the recoveries of the target phenolic compounds increase gradually with the inorganic salt concentration at the beginning and further decrease after reaching a maximum. Although interesting, this phenomenon has not been discussed, and so it can be hypothesized that at a certain concentration, some salts can start interacting with the phenolic molecules in the same way as they do with water at lower concentrations, and so this issue seems to be important to study.

Regarding natural compounds, a similar concentration-dependent behaviour to that reported by Wang et al. [51] was also documented by Fan et al. [59] during the extraction optimization of puerarin – a natural isoflavone glucoside – in $[C_4C_1im]$ Br/K_2CO_3 ABS, but not in the $[C_4C_1im]Br/K_2HPO_4$ ABS, which shows the importance of the salt used. The influence of the inorganic salt type and concentration on the extraction efficiency of anthraquinones in $[C_4C_1im][BF_4]$-based ABS was studied by Tan et al. [39]. It was shown that the gradual increase of the salt concentration, with Na_2SO_4, NaH_2PO_4 and $(NH_4)_2SO_4$, enhances the extraction efficiency towards the target compounds, whereas the introduction of additional quantities of $MgSO_4$ into the $[C_4C_1im][BF_4]$-based ABS leads to a decrease of extraction efficiency.

More detailed investigations on the effect of the inorganic salt have been performed by Cláudio et al. [57] on gallic acid partition in IL-based ABS. The authors tested the phase-forming ability of Na_2SO_4, K_3PO_4 and K_2HPO_4/KH_2PO_4 buffer solution by using a wide range of hydrophilic ILs. The three salts had been chosen in a way to ensure different pH values of the systems. It was found that the salts studied bring about a very different behaviour in partitioning the target compound, the latter suggesting that the choice of the inorganic salt is a dominant parameter in the extraction of phenolic compounds [57].

Salting-out agents of organic origin have been also employed to form extractive systems for phenolic compound recovery. For instance, Chen et al. [50] studied the partition of phenol into $[C_4C_1im][BF_4]$/glucose ABS. The authors showed that the increase of glucose concentration is favourable for increasing the distribution of phenol into the IL-rich phase and that the influence of glucose content is more significant than the IL used. Recently, Santos et al. [56] introduced another organic species, namely, potassium citrate, to induce the formation of IL-based ABS and studied the extraction of natural antioxidants such as eugenol and propyl gallate. The quantitative recovery of the two antioxidants into the IL-rich phase, achieved in this study, suggests that environmentally friendly salting-out agents can be also used to form effective IL-based ABS as extractive systems of phenolic compounds. A citrate-induced IL-based ABS was also introduced for the effective recovery of paracetamol from pharmaceutical wastes by Silva et al. [54].

In summary, inorganic salts as well as organic salts can be successfully employed to induce the formation of IL-based ABS. Salts containing $H_2PO_4^-$, PO_4^{3-}, CO_3^{2-} or SO_4^{2-} anions favour the partition of phenolic compounds into the IL-rich phase, whilst anions such as Cl^-, I^- and NO_3^- do not have influence on the extraction efficiency, even in hydrophobic IL/water biphasic systems. Additionally, the type and concentration of the salts used are of high importance, whilst the extraction efficiency can be manipulated by introduction of a certain salt and by changing its concentration. On the other hand, different salts can induce different pH values, which seem to be a significant factor towards the partition of phenolic compounds in IL-based ABS.

8.2.2 Ionic Liquid

It is known nowadays that imidazolium-based ILs can form aggregates in aqueous solutions, especially as the alkyl group attached to the imidazolium ring becomes longer, and a number of studies [65–70] demonstrated that the mesophase structure of IL-aqueous mixtures can be "tuned" by careful selection of the anion and adjustment of the water concentration, thereby "tailoring" the system for selective interactions with the solute of a particular interest. Taking into account that the IL-rich phase in ABS contains a considerable amount of water, hence different internal structures, one can expect different types of interactions, simultaneously dependent on the ion properties and water content. Therefore, the IL type seems to be an important factor in the partitioning of phenolic compounds.

A remarkable progress with respect to the establishment of IL structure influence on the extraction of phenolic compounds by IL-based ABS was achieved by Coutinho and co-authors. In a series of articles, the authors reported detailed studies on the partition of bisphenol A [52], vanillin [53], paracetamol [54], eugenol and propyl gallate [56] and gallic, vanillic and syringic acids [40, 57]. The broad range of the ILs chosen in these studies allowed the influence of their distinct properties on the partition of the compounds of interest to be assessed and some general conclusions to be drawn towards the extraction mechanism that takes place.

To evaluate the influence of the cation on the partition of vanillin, Cláudio et al. [53] used a series of chloride-based ILs, such as $[C_2C_1im]Cl$, $[C_4C_1im]Cl$, $[C_6C_1im]Cl$, $[C_7C_1im]Cl$, $[C_{10}C_1im]Cl$, $[(HO)^2C_2C_1im]Cl$, $[BzC_1im]Cl$ and $[(C_1=C_2)C_1im]$, and in order for the anion influence to be assessed, $[C_4C_1im]^+$ cation was combined with Cl^-, Br^-, $[C_1CO_2]^-$, $[CH_3SO_3]^-$, $[CF_3SO_3]$, $[C_1OSO_3]^-$ and $[N(CN)_2]^-$ anions. The extractions were performed at 298 K and K_3PO_4 was used as a salting-out agent, the latter ensuing a pH value of the system $ca.13$. The results obtained showed that the distribution of vanillin strongly depends on both cation and anion of the IL, since the partition coefficients ranged between 2.72 and 49.59 for similar ABS composition. Regarding the IL cation, the following order of extraction capacity was observed: $[C_6C_1im]Cl > [C_4C_1im]Cl > [BzC_1im]Cl > [C_7C_1im]Cl > [C_2C_1im]Cl \approx [(C_1=C_2)C_1im] >> [(HO)^2C_2C_1im]Cl > [C_{10}C_1im]$ Cl. To rationalize this order, the authors considered the chain length influence on the IL properties, such as free volume and surface tension, on the one hand, and coulombic and dispersive interactions, on the other hand, and concluded that the apparent best performance of $[C_6C_1im]Cl$ is due to the opposite direction of acting of these two contributing factors. Furthermore, the presence of different functional groups, such as phenyl, hydroxyl or allyl groups, in the alkyl side chain of the imidazolium cation does not improve the partition of vanillin, and this was attributed to the increased hydrophilicity and affinity of these ILs to water

[71]. Regarding the IL anion, it could be said that the partition of vanillin is favoured by the presence of anions with higher hydrogen-bond-accepting ability. This way, best partition coefficients were obtained with Cl^-, Br^- and $[N(CN)_2]^-$ anions [72]. In addition, the apparent extraction behaviour was explained by the salting-out/salting-in properties of the anions under study. It was deduced that the partition of vanillin into the IL-rich phase will be facilitated if anions with distinct salting-in properties are employed, since these low-charge-density ions favour the solute–IL interactions. In contrast, the high-charge-density ions, such as $[C_1CO_2]^-$ and $[CH_3SO_3]^-$, tend to form hydrated complexes, thus not interacting with vanillin, which in turn decrease the respective partition coefficients.

Similar explanation for the anion influence on the partition of gallic acid was given in a subsequent study by Cláudio et al. [57], albeit considerably lower partition coefficients were found in comparison to that obtained for vanillin at the same conditions. In this case, a preferred distribution of gallic acid into the IL-rich phase was realized only with two chloride-based ILs with long alkyl side chain cations, such as $[C_7C_1im]^+$ and $[C_8C_1im]^+$ (*EE%ca.* 89 and 74 %, respectively), which indicated that the mechanism of extraction is more complex than expected. Considering the pH value of the system (*ca.* 13), and based on the results obtained, the authors [57] assumed that the extraction of gallic acid can be attributed to the ability of the ILs under study to form micelles in presence of water. This way, they postulated that the driving force of the process is not related to the solvation of the corresponding conjugate base inside the micelles; instead, gallate reduces the repulsion between the imidazolium head groups and stabilizes the micelle, enhancing thus the formation of self-aggregating structures. This explanation seems not to be illogical since $[C_4C_1im][C_8OSO_3]$, also capable to form micelles, showed poorer performance in extracting gallate, most likely due to the unfavourable repulsive interactions between the two negatively charged species.

As it is described in Sect. 8.2.3, pH plays an important role in the partitioning of phenolic compounds in IL-based ABS. Considering this fact, Cláudio et al. [57] performed a subsequent study on the partition of gallic acid, but with Na_2SO_4 as a salting-out agent to induce the formation of IL-based ABS. It is noteworthy that different pH values ranging from 1.57 to 8.07 were obtained with different ILs. On the one hand, this fact is interesting itself, since Na_2SO_4 is a neutral salting-out agent and one can expect pH values of the system close to 7. On the other hand, this unusual behaviour allows IL-based ABS to be prepared with different pH values by using the same salt. However, the results obtained were consistent with the bellow discussed relationship pH–*pK*a extraction efficiency and suggests that in order for the higher distribution coefficients to be achieved, one should work at pH values close to the *pK*a of the compounds of interest. *EE%* higher than 80 % were obtained for all the systems studied, and this was attributed to the fact that the dominant species of gallic acid in the pH region close to 4.4 is neutral, the latter favouring the hydrogen bonding and hydrophobic interactions between the solute and IL. Exceptions were observed with the micelle-forming IL $[C_4C_1im][C_8OSO_3]$, which showed the lowest *EE%* (58 %), but this is consistent with the above-mentioned behaviour for this IL and suggests that the hydrogen-bonding ability plays the major role for the preferred partition of phenolic compounds into the IL-rich phase.

Considering these results, Coutinho and co-authors developed efficient methods for extraction of paracetamol from wastewaters [54], bisphenol A from human urine [52] and an elegant protocol for the recovery of natural antioxidants [40] by means of IL-based ABS.

8.2.3 pH

It is well-known that phenolic compounds are weak acids due to the possible resonance stabilization of the resulting anionic species in basic media. Therefore, one can expect that species of different polarities can exist at different pH values, and can play with that, in order for the most optimal conditions for phenolic compounds extraction to be established. Considering these facts, Wang et al. [51] studied the pH influence on the distribution coefficients of chlorinated phenols in $[C_4C_1im][BF_4]/NaH_2PO_4$ ABS. The partition into the IL-rich phase was found to be strongly dependent on the pH in the studied interval, between 1.5 and 9.0, increasing from the very beginning with the increase in pH and then decreasing in the same way, after reaching a maximum (*ca.* 90 % recovery) at about pH 4–4.5. A similar behaviour was also observed by Tan et al. [39] in the extraction of anthraquinones in $[C_4C_1im][BF_4]/Na_2SO_4$ ABS, and in this case, the target compounds were found to be distributed preferably to the IL-rich phase (*EE%* ca. 75 %) at pH 4 and not at pH 14, where the *EE%* was about 75 %, but in favour of the salt-rich phase. These results, however, brought about the development of a procedure for the efficient recovery of the target phenolic compounds by means of sequential liquid–liquid extractions at the optimized pH values for the forward and back extraction steps [40].

In support to the previously described observations, the *EE%* of gallic acid ($pKa = 4.4$) was found optimal at pH 3 and then to decrease gradually with increase of pH into $[C_4C_1im][CF_3SO_3]/Na_2SO_4$ ABS [57]. To explain these results and to understand better the influence of the pH on the extraction mechanism of phenolic compounds by IL-based ABS, Cláudio et al. [57] investigated the partition of gallic acid at different pH values, tailored by subsequent addition of NaOH. The results obtained showed that the higher the pH value is, the lower the partition coefficient into the IL-rich phase. Thus, since gallic acid exists as uncharged species at pH values below its *p*Ka, it preferably migrates to the IL-rich phase, whilst when the pH is higher than 4.4, the corresponding gallate, being a negatively charged and so more hydrophilic species, preferably partitions into the salt-rich phase. Similar explanation was given by Sheikhian et al. [60] to rationalize the partition behaviour of the phenolic azo dye 4-(2-pyridylazo)-resorcinol (PAR) in $[C_4C_1im]Br$/potassium-phosphate-salts ABS. The authors stressed that PAR can exist in three distinct forms – neutral (PAR), anionic (PAR$^-$) and di-anionic (PAR^{2-}) – each of them different in polarity ($pKa_1 = 3.1$, $pKa_2 = 5.6$, $pKa_3 = 11.9$, respectively [60]). This way, at high pH values (11.6), the extraction into the IL-rich phase decreases as the less hydrophobic species PAR^{2-} is present, the latter also indicating that the salting-out effect of PO_4^{3-} is not so effective at this pH and that the increased hydrophilic character of the di-anionic species hinders its partition into the IL-rich phase. The same trend of pH dependence was also documented by Noorashikin et al. [55], who studied the partition of parabens in $[C_4C_1im]Cl/K_3PO_4$ ABS in the pH range 2–12, but in contrast to the above results, the recovery of phenols was found to be independent in the low pH region (pH 2–4), showing a constant $EE\%$ of about 60 %, and then to increase gradually with the pH, reaching a maximum (about 95 % recovery) at pH 9, before going back to 75 % at pH 12. A plausible explanation of the observed change in the partition behaviour was given by the authors [55]. Considering the acid–base properties of the compounds under study, they postulated that at pH values below 3, parabens exist in a protonated form, which leads to an electrostatic repulsion with the positively charged IL cation, thus lowering the recovery. At pH 4–6.5 parabens exist as neutral species [73], at pH 7–9 are partly ionized and above pH 9 are negatively charged because the hydroxyl group is fully deprotonated. Thus, in the middle region, the recovery increases gradually with pH and achieves its maximum at pH 9, which is very close to the pKa values of the compounds under study (around 8.9). Consequently, the presence of functional groups that influence the relative polarity of the compounds to be extracted can be considered of an immense importance, and the pH influence on the $EE\%$ has to be always addressed.

The chemical structure of the ILs employed to form ABS is also important, and this can be rationalized by considering the results obtained by Silva at al. [54], who studied the partition of paracetamol in $[(C_2)_4N]Br$/potassium citrate buffer and $[(C_2)_4N]Br/K_3PO_4$ ABS at different pH values. The pH dependence on the $EE\%$ was found not so significant in this case, since for systems with similar component composition, very small deviations in the apparent $EE\%$, being between 90 and 95% in the pH range 5–13, were observed. However, it was deduced that the existence of paracetamol in its deprotonated form at pH values higher than its pKa value restricts its partition into the IL-rich phase to some extent, which can be attributed to the enhanced ionic interactions between the charged paracetamol and the components of the salt-rich phase.

Based on the above findings, Cláudio et al. [40] developed recently an elegant general strategy for the recovery of natural antioxidants, such as gallic, vanillic and syringic acids by means of IL-based ABS. Shortly, in the first step, $[C_4C_1im][CF_3SO_3]/Na_2SO_4$ ABS was employed to extract the compounds of interest from an aqueous solution into the IL-rich phase ($EE\%$ ca. 95 %); and after phase separation, another ABS composed of $[C_4C_1im][CF_3SO_3]/Na_2CO_3$ was used to achieve the back extraction ($EE\%$ ca. 99 %) of the natural products and to "wash" the IL-rich phase for its subsequent reutilization.

In sum, pKa values of the phenolic compounds play an important role towards their partition between two immiscible phases of different polarity. It can be postulated that the highest recoveries into the IL-rich phase of ABS can be achieved at pH values close to the pKa of the extracted compounds. Therefore, the presence of functional groups that influence the relative polarity of the compounds to be extracted can be considered of an immense importance, and the pH influence on the $EE\%$ has to be always taken into account.

8.2.4 Temperature

According to previous studies on IL-based ABS [74, 75], temperature can be considered to play an important role in the partitioning of molecules bearing different functional groups and hence offering different types of interactions. For instance, He et al. [74] reported that temperature has no significant influence on the partition behaviour of steroids, whereas Pei et al. [75] showed that it influences in a great manner the partition of proteins. Consequently, it can be suggested that the temperature dependence on the partition coefficients in IL-based ABS is simultaneously related to the solute properties and system composition; however, this issue has been addressed in a very limited extent.

A comprehensive investigation of the temperature influence on the distribution of a particular phenolic compound in IL-based ABS was first reported by Cláudio et al. [53]. The authors studied the partition of vanillin at different temperatures (in the range 288.15–328.15 K), and in order to assess the influence of the ion structure, ILs such as $[C_4C_1im]Cl$, $[C_4C_1im][CH_3SO_4]$, $[C_7H_7C_1im]Cl$ and $[(C_1,C_2)C_1im]Cl$ have been used to form ABS with K_3PO_4. The results obtained showed that the vanillin partition temperature dependence follows the same trend in all cases studied, but in a different extent. The observed partition coefficients were found to increase by changing temperature from 288 K to 298 K and then to decrease gradually with further increase in temperature. It is noteworthy that, compared to the remaining ILs, this effect was found less pronounced for the $[C_4C_1im][CH_3SO_4]$-based ABS and that the best results for all systems under study were obtained at 298 K. The latter suggests that the IL structure is an important factor and that both cation and anion contribute to the different partition behaviour of vanillin. In order to put some light on the mechanism of extraction and to clarify the driving forces of the process, Cláudio et al. [53] further performed thermodynamic analysis by using the van't Hoff relationship, $\ln K = -\Delta H/RT + \Delta S/R$. This way, by plotting lnK versus $1/T$, the thermodynamic parameters of transfer, such as standard molar Gibbs energy ($\Delta_{tr}G_m$), standard molar enthalpy ($\Delta_{tr}H_m$) and standard molar entropy ($\Delta_{tr}S_m$), were obtained and discussed. For the four systems studied, good linear relationships were obtained, which suggests that the enthalpy of transfer is independent on the temperature in the studied range between 298 and 328 K. Further, the negative values of $\Delta_{tr}H_m$ found in all systems indicated that the partition of vanillin from the salt-rich phase into the IL-rich phase is facilitated by the favourable vanillin-IL interactions, which were deduced to be controlled essentially by the anion. Moreover, the calculated negative $\Delta_{tr}G_m$ indicated that the process is endothermic, the latter being supported by the spontaneous and preferential distribution of vanillin into the IL-rich phases, as indicated by $K > 1$ for all systems studied.

Consistent results with the above discussed were obtained later by Tan et al. [39], during the extraction optimization of aloe anthraquinones (AQs) in $[C_4C_1im][BF_4]/Na_2SO_4$ ABS. In this case, the temperature was considered as a factor of high relevance, since the phase-forming ability of the IL is known to decrease with the increase in temperature, thus altering the composition of the two phases [76]. The AQ partition was studied in the range 298–323 K, and the partition coefficients were found to decrease gradually with the increase of temperature, and this was accompanied with no significant change of the phase–volume ratio. Thus, the temperature influence can be rather attributed to the disruption of the weak solute–IL interactions, responsible for the favourable partition into the IL-rich phase, than to the contraction of the two-phase region. Therefore, the authors postulated that such processes should be conducted at the lowest temperature possible. In addition, by conducting extractions at low temperatures, one can avoid the solute degradation, since it is known that phenolic species tend to degrade at high temperatures [1, 77] and also to suppress the hydrolysis of fluorinated anions such as $[BF_4]^-$ and $[PF_6]^-$ [78] if ILs composed of them are used.

It is noteworthy that the above discussed results are not consistent with these reported for the phenolic compounds partitioned in hydrophobic IL/water systems [62, 63], but a plausible explanation on this difference has not been given yet.

8.3 Conclusions

ILs, mainly due to the versatility of possible ion combinations and fine-tunable unique physicochemical properties, have received the attention of the scientific community in the recent years, and as a result of the extensive research conducted, ILs were introduced as substituents of volatile, flammable and toxic organic solvents in diverse processes. For instance, IL-based ABS were successfully employed instead of VOCs in the widely applied conventional liquid–liquid extraction processes, mainly for purification purposes. IL-based ABS have proved to be efficient systems for the recovery of a broad range of value-added chemicals, including natural products, in majority of the cases providing simultaneous enrichment, separation, purification and isolation of the species of interest.

The successful extraction of a certain molecule with IL-based ABS depends both on the physicochemical properties of the compound to be extracted and on the constituents of the system. Regarding natural phenolic compounds, the inorganic salt used to induce the formation of IL-based ABS has proved to be of extreme importance, since the extraction efficiency can be manipulated by the introduction of a certain salt and by changing its concentration. Moreover, different salts were found to induce different pH values of the systems, which seem to be the most significant factor towards the partition of phenolic compounds within IL-based ABS.

Both IL cation and anion structure and their properties are also important factors. As a rule, it can be suggested that a combination of between hydrophobic cations, which lower the surface tension of the IL-rich phase, and anions with prominent hydrogen-bond-accepting character favours the partition of phenolic compounds towards the IL-rich phase. It can be also suggested that the extraction of phenolic compounds with IL-based ABS has to be conducted at the lowest temperature possible, since the increase in temperature results in a decrease of the extraction efficiency, the latter being attributed to the disruption of the weak solute–IL interactions responsible for the favourable partition into the IL-rich phase. In addition, by conducting extractions at low temperatures, one can avoid the solute degradation, since phenolic species tend to degrade at high temperatures and also to suppress the hydrolysis of IL fluorinated anions, such as $[BF_4]^-$ and $[PF_6]^-$.

The pKa values of the phenolic compounds play the most important role towards their partition between the two phases of IL-based ABS. The highest recoveries into the IL-rich phase can be achieved at pH values close to the pKa of the extracted compounds, and the presence of functional groups which influence the relative polarity of the solutes has to be always taken into account. It is noteworthy that the observed pH–pKa extraction efficiency dependence allows controlled partition of phenolic compounds by variation of the system composition or its pH value, thus broadening the scope of IL-based ABS applications. Indeed, as summarized in this chapter, IL-based ABS were shown to be highly effective extractive platforms for phenolic compounds. Moreover, IL-based ABS can be considered as sustainable and cost efficient due to the possibility for their successful reuse in several extractive cycles without loss of performance.

In conclusion, the knowledge gained to date for IL-based ABS shows their high potential to be employed as extractive media to recover natural phenolic compounds and represents a promising basis for the development of sustainable industrial methods. However, IL-based ABS have been successfully employed mainly on a laboratory scale to extract model compounds, and so, there is the need for a further transfer of this knowledge into industrial scale processes. To this end, more efforts towards the development of procedures for the separation/fractionation of complex mixtures and the assessment of the overall process' costs are necessary. Finally, more easily available and inexpensive ILs with improved properties from an environmental standpoint, and the development of efficient procedures for ILs recovery and recycling, are also crucial factors which have to be taken into account.

Acknowledgements The authors would like to acknowledge the support of the National Science Fund of Bulgaria at the Ministry of Education and Science (Project DFNI T02/23) and the contribution of COST Action CM1206-Exchange on Ionic Liquids.

References

1. Dai J, Mumper RJ (2010) Plant phenolics: extraction, analysis and their antioxidant and anticancer properties. Molecules 15:7313–7352
2. Mansfield JW (2000) The role of phytoalexins and phytoanticipins. In: Slusarenko AJ, Fraser RSS, van Loon LC (eds) Mechanisms of resistance to plant diseases. Kluwer Academic Publishers, Dordrecht, pp 325–370
3. Fresco P, Borges F, Diniz C et al (2006) New insights on the anticancer properties of dietary polyphenols. Med Res Rev 26:747–766
4. Kamat CD, Gadal S, Mhatre M et al (2008) Antioxidants in central nervous system diseases: preclinical promise and translational challenges. J Alzheimers Dis 15:473–493
5. Zhao B (2009) Natural antioxidants protect neurons in Alzheimer's disease and Parkinson's disease. Neurochem Res 34:630–638
6. Fresco P, Borges F, Marques MP et al (2010) The anticancer properties of dietary polyphenols and its relation with apoptosis. Curr Pharm Des 16:114–134
7. Benfeito S, Oliveira C, Soares P et al (2013) Antioxidant therapy: still in search of the 'magic bullet'. Mitochondrion 13:427–435
8. Fais A, Corda M, Era B et al (2009) Tyrosinase inhibitor activity of coumarin-resveratrol hybrids. Molecules 14:2514–2520
9. Serafim TL, Carvalho FS, Marques MP et al (2011) Lipophilic caffeic and ferulic acid derivatives presenting cytotoxicity against human breast cancer cells. Chem Res Toxicol 24:763–774
10. Viña D, Matos MJ, Ferino G et al (2012) 8-Substituted 3-arylcoumarins as potent and selective MAO-B inhibitors: synthesis, pharmacological evaluation, and docking studies. Chem Med Chem 7:464–470
11. Garrido J, Gaspar A, Garrido EM et al (2012) Alkyl esters of hydroxycinnamic acids with improved antioxidant activity and lipophilicity protect PC12 cells against oxidative stress. Biochimie 94:961–967
12. Teixeira J, Silva T, Benfeito S et al (2013) Exploring nature profits: development of novel and potent lipophilic antioxidants based on galloyl-cinnamic hybrids. Eur J Med Chem 62:289–296
13. Miliovsky M, Svinyarov I, Mitrev Y et al (2013) A novel one-pot synthesis and preliminary biological activity evaluation of *cis*-restricted polyhydroxy stilbenes incorporating protocatechuic acid and cinnamic acid fragments. Eur J Med Chem 66:185–192
14. Catto M, Pisani L, Leonetti F et al (2013) Design, synthesis and biological evaluation of coumarin alkylamines as potent and selective dual binding site inhibitors of acetylcholinesterase. Bioorg Med Chem 21:146–152
15. Mura F, Silva T, Castro C et al (2014) New insights into the antioxidant activity of hydroxycinnamic and hydroxybenzoic systems: spectroscopic, electrochemistry, and cellular studies. Free Radic Res 48:1473–1484
16. Azmir J, Zaidul ISM, Rahman MM et al (2013) Techniques for extraction of bioactive compounds from plant materials: a review. J Food Eng 117:426–436
17. Bucar F, Wube A, Schmid M (2013) Natural product isolation – how to get from biological material to pure compounds. Nat Prod Rep 30:525–545
18. Anastas PT, Kirchhoff MM (2002) Origins, current status, and future challenges of green chemistry. Acc Chem Res 35:686–694
19. Bogdanov MG (2014) Ionic liquids as alternative solvents for extraction of natural products. In: Chemat F, AbertVian M (eds) Alternative solvents for natural products extraction, green chemistry and sustainable technology. Springer, Berlin/Heidelberg, pp 127–166. doi:10.1007/978-3-662-43628-8_7
20. Passos H, Freire MG, Coutinho JAP (2014) Ionic liquid solutions as extractive solvents for value-added compounds from biomass. Green Chem 16:4786–4815
21. Freemantle M (2009) An introduction to ionic liquids. RSC Publishing, Cambridge

22. Smiglak M, Pringle JM, Lu X et al (2014) Ionic liquids for energy, materials, and medicine. Chem Commun 50:9228–9250
23. Bogdanov MG, Kantlehner W (2009) Simple prediction of some physical properties of ionic liquids: the residual volume approach. Z Naturforsch B 64:215–222
24. Bogdanov MG, Iliev B, Kantlehner W (2009) The residual volume approach II: simple prediction of ionic conductivity of ionic liquids. Z Naturforsch B 64:756–764
25. Bogdanov MG, Petkova D, Hristeva S et al (2010) New guanidinium-based room-temperature ionic liquids. Substituent and anion effect on density and solubility in water. Z Naturforsch B 65:37–48
26. Bogdanov MG, Svinyarov I, Kunkel H et al (2010) Empirical polarity parameters for hexaalkylguanidinium-based room-temperature ionic liquids. Z Naturforsch B 65:791–797
27. Hallett JP, Welton T (2011) Room-temperature ionic liquids: solvents for synthesis and catalysis. 2. Chem Rev 111:3508–3576
28. Parvulescu VI, Hardacre C (2007) Catalysis in ionic liquids. Chem Rev 107:2615–2665
29. Opallo M, Lesniewski A (2011) A review on electrodes modified with ionic liquids. J Electroanal Chem 656:2–16
30. Ho TD, Zhang C, Hantao LW et al (2014) Ionic liquids in analytical chemistry: fundamentals, advances, and perspectives. Anal Chem 86:262–285
31. Bogdanov MG, Svinyarov I (2013) Ionic liquid-supported solid–liquid extraction of bioactive alkaloids. II. Kinetics, modeling and mechanism of glaucine extraction from *Glauciumflavum* Cr. (Papaveraceae). Sep Purif Technol 103:279–288
32. Carneiro AP, Rodriguez O, Macedo EA (2014) Separation of carbohydrates and sugar alcohols from ionic liquids using antisolvents. Sep Purif Technol 132:496–504
33. Jiao J, Gai QY, Fu YJ et al (2013) Microwave-assisted ionic liquids treatment followed by hydro-distillation for the efficient isolation of essential oil from *Fructusforsythiae* seed. Sep Purif Technol 107:228–237
34. Bogdanov MG, Svinyarov I, Keremedchieva R et al (2012) Ionic liquid-supported solid–liquid extraction of bioactive alkaloids. I. New HPLC method for quantitative determination of glaucine in *Glauciumflavum* Cr. (Papaveraceae). Sep Purif Technol 97:221–227
35. Cláudio AFM, Ferreira AM, Freire MG et al (2013) Enhanced extraction of caffeine from Guaraná seeds using aqueous solutions of ionic liquids. Green Chem 15:2002–2010
36. Ressmann AK, Zirbs R, Pressler M et al (2013) Surface-active ionic liquids for micellar extraction of piperine from black pepper. Z Naturforsch B 68:1129–1137
37. Zirbs R, Strassl K, Gaertner P et al (2013) Exploring ionic liquid–biomass interactions: towards the improved isolation of shikimic acid from star anise pods. RSC Adv 3:26010–26016
38. Zhu S, Ma C, Fu Q et al (2013) Application of ionic liquids in an online ultrasonic assisted extraction and solid-phase trapping of rhodiosin and rhodionin from *Rhodiolarosea* for UPLC. Chromatographia 76:195–200
39. Tan Z, Li F, Xu X (2012) Isolation and purification of aloe anthraquinones based on an ionic liquid/salt aqueous two-phase system. Sep Purif Technol 98:150–157
40. Cláudio AFM, Marques CFC, Boal-Palheiros I et al (2014) Development of back-extraction and recyclability routes for ionic-liquid-based aqueous two-phase systems. Green Chem 16:259–268
41. Tonova K, Svinyarov I, Bogdanov MG (2014) Hydrophobic 3-alkyl-1-methylimidazolium saccharinates as extractants for L-lactic acid recovery. Sep Purif Technol 125:239–246
42. Gutowski KE, Broker GA, Willauer HD et al (2003) Controlling the aqueous miscibility of ionic liquids: aqueous biphasic systems of water-miscible ionic liquids and water-structuring salts for recycle, metathesis, and separations. J Am Chem Soc 125:6632–6633
43. Tan ZJ, Li FF, Xu XL et al (2012) Simultaneous extraction and purification of aloe polysaccharides and proteins using ionic liquid based aqueous two-phase system coupled with dialysis membrane. Desalination 286:389–393

44. Tonova K (2012) Separation of poly- and disaccharides by biphasic systems based on ionic liquids. Sep Purif Technol 89:57–65
45. Deive FJ, Rodríguez A, Pereiro AB et al (2011) Ionic liquid-based aqueous biphasic system for lipase extraction. Green Chem 13:390–396
46. Marques CFC, Mourão T, Neves CMSS et al (2013) Aqueous biphasic systems composed of ionic liquids and sodium carbonate as enhanced routes for the extraction of tetracycline. Biotechnol Prog 29:645–654
47. Freire MG, Neves CMSS, Marrucho IM et al (2010) High-performance extraction of alkaloids using aqueous two-phase systems with ionic liquids. Green Chem 12:1715–1718
48. Freire MG, Teles ARR, Canongia Lopes JN et al (2012) Partition coefficients of alkaloids in biphasic ionic-liquid-aqueous systems and their dependence on the Hofmeister series. Sep Sci Technol 47:284–291
49. Zhang D, Deng Y, Chen J (2010) Enrichment of aromatic compounds using ionic liquid and ionic liquid-based aqueous biphasic systems. Sep Sci Technol 45:663–669
50. Chen Y, Meng Y, Yang J et al (2012) Phenol distribution behavior in aqueous biphasic systems composed of ionic liquid-carbohydrate-water. J Chem Eng Data 57:1910–1914
51. Wang L, Zhu H, Sun YT et al (2011) Determination of trace chlorophenols endocrine disrupting chemicals in water sample using [Bmim]BF4- NaH2PO4 aqueous two-phase extraction system coupled with high performance liquid chromatography. Chin J Anal Chem 39:709–712
52. Passos H, Sousa ACA, Pastorinho M et al (2012) Ionic-liquid-based aqueous biphasic systems for improved detection of bisphenol A in human fluids. Anal Methods 4:2664–2667
53. Cláudio AFM, Freire MG, Freire CSR et al (2010) Extraction of vanillin using ionic-liquid-based aqueous two-phase systems. Sep Purif Technol 75:39–47
54. e Silva FA, Sintra T, Ventura SPM et al (2014) Recovery of paracetamol from pharmaceutical wastes. Sep Purif Technol 122:315–322
55. Noorashikin MS, Mohamad S, Abas MR (2014) Extraction and determination of parabens in water samples using an aqueous two-phase system of ionic liquid and salts with beta-cyclodextrin as the modifier coupled with high performance liquid chromatography. Anal Methods 6:419–425
56. Santos JH, e Silva FA, Ventura SPM et al (2014) Ionic liquid-based aqueous biphasic systems as a versatile tool for the recovery of antioxidant compounds. Biotechnol Prog 31:70–77
57. Cláudio AFM, Ferreira AM, Freire CSR et al (2012) Optimization of the gallic acid extraction using ionic-liquid-based aqueous two-phase systems. Sep Purif Technol 97:142–149
58. Sánchez-Rangel JC, Jacobo-Velázquez DA, Cisneros-Zevallos L et al (2014) Primary recovery of bioactive compounds from stressed carrot tissue using aqueous two-phase systems strategies. J Chem Technol Biotechnol 91:144–154
59. Fan JP, Cao J, Zhang XH et al (2012) Extraction of puerarin using ionic liquid based aqueous two-phase systems. Sep Sci Technol 47:1740–1747
60. Sheikhian L, Akhond M, Absalan G (2014) Partitioning of reactive red-120, 4-(2-pyridylazo)-resorcinol, and methyl orange in ionic liquid-based aqueous biphasic systems. J Environ Chem Eng 2:137–142
61. Ribeiro BD, Coelho MAZ, Rebelo LPN et al (2013) Ionic liquids as additives for extraction of saponins and polyphenols from mate (*Ilex paraguariensis*) and tea (*Camellia sinensis*). Ind Eng Chem Res 52:12146–12153
62. Yu YY, Zhang W, Cao SW (2007) Extraction of ferulic acid and caffeic acid with ionic liquids. Chin J Anal Chem 35:1726–1730
63. Fan J, Fan Y, Pei Y et al (2008) Solvent extraction of selected endocrine-disrupting phenols using ionic liquids. Sep Purif Technol 61:324–331
64. Shahriari S, Neves CMSS, Freire MG et al (2012) Role of the Hofmeister series in the formation of ionic-liquid-based aqueous biphasic systems. J Phys Chem B 116:7252–7258
65. Cammarata L, Kazarian SG, Salter PA et al (2001) Molecular states of water in room temperature ionic liquids. Phys Chem Chem Phys 3:5192–5200

66. Gaillon L, Sirieix-Plénet J, Letellier P (2004) Volumetric study of binary solvent mixtures constituted by amphiphilic ionic liquids at room temperature (1-alkyl-3-methylimidazolium bromide) and water. J Solut Chem 33:1333–1347
67. Sirieix-Plénet J, Gaillon L, Letellier P (2004) Behaviour of a binary solvent mixture constituted by an amphiphilic ionic liquid, 1-decyl-3-methylimidazolium bromide and water: potentiometric and conductimetric studies. Talanta 63:979–986
68. Inoue T, Dong B, Zheng L-Q (2007) Phase behavior of binary mixture of 1-dodecyl-3-methylimidazolium bromide and water revealed by differential scanning calorimetry and polarized optical microscopy. J Colloid Interface Sci 307:578–581
69. Bhargava BL, Klein ML (2009) Formation of micelles in aqueous solutions of a room temperature ionic liquid: a study using coarse grained molecular dynamics. Mol Phys 107:393–401
70. Bhargava BL, Yasaka Y, Klein ML (2011) Computational studies of room temperature ionic liquid–water mixtures. Chem Commun 47:6228–6241
71. Neves CMSS, Ventura SPM, Freire MG et al (2009) Evaluation of cation influence on the formation and extraction capability of ionic-liquid-based aqueous biphasic systems. J Phys Chem B 113:5194–5199
72. Cláudio AFM, Swift L, Hallett JP et al (2014) Extended scale for the hydrogen-bond basicity of ionic liquids. Phys Chem Chem Phys 16:6593–6601
73. Angelov T, Vlasenko A, Tashkov W (2007) HPLC determination of pKa of parabens and investigation on their lipophilicity parameters. J Liq Chromatogr Relat Technol 31:188–197
74. He C, Li S, Liu H et al (2005) Extraction of testosterone and epitestosterone in human urine using aqueous two-phase systems of ionic liquid and salt. J Chromatogr A 1082:143–149
75. Pei Y, Wang J, Wu K et al (2009) Ionic liquid-based aqueous two-phase extraction of selected proteins. Sep Purif Technol 64:288–295
76. Zafarani-Moattar MT, Hamzehzadeh S (2009) Phase diagrams for the aqueous two phase ternary system containing the ionic liquid 1-butyl-3-methylimidazolium bromide and tri-potassium citrate at T = (278.15, 298.15, and 318.15) K. J Chem Eng Data 54:833–841
77. Liazid A, Palma M, Brigui J et al (2007) Investigation on phenolic compounds stability during microwave-assisted extraction. J Chromatogr A 1140:29–34
78. Freire MG, Neves CMSS, Marrucho IM et al (2010) Hydrolysis of tetrafluoroborate and hexafluorophosphate counter ions in imidazolium-based ionic liquids. J Phys Chem A 114:3744–3749

Chapter 9
Extraction of Metals with ABS

Isabelle Billard

Abstract The extraction of metallic ions by the use of mixtures of water and ILs, possibly completed by mineral acids, salts, and/or extracting agents, leading to ionic-liquid-aqueous biphasic systems (IL-ABS), is critically reviewed. In this chapter, the extraction performance induced by temperature or concentration stimulus is considered. First, the IL-ABS basic physicochemical properties are recalled, highlighting those of interest to metal extraction. Then, the main results are presented and discussed for systems ordered and categorized by IL types. Various extensions of the notion of IL-ABS are given and briefly discussed as an advocacy in favor of a continuous line between "real" ABS comprising ILs and other liquid-liquid extraction systems including one IL.

Keywords Ionic liquids • Metal extraction • Aqueous biphasic systems • Stripping • Crossed ionic solubilities • Liquid-liquid extraction

9.1 Introduction

The extraction of metals from aqueous phases is a huge problem connected to a large number of industrial applications: it is important not only for ore exploitation or for recycling of precious metals (found in "urban wastes" or in used nuclear fuels) but also in view of polluted water characterization, to name a few. Furthermore, it concerns all metals, from Cs and Sr to *d* elements, such as Cu, Co, Ni, Zn, Pb, or Hg, as well as Au, Pt, Ir, lanthanides, and finally 5*f* metals, such as U, Pu, or Cm. Most of the time, the metal to be extracted is not the only element present in the aqueous feed, which often contains other metals and large amounts of mineral acids, such as nitric or hydrochloric acid, that have been used to dissolve the ores or the wastes to be recycled. As a consequence, metals in such aqueous acidified solutions exist as ions, either cations or anions, depending on their specific affinity

I. Billard (✉)
Université Grenoble Alpes, LEPMI, Grenoble F-38000, France

CNRS, LEPMI, Grenoble F-38000, France

LEPMI, 1140, rue de la Piscine, Saint Martin d'Hères 38402, France
e-mail: Isabelle.billard@lepmi.grenoble-inp.fr

M.G. Freire (ed.), *Ionic-Liquid-Based Aqueous Biphasic Systems*, Green Chemistry and Sustainable Technology, DOI 10.1007/978-3-662-52875-4_9

toward the counter-anion of the acid and concentration of the latter. In the case of water depollution, samples may contain organic compounds, possibly bacteria or other biological entities that all may also be complexed to the metal ions. Needless to say, high extraction efficiencies and selectivity for the metal of interest are requested, together with cost-effectiveness, protocol easiness, and limited operation times. To solve this problem, or better to say, these problems, liquid-liquid extraction has emerged years ago as the best industrial process ever [1], and since the industrial implementation of the PUREX process, first designed to extract U and Pu in the 1940s, liquid-liquid extraction of metals has successfully achieved all goals cited above.

Typical liquid-liquid extractions from aqueous phases are based on the use of volatile organic compounds (VOCs), such as chloroform, dichloromethane, kerosene, etc., as the second liquid phase, to which the metallic entities are extracted. Therefore, and in addition to costs linked to the design of always better extracting agents, one should consider, under the light of the European REACH regulation, for example, environmental costs. Under such a frame, liquid-liquid extraction is now looking for an ecologic revival.

After the work of Dai and co-workers [2] evidencing an increase up to four orders of magnitude in Sr(II) extraction efficiency by replacing traditional solvents by some ionic liquids (ILs), these new solvents have offered fantastic opportunities as VOC replacements in view of liquid-liquid extraction of metals, and there are now numerous academic studies that have demonstrated their advantages in terms of extraction efficiency of metals [3–5] and one critical review [6]. Interestingly enough, ILs have also rapidly proved to provide even better uses than mere VOC replacement solvents. Actually, they can also act as pure liquid phases, i.e., without the use of any additional extracting agent as usually needed when using molecular solvents. Depending on systems and metals, such a surprising result can be obtained either with very simple and "classical" (i.e., commercially available) ILs [4] or with specifically synthesized ILs, the so-called task-specific ILs (TSILs), bearing complexing patterns such as phosphine oxide, amides, crown ether, or calixarenes [7–11]. Such innovative liquid-liquid extraction procedures, however, correspond to higher costs, renewed synthetic studies, and collateral problems linked to the high viscosity of ILs among other problems.

In parallel to these evolutions of liquid-liquid extraction studies and practices, and in a rather disconnected way, the academic community has developed and brought to industrial applications the concept of aqueous biphasic systems (ABS), mainly for extraction/separation of biological molecules and a large variety of compounds (see other chapters of this book on this point). Although it may now appear evident that the use of ABS is not restricted to organic or biological samples, the idea of bringing the knowledge gained on ILs and on ABS in view of metal extraction has emerged only quite recently. To our opinion, this field of research will soon blossom, and the rather limited number of papers available at the moment is just the beginning of a topic soon to expand tremendously.

In this chapter, Sect. 9.2 is devoted to notations and definitions of ILs and IL-based ABS concepts. Properties of IL-based ABS of interest to metal extraction are also recalled and discussed in details. Next, in Sect. 9.3, papers on metal extraction by IL-ABS are critically reviewed. Literature is ordered based on IL types and not on the type of metal being extracted; however, the reader can find in Table 9.1 a summary where the key entrance is metallic elements. For each case, we first recall why the system under review falls in the scope of the chapter, and we thereafter present and discuss the extraction results and other data of relevance to the understanding of the fascinating IL-based ABS. Section 9.4, by gathering data dealing with systems close to IL-based ABS, hopefully demonstrates to the reader that there is actually a continuous line linking all these systems.

9.2 IL-Based ABS: Definitions, Notations, and Main Properties

9.2.1 *Definitions and Notations*

Although all chapters of this book deal with IL-ABS, it is very important to precisely define what this term means, in order to border the scope of this chapter. As far as ILs are concerned, we will comply with the usual definition stating that an IL is a salt of which the melting temperature, under normal pressure conditions, is equal or below 100 °C. With regard to IL basic properties, the reader is referred to excellent books and reviews on this topic [30–34]. ILs' cations and anions will be noted as [Cat$^+$] and [Ani$^-$], respectively, while IL will be given as [Cat$^+$][Ani$^-$], in order to highlight the ionic character of such solvents. As regards the most typical IL family, imidazolium cations will be denoted as [C$_n$C$_m$im]$^+$ for *n*-alkyl-*m*-alkylimidazolium, while the anion [(CF$_3$SO$_2$)$_2$N]$^-$ (bis(trifluoromethylsulfonyl) imide) has been by far the most investigated. All other compounds cited in this chapter are displayed in the chemical chart provided at the end. All ILs are salts but not all salts are ILs, depending on their melting temperature. Mineral acids and salts will be denoted according to the IL notation adopted above, e.g., [H$^+$][Cl$^-$], [Co^{2+}][Cl$^-$]$_2$, or [Na$^+$]$_2$[CO$_3^{2-}$].

By contrast, defining an ABS is somewhat more difficult. One general definition can be found in a previous review on ABS [35] as: "ABS consist of two immiscible aqueous-rich phases based on polymer-polymer, polymer-salt or salt-salt combinations. Although both solutes are water-soluble, they separate into two coexisting phases above a given concentration: one of the aqueous phases will be enriched in one of the solutes while in the other phase there is prevalence for the second polymer or salt." It thus could be easily inferred from this definition that the IL-ABS of concern in this book are simply those ABS composed of (at least) one IL, whatever the type of the other solute. However, the definition above raises more questions. First, one should define what an "aqueous-rich phase" means as the scale on which this notion is based is not specified. Usual scales are molar, weight, or volume percentages, but other scales are also meaningful (see below). Considering

Table 9.1 List of the metallic ions extracted by the use of IL-ABS. Elements are listed in alphabetical order and are displayed if $D_{X,\ell}$ is ≥ 1

Metal, oxidation number	IL-ABS	Ref.
Ag(I)	$H_2O\&[C_1C_6im^+][BF_4^-]\&[Na^+][PF_6^-]\&[H^+][NO_3^-]$	[12]
	$H_2O\&[C_1C_4im^+][Cl^-]\&[H^+][NO_3^-]\&[K^+]_2[HPO_4^{2-}]$	[13]
Au(III)	$H_2O\&[C_1C_8im^+][Cl^-]\&[H^+][Cl^-]$	[14]
	$H_2O\&[C_1C_8im^+][Br^-]\&[H^+][Br^-]$	[14]
	$H_2O\&[C_8pyr^+][Br^-]\&[H^+][Br^-]$	[14]
Ce(IV)	$H_2O\&[C_1C_8im^+][Cl^-]\&[H^+][NO_3^-]$	[15]
Cd(II)	$H_2O\&[TBA^-][Br^-]\&2[NH_4^+][SO_4^{2-}]$	[16]
Co(II)	$H_2O\&$Girard-IL	[17]
	$H_2O\&[P_{444}E_i^+][DEHP^-]$ (i = 1,2,3)	[18]
Cr(III)	$H_2O\&$Girard-IL	[17]
Cr(IV)	$H_2O\&[TBA^-][Br^-]\&2[NH_4^+][SO_4^{2-}]$	[19]
Cu(II)	$H_2O\&[Hbet^+][Tf_2N^-]\&$bet	[16]
	$H_2O\&$Girard-IL	[17]
	$H_2O\&[P_{444}E_i^+][DEHP^-]$ (i = 1,2,3)	[18]
Dy(III)	$H_2O\&[Hbet^+][Tf_2N^-]\&$bet	[16]
Er(III)	$H_2O\&[Hbet^+][Tf_2N^-]\&$bet	[16]
Fe(III)	$H_2O\&[C_1C_4im^+][FeCl_4^-]$	[20]
Ga(III)	$H_2O\&[Hbet^+][Tf_2N^-]\&$bet	[16]
Ho(III)	$H_2O\&[Hbet^+][Tf_2N^-]\&$bet	[16]
In(III)	$H_2O\&[Hbet^+][Tf_2N^-]\&$bet	[16]
Ir(IV)	$H_2O\&[C_1C_1C_8pyrro^+][Br^-]\&[H^+][Cl^-]$	[21]
La(III)	$H_2O\&[Hbet^+][Tf_2N^-]\&$bet	[16]
Mn(II)	$H_2O\&[Hbet^+][Tf_2N^-]\&$bet	[16]
Nd(III)	$H_2O\&[Hbet^+][Tf_2N^-]\&$bet	[16, 22]
	$H_2O\&[chol^+][Tf_2N^-]\&[chol^+][hfac^-]\&[H^+][NO_3^-]$	[23]
Ni(II)	$H_2O\&[Hbet^+][Tf_2N^-]\&$bet	[16]
	$H_2O\&$Girard-IL	[17]
	$H_2O\&[P_{44414}^+][Cl^-]\&[Na^+][Cl^-]$	[24]
	$H_2O\&[P_{444}E_i^+][DEHP^-]$ (i = 1,2,3)	[18]
Pb(II)	$H_2O\&[TBA^+][Br^-]\&[NH_4^+]_2[SO_4^{2-}]$	[16]
Pd(II)	$H_2O\&[Hbet^+][Tf_2N^-]\&[H^+][NO_3^-]$	[25]
	$H_2O\&[chol^+][Tf_2N^-]\&[H^+][NO_3^-]$	[25]
Pr(III)	$H_2O\&[Hbet^+][Tf_2N^-]\&$bet	[16, 22]
Pt(IV)	$H_2O\&[C_1C_8pyrro^+][Br^-]\&[H^+][Cl^-]$	[26]
Rh(III)	$H_2O\&[Hbet^+][Tf_2N^-]\&[H^+][NO_3^-]$	[25]
Ru(III)	$H_2O\&[Hbet^+][Tf_2N^-]\&[H^+][NO_3^-]$	[25]
	$H_2O\&[chol^+][Tf_2N^-]\&[H^+][NO_3^-]$	[25]
Sc(III)	$H_2O\&[Hbet^+][Tf_2N^-]\&$bet	[16]
	$H_2O\&[P_{444}C_1COOH^+][Cl^-]\&[Na^+][Cl^-]$	[27]
Tc(VII)	$H_2O\&[C_1C_4im^+][Cl^-]\&[K^+]_3[PO_4^{3-}]$	[28]
	$H_2O\&[C_1C_4im^+][Cl^-]\&[K^+]_2[HPO_4^{2-}]$	[28]
	$H_2O\&[C_1C_4im^+][Cl^-]\&[K^+]_2[CO_3^{2-}]$	[28]

(continued)

Table 9.1 (continued)

Metal, oxidation number	IL-ABS	Ref.
U(VI)	H_2O&[Hbet$^+$][Tf$_2$N$^-$]&[H$^+$][NO$_3^-$]	[29]
Y(III)	H_2O&[Hbet$^+$][Tf$_2$N$^-$]&bet	[16]
Zn(II)	H_2O&[Hbet$^+$][Tf$_2$N$^-$]&bet	[16]
	H_2O&[P$_{444}$E$_i^+$][DEHP$^-$] (i = 1,2,3)	[18]

the very large molar *per* liter value of water (55 moles *per* liter) as compared to other liquid compounds, such as ILs (typically in the range of 2–8 moles *per* liter), while ILs' molar weights are usually at least one order of magnitude above the 18 g·mol^{-1} of water, it is clear that the molar scale favors water, while weight percentage highlights ILs' contribution. In other words, for a biphasic sample containing water and IL, either the terms water-rich or IL-rich phase might be appropriate, depending on the chosen scale. Similarly, the notions of "enrichment" or "prevalence" of a solute in one of the phases should be accompanied by the indication of a scale. Second, the biphasic/monophasic changes that are the signature of ABS can also occur under a temperature stimulus (thermomorphic behavior) for a fixed chemical composition, a very important point to add to the definition cited above: systems displaying either an upper critical solution temperature, UCST (above this temperature, the system is monophasic, whatever its composition) or a lower critical solution temperature, LCST (temperature below which the system is monophasic whatever its composition) exist. An alternative way to induce a change from monophasic to biphasic regimes and vice versa, by bubbling of CO_2 and N_2 gases, was demonstrated by Kohno et al. [36]. However, this is in fact due to changes in the pH of the sample, because CO_2 converts in [CO_3^{2-}] anions, while N_2 bubbling expels CO_2. Thus, we should categorize this state change as a concentration-induced change, which is however more easy to perform and more reversible than addition of salts/dilution that always varies concentrations and volumes to large extents.

Based on these remarks, we will therefore adopt a slightly different and somehow more restrictive definition of IL-based ABS for this chapter only. Under our understanding, IL-based ABS are systems which comply with the following five criteria:

1. They contain water.
2. They contain at least one IL compound.
3. They do not contain any molecular compound, apart from acidic or zwitterionic ones.
4. They may contain other compounds in agreement with the above three criteria.
5. Within the temperature range from solidification to boiling or decomposition of the sample (under normal pressure), they present a change from monophasic to biphasic state by an increase or decrease of either temperature or concentration of at least one of their components.

Such a definition calls for some comments. First, by excluding most of the molecular compounds, we set apart systems containing polymers. This is clearly a restrictive choice of the author of this chapter as compared to the general policy of this book (in particular, see Chap. 4), but it was made in order to limit this review chapter to systems without usual neutral extracting agents, such as tributylphosphate, calixarenes, or monoamides, to name a few. Second, it is important to note that this definition does not rely on notion of water-rich phase or any other type of whatsoever-rich phases. We are strongly in favor of using solely the terms upper and lower phases for describing the biphasic liquid state of ABS and IL-ABS. We will use subscripts "u" and "ℓ" to denote them, while the term "initial" and its corresponding subscript "in," together with the terms "IL phase" and "aqueous phase," will refer to initial quantities and initial stock solutions, i.e., prior to mixing of the IL-ABS components. Another important point to be noted from this definition is that we consider mixtures of water and one IL to be IL-ABS, provided these present a UCST or an LCST phenomenon. This is in clear contrast with the definition given by Freire and co-workers [35] which limits ABS to a mixture of three compounds at minimum.

As regards notation of IL-ABS of concern in this chapter, this is of importance although no consensus can be found at the moment in the literature. We will list all chemicals under the chemical form they have been used for sample preparation, starting with water, each of them separated by symbol "&," for example, $H_2O\&[H^+][E^-]\&[H^+][X^-]\&[K^+]_3[PO_4^{3-}]\&[C_1C_4im^+][Cl^-]$, where $[H^+][E^-]$ is an acidic extractant and $[H^+][X^-]$ is a (mineral) acid. In case of necessity, quantities will be indicated in parenthesis after each component. Whenever possible, and in view of practical easiness (preparation protocol, for instance), we will use the mole amount of each compound, because interactions responsible for mono- to biphasic changes are clearly occurring at a molecular/ionic level. However, depending on the information available in each experimental section of the papers to be discussed, we may use molar scale or any other scale.

9.2.2 Properties of Interest of IL-ABS

9.2.2.1 Two-Component Systems

IL-ABS obviously offer limited possibilities to play with as compared to systems containing more than two components, but there are nonetheless important points to be recalled in view of the review to follow in Sect. 9.3. The easiest way to represent the thermodynamic behavior of an IL-ABS simply composed of water and one IL is through the plot of the partial miscibility curve, as schematically represented in Fig. 9.1a for a UCST-type system: the two-phase region is located inside the envelop and the UCST value is indicated on the y-axis. Conversely, Fig. 9.1b shows the schematic of an LCST-type IL-ABS. For systems solely composed of water and one IL, a 2D plot is sufficient to fully describe their mono-/biphasic

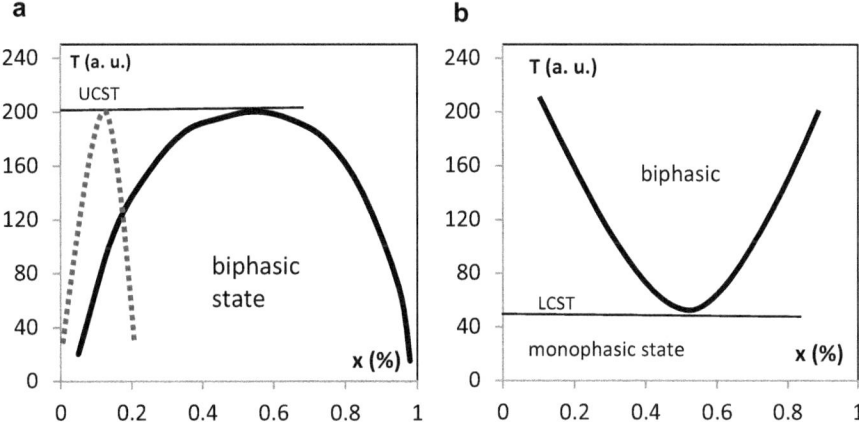

Fig. 9.1 (a) Schematic plot of a miscibility curve with UCST behavior for a hypothetical system $H_2O\&[Cat^+][Ani^-]$. Solid line: plot under the mass fraction scale. Dashed line: plot under the mole fraction scale. (b) Schematic plot of a miscibility curve with LCST behavior for a hypothetical system $H_2O\&[Cat^+][Ani^-]$

behavior, making use of the % scale (either in mass, volume, or mole). For example, the system $H_2O\&[C_1C_4im^+][BF_4^-]$ displays a UCST equal to ca. 5 °C while for a water amount equal to 20 wt%, the temperature at which this system turns from mono- to biphasic, critical temperature (T_c), is equal to −4 °C [37, 38]. Note that a clear isotopic effect is observed on UCST temperatures, as deuterated water induces an increase of the transition temperature of ca. 3.5 °C at maximum [37]. Other examples of larger variations (up to 25 °C) in the turnover temperature can be found for phosphonium-based ILs and water mixtures [39]. Increases of a few degrees in UCST can also be obtained by applying pressure up to 800 bar [37].

The rather symmetrical plot displayed in Fig. 9.1a (solid line) is only obtained for x-axis based on mass fractions. Changing the mole fraction leads to a distorted plot with a curve now shifted to the left-hand side of the diagram. This is clearly shown in the work of Rebelo et al. [37] for the $H_2O\&[C_1C_4im^+][BF_4^-]$ system and is also schematically illustrated as a dashed line in Fig. 9.1a. This effect is the graphical translation of what has been written above concerning the scales applied to define "water-rich" phases. Other illustrations of the distortions that can be obtained by changing the scale in such plots can be found elsewhere [40, 41], and a very interesting discussion on the best way to plot such thermodynamic data can be found in the paper by Wagner et al. [42]. In particular, it is clearly said that the mole, mass, or volume fraction scales have equal validity for plotting miscibility curves. Deeper thermodynamical considerations would be in favor of the volume fraction, but depending on the technique used to determine the state changes, other scales may be used, for example, the refractive index or the Lorentz function of this number [42].

According to the excellent work of Kohno and co-workers [43], who studied the transition behavior of 48 IL water systems (ILs based on phosphonium or ammonium cations), any system of the type $H_2O\&[Cat^+][Ani^-]$ may display one of the three following behaviors as a function of temperature (T):

1. Totally miscible in the whole T range from 0 to 100 °C
2. Totally immiscible within this T range
3. Change from mono- to biphasic state, either with UCST or LCST in that T range

ILs belonging to the first category are said "real" hydrophilic ILs, while those from the second category are qualified "real" hydrophobic ILs. This way to categorize systems also calls for some comments, as it is clear that the T range 0–100 °C has been chosen on the basis of water behavior. However, should a system $H_2O\&[Cat^+][Ani^-]$ be totally miscible whatever its composition in the range 0–100 °C, this does not preclude it to experience a change from mono- to biphasic state either above 100 °C (LCST behavior) or below 0 °C (UCST behavior). On the other hand, "totally immiscible" is merely a question of detection limits and personal feeling: as ILs are composed of ions, interactions with water molecules unavoidably occur, leading to mixing, even if very limited.

For the six systems studied by these authors which display a transition from mono- to biphasic state within the range 0–100 °C, the so-called (as written by the authors) IL-rich phase of the biphasic state contains from 4 to 22 H_2O entities per IL cation-anion pair [43], so it has to be better considered as a water-rich phase under the molar percentage scale. Such systems are thus clearly IL-ABS under the definition of Freire and co-workers [35]. The system $H_2O\&[chol^+][Tf_2N^-]$, which is of interest to this chapter [44], also clearly illustrates this point. $[chol^+][Tf_2N^-]$ melts at T = 30 °C and the system $H_2O\&[chol^+][Tf_2N^-]$ has a UCST at 72.1 °C. The authors comment that "one should use the terms hydrophobic and hydrophilic ionic liquids with caution," a statement perfectly in line with the work of Kohno and co-workers [43] discussed above. A comprehensive study of the $H_2O\&[chol^+][Tf_2N^-]$ system in terms of the Ising theory can be found in the literature [44]. Other works also highlight the large proportion of water remaining in the lower phase of IL-ABS under their biphasic states: see [45] for ILs based on phosphonium and dicarboxylate (still 7.5 H_2O entities per IL pair at the transition temperature) and refer to [46] for a detailed study of the liquid structure close to the transition temperature.

9.2.2.2 Three-Component Systems and Above

An impressive list of three-component systems together with detailed comments can be found in the review [35] by Freire and co-workers and in other previous publications of this group [47, 48], so we will limit ourselves to stress a few points of decisive interest to our review chapter. In particular, the reader is referred to the work of Merchuk et al. [49], which clearly sets the experimental basis of such phenomenon although systems here studied are not IL-ABS (for kinetic aspects, see also, from the same author, Ref. [50]). The change from mono- to biphasic state by addition of a salt is also referred to "salt-in/salt-out effect" and salts are qualified kosmotropic/chaotropic, respectively. For these IL-ABS containing at least three different components, for example, $H_2O\&[C_1C_4im^+][Cl^-]\& [K^+]_3[PO_4^{3-}]$ as studied (among others) by Gutowski and co-authors [51], a ternary phase diagram would be necessary, but authors most of the times limit themselves to orthogonal phase diagrams, displaying the mutual coexistence curve, called binodal curve, which delineates the regions of the composition diagram for which the system is either mono- or biphasic, at a fixed temperature. In such graphical representation, limits of the tie lines give the exact compositions of the two phases, for a given global composition of the system and fixed T. Typical binodal curves and tie lines are schematically presented in Fig. 9.2. Note that atypical binodal curves were also found in the literature [52].

It is of tremendous importance for the rest of this chapter to be aware that, starting from the simplest IL-ABS we defined as $H_2O\&[Cat^+][Ani^-]$, addition of any compound may impact on their thermomorphic behavior. Apart from addition of common salts such as potassium phosphate, sodium sulfate, or many others as illustrated in [35], striking examples of interest to this chapter concern the addition of either extracting agent (in view of enhancement of the extraction efficiency), mineral acids (in order to avoid hydrolysis/precipitation of the metallic ion), or metallic salts themselves. For example, adding betaine to $H_2O\&[Hbet^+][Tf_2N^-]$ induces a decrease of the UCST from 55.5 °C (no betaine) to less than 40 °C (28.8 wt% of betaine) [22]. The addition of $[H^+][Cl^-]$ increases the UCST (1 M added, increase of $\approx$12.5 °C; 2 M added, increase of $\approx$20 °C) [16, 53], while 1 M $[H^+][ClO_4^-]$ leads to a decrease of ca. 26 °C [53] as compared to $H_2O\&[Hbet^+][Tf_2N^-]$ (UCST at 55 °C). Also of tremendous relevance in view of metal extraction is the reported decrease of UCST of ca. 20 °C by addition of Nd(III) to $H_2O\&[Hbet^+][Tf_2N^-]$ [54], which is accompanied by an increase of the overall mutual solubilities of water and IL. Other

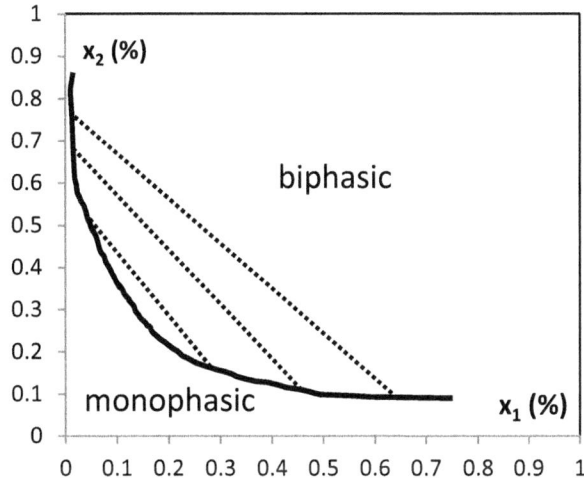

Fig. 9.2 Schematic plot of a binodal curve for an IL-ABS system with three components: water, IL, and a salt. *Dotted lines* are *tie lines*

examples of changes in the UCST or LCST values by additions of various metallic salts ([Cs$^+$][Cl$^-$], [Li$^+$][Cl$^-$], [Na$^+$][ClO$_4$$^-$], [Na$^+$][NO$_3$$^-$], etc.) can be found in the literature [53]. Obviously, the comprehensive experimental study of binodal curves is impossible for systems of general formulation H$_2$O&[Cat$^+$][Ani$^-$]&[E]&[H$^+$][X$^-$] to be reviewed in this chapter because most extraction studies involve batch experiments with variations of the acid concentration and/or of the extractant in rather large ranges. We thus commend the attempts made to derive intuitive prediction models of such phenomena [53, 55]. However, *stricto sensu*, should we limit this review to publications dealing with systems fully complying with the definition of IL-ABS we set and in particular with the fifth criterion, the number of papers to review would be dramatically limited, in part because of a lack of data on their mono/biphasic behavior. Considering these problems of classification, although we rely on our definition of IL-ABS as stated above, papers under review in this chapter have been selected because (unless otherwise specified, see Sect. 9.4) a subpart of the full system under study is known to display a change from monophasic to biphasic by either temperature or concentration variation.

Another very important point to be recalled all times is the differences in distribution ratios of all chemical entities between the two phases. This has been detailed above for water molecules but is also true for any of the other chemical moieties of the system. As clearly demonstrated in Gutowski's work for H$_2$O& [C$_1$C$_4$im$^+$][Cl$^-$]&[K$^+$]$_3$[PO$_4$$^{3-}$] [51], the distribution ratios for [C$_1$C$_4$im$^+$] and [Cl$^-$] are not perfectly equal, pointing to the independent behavior of each ion. In other words, it is not because [C$_1$C$_4$im$^+$] ions are introduced together with [Cl$^-$] ions that they will perfectly distribute similarly between the two phases. This should not be overlooked. Thus, apart from the metallic entity, which is hopefully liable to massive transfer from one phase to the other, any other component of the system may be considered under the light of its extraction efficiency. This, however, necessitates the precise definition of "extraction" in such systems. Again, no real consensus exists at the moment, so we will apply a rather general, basic concept: extraction of species X means that X distributes between the two phases. On a quantitative basis, most papers refer to distribution ratios of species X, which make reference to the concentration of that species in both the "aqueous" and the "IL" phases. On the basis of what has been discussed above, the terms upper and lower phase would be more meaningful. Furthermore, changes in the overall water content depending on T and chemical composition impose extensive experimental measurements of phase volumes and/or masses in order to apply the traditional distribution definition [22]. Therefore, in the following, whenever possible, we will better use the following expressions instead:

Extraction efficiencies:

$E_{X,u}$ = 100*amount of X in the upper phase (in mole)/initial total amount of X (in mole)

$E_{X,\ell}$ = 100*amount of X in the lower phase (in mole)/initial total amount of X (in mole)

Distribution ratios:

$D_{X,u}$ = amount of X in the upper phase (in mole)/amount of X in the lower phase (in mole)

$D_{X,\ell}$ = amount of X in the lower phase (in mole)/amount of X in the upper phase (in mole)

In the next sections, the main topic of this chapter is attempted in more detail, by first critically discussing results on metal extraction by the use of "real" IL-ABS. Systems experiencing T stimulus only will be reviewed first (Sects. 9.3.1, 9.3.2, and 9.3.3), starting from very simple systems (only two components) to systems including extractant and/or acid in addition to water, IL, and metallic ions. Second, systems for which concentration changes have been applied, at a fixed temperature, will be presented (Sects. 9.3.4, 9.3.5 and 9.3.6). Section 9.3.6 discusses systems for which T and concentration stimulus are applied at the same time, while Sect. 9.3.8 summarizes results about metal stripping and recovery of ILs. Section 9.4 extends the topic to "borderline" systems.

9.3 IL-ABS for Metallic Ion Extraction

9.3.1 The Ionic Liquid Already Contains the Metallic Ion to Be Extracted

Although there is, to the best of our knowledge, only a single paper [20] falling in this category, we found the results obtained of high interest. In particular, the number of compounds is limited to two, water and one IL, but nevertheless contains everything needed for metal extraction. In this respect, this can be considered as a case study and an emblematic system.

Several ILs exist that contain a metallic ion, most often as the anionic part of the IL [56–58] but some (rare) examples can be found with a cationic metallic entity instead. Examples with both cationic and anionic components being metallic entities can also be found [59]. For most of these metal-containing ILs (Mn, Fe, Co, Gd, Dy, etc.), magnetic properties are mostly studied and very limited knowledge is found on their thermomorphic behavior in the literature, the only example being for $H_2O\&[C_1C_4im^+][FeCl_4^-]$ (glass transition of the IL at $T = -88\ °C$ [56]) which displays an LCST behavior [20, 60]. For 25 wt% of IL, the change from mono- to biphasic state occurs at $T_c = 50\ °C$, while at 20 wt% of IL, changes occur at $T_c = 90\ °C$. Below 20 wt% of IL, T is above 100 °C, and above 30 wt%, T_c is below 10 °C, so the authors were not able to determine the complete miscibility curve [20]. Note that similar changes occur for $H_2O\&[C_1C_{12}im^+][FeCl_4^-]$ but we could not find any indication of melting point for this imidazolium compound.

Experiments have been performed by mixing $[C_1C_nim^+][FeCl_4^-]$ and water for different chemical compositions, heating above $T = 70\ °C$ to reach, whenever possible, the biphasic state. Characterization of the "oily liquid recovered" after phase separation shows that this is indeed the original IL, $[C_1C_n\text{im}^+][\text{FeCl}_4^-]$. Iron distribution between the upper and lower phases has been followed by ICP-OES and evidences a rather identical presence of Fe(III) in both phases: $E_{Fe,u} = 56\%$ ($n = 4$) and 61% (n = 12). pH measurements indicate that Fe(III) ions undergo strong hydrolysis but no exact determination of the species found in the upper phase has been achieved. Surprisingly, the water distribution between the two phases seems to be very different from that of Fe(III), as no water IR signature could be found in the "oily" phase. Although these two systems may not come to an application, because of probable formation of hydrolyzed products, and because distribution ratios are too close to 1, the interest here rests on the so to say ideal number of components.

Interestingly, $[C_1C_n\text{im}^+][\text{FeCl}_4^-]$ ($n = 4$, 12) are prepared by simply mixing $[C_1C_n\text{im}^+][\text{Cl}^-]$ and the salt $\text{FeCl}_3.6\text{H}_2\text{O}$ [20]. We can therefore consider that the two-component IL-ABS $\text{H}_2\text{O}\&[C_1C_{12}\text{im}^+][\text{FeCl}_4^-]$ discussed above is equivalent to $\text{H}_2\text{O}\&[C_1C_{12}\text{im}^+][\text{Cl}^-]\&[\text{Fe}^{3+}]_3[\text{Cl}^-]$, which is clearly an IL-ABS according to our definition but contains three individual components.

We would like also to mention another very unusual way to simulate a phase separation in such systems: as the IL under study possesses magnetic properties, the monophasic state can suffer very large concentration gradients by simply approaching a strong magnet close to the test tube. In this case, no meniscus is visible, so this does not correspond *per se* to a change from mono- to biphasic state, but this is a brilliant use of the IL properties [60].

9.3.2 IL-ABS with Betainium or Choline Cations and the [Tf₂N⁻] Anion and Other Related Systems

Taken alone as a dry compound, $[\text{Hbet}^+][\text{Tf}_2\text{N}^-]$ is a solid at room temperature and melts at ca. 54–57 °C [54, 58]. Thus, this compound fits within the IL category, being also highly hygroscopic as it takes up to ca. 13 wt% of water at room temperature [22], is a highly viscous liquid, and, conversely, dissolves up to 15 wt% in water for equal volumes of water and IL put into contact at room temperature [16]. $\text{H}_2\text{O}\&[\text{Hbet}^+][\text{Tf}_2\text{N}^-]$ displays a UCST at ca. 55.5 °C [54, 58]. Changes in initial pH values of the aqueous phase also induce changes from mono- to biphasic state [58], but this is not really a reversible process because neutralization of $[\text{Li}^+][\text{OH}^-]$, $[\text{Na}^+][\text{OH}^-]$, or $[\text{K}^+][\text{OH}^-]$ by any acid dilutes all the components and adds other ions to the system. Detailed partial miscibility curves, together with other physicochemical properties of general interest (viscosity, density, thermal analysis, etc.) for $\text{H}_2\text{O}\&[\text{Hbet}^+][\text{Tf}_2\text{N}^-]$, can be found in [54] and [58].

The large water uptake and large solubilization of $[\text{Hbet}^+][\text{Tf}_2\text{N}^-]$ in water render the terms "IL phase" and "aqueous phase," used in the papers under discussion, rather unsuitable. Although no information of densities is given, we will consider that the lower phase corresponds to the so-called IL (rich) phase of the papers, and, conversely, the upper phase is ascribed to the "aqueous phase."

Extraction experiments by the use of $H_2O\&[Hbet^+][Tf_2N^-]$ concern U(VI) [29]; Pd(II), Rh(III), and Ru(III) [25]; or Sc(III) and Fe(III) [61]. By addition of a complexing agent, betaine, the system becomes $H_2O\&[Hbet^+][Tf_2N^-]\&$bet, and extraction concerns Cu(II), Y(III), Dy(III), Er(III), Ho(III), La(III), Pr(III), Nd(III), Ga(III), In(III), and Sc(III) [16]. The extraction data in [22] are very similar and are limited to Nd(III) alone. Experiments have been performed under two very different protocols: the first one relates to traditional experimental procedures, as the aqueous and IL phases are contacted by vigorous shaking at room temperature, thus under the biphasic state [25, 29]. Although the existence of a thermomorphic phenomenon is clearly acknowledged by the authors, they did not take advantage of it. In the second extraction protocol, the samples were heated above the UCST (at 60 °C) and then cooled back to room temperature [16]. The effect of some technical aspects (the temperature at which the system is cooled below the UCST to obtain phase separation, the duration of the heating part of the protocol above UCST, the centrifugation for biphasic settlement, etc.) has been studied, but no comparison between the two protocols was made. As a matter of fact, comparison cannot be made by us between the two protocols based on these publications, because the metals under study are different and some of the systems investigated do not contain betaine, while others do. Finally, in a recent paper, Onghena and co-workers [61] do compare the two protocols, in view of Sc(III) and Fe(III) extraction, by the use of $H_2O\&[Hbet^+][Tf_2N^-]$.

The potential interest of reaching the homogeneous stage of the system, and then of inducing phase separation by decreasing the temperature below the UCST, is a question of both fundamental and industrial nature. As explained in [16], the homogeneous state may offer great advantages, because of beneficial kinetic effect onto the extraction process, but the physicochemical reasons behind this are rather entangled. In particular, Onghena and co-workers note that stirring speeds up heat homogenization [61] and a higher temperature lowers viscosity [22]. At the same time, it is well-known that stirring, whatever the temperature, accelerates extraction by a simple mechanical effect. Therefore, reaching the monophasic state for better mixing and finally performing phase separation at room temperature by a cooling process or simply using vigorous stirring below UCST actually gave the same very efficient extraction of Sc(III) (E $\approx$ 100 %). Conversely, as pinpointed in [61], heating above 55 °C has its own cost, so a minute balance of pros and cons should be made in view of industrial applications, but this is out of the scope of this review. Therefore, the burst of enthusiasm about the monophasic stage advantages found in [22] is somehow dampen in a more recent publication [61]. We are convinced that performing extraction by the use of the UCST limit (above and below) accelerates the kinetics of the extraction, but it is our understanding that it does not enhance the thermodynamical aspect of the extraction and thus the equilibrium E and D values. A detailed study of these two aspects would require choosing a system for which extraction is not quantitative, in order to accurately rate the possible impacts of the experimental parameters.

In the absence of extractant or any additional compound, the extraction of Ln ions is said negligible by the use of H_2O&[Hbet$^+$][Tf$_2$N$^-$] [16]. No indication of means to prevent metal hydrolysis could be found in this work. In order to modify the extraction of metallic ions, two chemicals have been studied, betaine and nitric acid. The zwitterionic betaine (13 wt%) efficiently extracts a large variety of elements toward the lower phase, from Cu(II) ($D_{Cu,\ell} \approx 30$) to Sc(III) ($D_{Sc,\ell} \approx 600$) [16]. The lanthanides Dy, Er, Ho, La, Pr, and Nd all display $D_{Ln,\ell} \approx 10$, Mn, Ni, and Zn roughly distributed equally between the two phases ($D_{X,u} \approx 1.7$ for the three elements), while Ag mainly remains in the upper phase ($D_{X,u} \approx 6.7$). It is interesting to note that the betaine compound has been dissolved in the aqueous phase before contact with the IL phase, which is a rather unusual procedure in traditional liquid-liquid extraction. The Nd distribution ratio increases as a function of betaine concentration, but the authors note that betaine needs to be in very large excess as compared to the metallic ion in order to get high efficiencies (ratio of 200, $E_{Nd,\ell} = 99\%$). As already pinpointed in Sect. 9.2, addition of large amounts of betaine has a strong impact onto the UCST. On the other hand, nitric acid also modifies significantly the extraction efficiencies (in the absence of betaine) for other metallic ions: Pd(II), Rh(III), and Ru(III), with $E_{X,\ell}$ values ranging from 100 to 95%, 70 to 40%, and 40 to 20%, respectively, as nitric acid amount is increased [25]. Such values lead to separation factors for Pd/Ru and Pd/Rh in the range 50–5000. Similar decrease was observed for U(VI), with values $E_{U,\ell}$ from 62 to 0% as nitric acid concentration varies from 10^{-2} to 2 M. Therefore, for U(VI), Pd(II), Rh(III), and Ru(III), a compromise should be found between high $E_{X,\ell}$ values and troubles arising from hydrolysis, so the authors suggest $[H^+]_{init} \approx 0.3$ M [25]. Finally, $E_{X,\ell}$ values for Al(III) and Na(I) remain negligible in the whole nitric acid range investigated [29].

A mechanism has been proposed for Pd(II) extraction in the absence of betaine, but the presence of nitric acid [25, 29] is also suggested for U(VI) extraction [29], based on an ion exchange between the upper and lower phase as

$$Pd^{2+}_u + 2Hbet^+_\ell \rightleftarrows 2H^+_u + [Pd(bet)_2]^{2+}_\ell \quad (9.1)$$

Note that in the work of Sasaki et al. [25], subscripts in the chemical equilibrium proposed refer to the "aqueous" and "IL" phase so we have just changed these to the "u" and "ℓ" subscripts we refer to. This equation is in line with the observed decrease in $D_{Pd,\ell}$ as the initial nitric acid concentration in the system is increased, because H^+ increase would disfavor the formation of $[Pd(bet)_2]^{2+}$ in the lower phase, by enhancing the protonation of bet into Hbet$^+$ (p$K_a = 1.83$). As no measurement of the equilibrium pH of the upper phase has been done, this plausible explanation remains nevertheless to be firmly assessed.

In the presence of betaine and the absence of nitric acid, the Binnemans group [22] independently proposed a mechanism to describe their data on Nd(III) extraction, based on the following extraction equilibrium:

$$Nd^{3+}{}_u + n\ bet_u + 3Tf_2N^-{}_u + x\ H_2O_u \rightleftarrows Nd(bet)_n(H_2O)_x(Tf_2N)_{3\ell} \quad (9.2)$$

Here again, instead of species in aqueous and organic phases, we have assigned to the upper and lower phases. Based on a slope analysis of D as a function of the initial betaine concentration, the authors [22] suggested the extracted species to be $[Nd_2(bet)_3(H_2O)_x]^{3+}$, with the addition that "electrical neutrality can be achieved by [Tf_2N^-] or nitrate ions." This would require six negatively charged species to join the structure, as the exact charge of the complex is formally +6, because betaine is a zwitterion, therefore an overall neutral entity.[1] Additionally, experimental results performed with $[Pr^{3+}][Br^-]_3$, $[Nd^{3+}][NO_3^-]_3$, and $[Nd^{3+}][Cl^-]_3$ all point to an ion exchange mechanism, as suggested by the authors [22], without involvement of the metal counter-anion.

It is our opinion that although the systems studied by the Japanese [25] and the Belgium [16] groups differ slightly, one single mechanism should be able to describe both data sets. In particular, we found the protonation process of betaine as acidity is increased, as discussed in [25], a reasonable assumption. Furthermore, there is no evidence of extraction of a Nd neutral species, and the global charge of the Nd complex remains unknown at the moment. We thus propose an alternative mechanism as

$$Nd^{3+}{}_u + x\ H_2O_u + 3\ Hbet^+{}_\ell \rightleftarrows \left[Nd(bet)_3(H_2O)_x\right]_\ell + 3\ H^+{}_u \quad (9.3)$$

Such a mechanism is formally identical to Eq. 9.2, and we think it should be valid for any of the other Ln(III) ions.

An impressive demonstration of an entire recycling process is given in [62], based on the thermomorphic system $H_2O\&[Hbet^+][Tf_2N^-]$, and applied to the recycling of rare earths from NdFeB magnets. Real magnets were first roasted, thus turned to oxides and milled to generate particles of size in the range 6–310 μm. The thermomorphic system $H_2O\&[Hbet^+][Tf_2N^-]$ (operating temperature: 80 °C) easily dissolves such particles (except for a small residue, depending on the size of the particles), and by cooling the monophasic system down to room temperature, iron is mainly present in the lower phase ("IL phase"), while the rare earths (Nd, Dy) and Co are mainly present in the upper phase. Transferring iron from the lower to the upper phase can be achieved by addition (1 M) of potassium salts of [Cl^-] and [$C_2O_4^{2-}$], while [ClO_4^-] and [NO_3^-] counter-anions only increase the affinity of Nd, Dy, and Co for the upper phase, without modifying the iron preference for the lower phase. Precipitation of the rare earths and cobalt with oxalic acid followed by a treatment with aqueous ammonia (which dissolves cobalt oxalate only) and

---

[1]Therefore Eq. 9.2 is not balanced in charge.

calcination of the purified rare-earth oxalates allowed the recovery of rare earths as very pure oxides (purity > 99.9 wt%) ready for the production of new magnets. In addition of being the first demonstration of a fully integrated process from milling of used magnets until production of new pure oxides by the use of IL-ABS, this study makes benefit of two properties of the $H_2O\&[Hbet^+][Tf_2N^-]$ system: its thermomorphic behavior and its ability to dissolve oxides [58, 63]. The latter aspect avoids the classical leaching step in acidic solutions, at the expense of a possibly costly roasting procedure.

The system $H_2O\&[chol^+][Tf_2N^-]$ is closely related to $H_2O\&[Hbet^+][Tf_2N^-]$. The compound $[chol^+][Tf_2N^-]$ is an IL (melting temperature = 30 °C) and displays a UCST at 72 °C [44]. At room temperature and for equal volumes of water and $[chol^+]$ $[Tf_2N^-]$, 10 wt% of water dissolves in the lower phase, while 12 wt% of IL is found in the upper phase, a situation very similar to the $H_2O\&[Hbet^+][Tf_2N^-]$ systems. Other similarities can be observed, such as small changes of the UCST by addition of either $[H^+][NO_3^-]$ or extractant (increase of ca. 1 °C at low concentration of added compound, then decrease of a few degrees). Changes are also observed in the UCST value by addition of $[Nd^{3+}][NO_3^-]_3$ [23].

The same two groups have investigated different protocols and different metallic ions. Binnemans and co-workers [23] studied the system $H_2O\&[chol^+][Tf_2N^-]\&$ $[chol^+][hfac^-]\&[H^+][NO_3^-]\&[Nd^{3+}]_3[NO_3^-]$. The compound $[chol^+][hfac^-]$ is an IL but here it is used as an extractant, with the additional advantage that it is less volatile than Hhfac. No experiment was performed by this group in the absence of extractant, and a low nitric acid concentration was used to prevent Nd hydrolysis ($pH_{init} = 2$) [23]. The extraction was performed by heating the two phases above the UCST (T = 80 °C). By contrast, Sasaki and co-workers focused on the impact of large initial nitric acid concentrations, without any extractant, and extraction of Pd (II), Ru(III), and Rh(III) was performed under the biphasic state (T = 25 °C) [25].

As previously discussed for the $H_2O\&[Hbet^+][Tf_2N^-]$ system, Nd(III) extraction experiments demonstrate the advantage of mixing the two phases above the UCST, in terms of kinetics but identical E values are obtained under the two regimes (monophasic or biphasic) [23]. Transfer of Nd to the lower phase increases as a function of extractant, and $E_{Nd,\ell}$ reaches nearly 100 % above $[chol^+][hfac^-] =$ 60 mmol.kg^{-1}.[2] Increase in the extractant concentration leads to a concomitant increase of the pH of the upper phase, which is ascribed by the authors to the protonation of the anion $[hfac^-]$. Interesting tests were performed in order to determine the maximum loading capacity of the extracted species, which is estimated at 43 mmol.kg^{-1}. Above this value, a precipitate appears. On the other hand, addition of $[H^+][NO_3^-]$ alone (no extractant) does not show a large impact onto Pd (II) and Rh(III) extraction in the range of initial acidities 0.3–2 M, as all D values remain approximatively constant and in the range 0.1–5 [25].

Cationic exchange is proposed by the two groups [23, 25] to account for their data. Sasaki et al. [25] describe the extraction process according to

---

[2]This concentration corresponds to the initial concentration in the single IL phase.

$$Pd^{2+}{}_u + 2\ chol^+{}_\ell \rightleftarrows Pd^{2+}{}_\ell + 2\ chol^+{}_u \qquad (9.4)$$

Cationic exchange is also proposed for the Nd experiments in the presence of extractant, but no equilibrium is written, mainly because the exact stoichiometry of the extracted species cannot be determined. Some experimental evidence [23] would be in favor of four [hfac⁻] entities per Nd ion, but the protonation of the [hfac⁻] introduces a bias in the slope analysis, thus hampering a firm assessment. However, addition of [chol⁺][Cl⁻] to the upper phase induces a decrease in $E_{Nd,\ell}$ (from 70 to 52 % in the range 0–70 mmol.kg^{-1} of [chol⁺][Cl⁻]). This is ascribed to [chol⁺] migration toward the upper phase upon Nd(III) transfer to the lower phase. Based on these experimental results and on the detailed discussion found in [23], we therefore suggest a possible extraction mechanism as

$$Nd^{3+}{}_u + 4hfac^-{}_\ell + 3\ chol^+{}_\ell \rightleftarrows \left[Nd(hfac)_4\right]^-{}_\ell + 3\ chol^+{}_u \qquad (9.5)$$

Note that Eqs. 9.1, 9.2, 9.4, and 9.5 are all based on cationic exchange between the metallic and the IL's cations. Cationic exchange could be further supported by analytical examination of the upper and lower phase compositions before and after metal extraction.

The cholinium cation is nontoxic and biodegradable and has been recommended in a study on antibiotic extraction as a greener approach to extraction [48]. Unfortunately, the IL compound [chol⁺][hfac⁻] is not stable upon heating and storage: 14 days of storage reduces the extraction efficiency to less than a half of the initial efficiency; a tendency is also observed if stored at ca. 6 °C, and the compound is known to decompose upon heating [23]. These facts render the system rather limited for industrial applications.

Some narrow studies have been performed with the related system H₂O& [TMPA⁺][Tf₂N⁻]&[H⁺][NO₃⁻]: U(VI) is poorly extracted [29] and so are Pd(II), Rh(III), and Ru(III) [25].

Thermomorphic behaviors of carboxyl-functionalized ILs based either on the morpholinium cation or betaine derivatives [63], with UCST equal to 52 °C (morpholinium cation) and 55 °C or 64 °C, have been reported. Metal extraction experiments were not performed but the authors easily dissolved many metal oxides and hydroxides in such ILs. So, using these ILs for IL-ABS-based metal extraction is quite a reasonable perspective.

9.3.3 Use of Ionic Liquid Analogues of Girard's Reagents

Blesic and co-workers performed the synthesis of a series of new ionic liquid compounds, based on the classical [Tf₂N⁻] anion and several variations upon the Girard's reagent pattern [17]. Although the Girard's reagents have melting points around ca. 200 °C and are thus not ionic liquids, the six ionic liquid versions of these salts display melting points in the range from −55 to −36 °C. Their mixtures with water behave as USCT systems, with T_c values varying from ca. 5 to 95 °C depending on the IL nature and relative proportions of water and ILs. Preliminary extraction experiments have been performed by contacting aqueous solution of either Ni(II), Cu(II), Co(II), or Cr(III) ions (initial aqueous concentration, ca. 10 mM; counter-anions, [$CH_3CO_2^-$]) with one of the Girard ILs. The authors took advantage of the USCT phenomenon and performed extraction by heating above the USCT value (T = 60 °C) and then cooled down the samples below (T = 20 °C) to derive the D values. No indication could be found in the paper on the respective volumes of the aqueous and IL phases put into contact so we thus just recall the D values as published. All four metallic ions are individually extracted, quantitatively for Ni and Cu (D values at ca. 780 and $> 2 \times 10^4$, respectively), very well for Co (D = 36) and much less for Cr(III) (D ≈ 5) [17].

9.3.4 Systems Using Tetrabutylammonium Bromide

Three publications (two of them by the same group) concern the system H_2O & [TBA^-][Br^-] & [NH_4^+]$_2$[SO_4^{2-}], where [TBA^-][Br^-] is at the upper limit of the definition we set for an IL (T_{melt} = 100 °C) and in which extractions were performed at a constant temperature by addition of one salt in order to induce the mono-/biphasic state change.

In a first step, Akama and co-workers have examined the chemical conditions leading to the ABS behavior and observed that several quaternary ammonium salts and other usual salts ([Na^+][Cl^-], [Na^+][NO_3^-], or [NH_4^+][Cl^-]) also enable the mono-/biphasic changes, but concluded that the combination of [TBA^+][Br^-] and [NH_4^+]$_2$[SO_4^{2-}] is the most suitable one [64]. For example, a two-phase state is obtained at T = 20 °C for [TBA^+][Br^-] concentration in the range 0.3–1 M (5 mL of solution), together with [NH_4^+]$_2$[SO_4^{2-}] from 0.1 to 1.7 g. Starting from an initial monophasic aqueous solution volume of 5 mL, variation of the concentrations of the two salts induces large changes in the volume of the upper phase, from 0 mL (monophasic state) to 4 mL (biphasic state). Note that such a plot (volume of upper phase as a function of global composition) is another way to display the information related to the miscibility curve, although less convenient. The authors [64] have also determined the amounts of the main ions composing the IL-ABS, namely, [TBA^+], [NH_4^+], [SO_4^{2-}], and [Br^-], in the upper and lower phase (see Table 9.2). Their general experimental procedure indicates adjustment of the initial pH value at pH = 3, but unfortunately they do not comment on the equilibrium pH values to be observed in the upper and lower phases, although H^+ ions are known to distribute in a complex way in other water/IL biphasic systems [65]. Apart from uncertainties in the analytical method, this could be part of the reason explaining the observed charge balance discrepancy in the upper phase and the 10 % difference in the measured and expected values for [NH_4^+] amounts. As a consequence, calculated $E_{X,u}$ and $E_{x,\ell}$ values we have added in Table 9.2 should be considered with caution but these data

Table 9.2 Amounts (in mmol) of the various ions composing the IL-ABS H$_2$O&[TBA$^+$][Br$^-$]& [NH$_4^+$]$_2$[SO$_4^{2-}$] in the upper and lower phase, as measured by Akama and co-workers [64] and as calculated from the initial amounts used. Corresponding values of E$_{X,u}$ and E$_{X,\ell}$ as calculated from these data (values of E are displayed only if above 50%)

	[TBA$^+$]	[Br$^-$]	[NH$_4^+$]	[SO$_4^{2-}$]
Upper phase (mmol)	2.5	1.0	0.9	0.5
Lower phase (mmol)	0.6	1.8	17.0	7.9
Total measured (mmol)	3.1	2.8	17.9	8.4
Total expected (mmol)	3.0	3.0	16.6	8.3
E$_{X,u}$ (%)	83.3			
E$_{X,\ell}$ (%)		60.0	102.4	95.2

would indicate, as previously stressed, that although [TBA$^+$] and [Br$^-$] have been introduced in identical molar amounts, owing to their identical charges, they distribute quite differently between the two phases: the equilibrium amount (in mmol) of [TBA$^+$] in the lower phase is three times less than that of [Br$^-$].

In a second step, the same group of authors have investigated the transfer from one phase to another of Cd(II), Pb(II), Co(II), Cu(II), Fe(III), and Zn(II) in [64] and Cr(VI) and Cr(III) (introduced as potassium salt) [19]. They have studied, in view of Cd(II) extraction, the impact of parameters such as sample volume, amount of metallic ion, and initial pH value of the aqueous phase in the presence of interfering metallic ions (Co(II), Cu(II), Fe(III), and Zn(II)). Note that these authors [64] use the terms "upper" and "lower" phases exactly as we do and indicate densities to be 1.05 and 1.14 at T = 20 °C, respectively. They indicate that low pH values (below 0.5) render the IL-ABS quite unstable, and their data show a strong dependence of individual extractions on the initial pH values (E$_{Cd,\ell}$ = 100 % for pH > 2; E$_{Pb,\ell}$ > 80 % for pH < 4). Chemical conditions for the efficient and selective extraction of Cd(II) (Cd recovery: above 90 % and up to 99 %) in the presence of large amounts of Zn(II) have been determined. In case of Cr(VI) extraction, the initial pH value was shown to be a decisive parameter, while Cr(III) was never extracted to the upper phase in significant amounts whatever the chemical conditions investigated [64]. We hypothesize that the difference in extraction ability from Cr(III) to Cr(IV) is mainly due to differences in charge and charge densities of the entities under study that modify their solvation abilities. Tests have been performed in order to discriminate Cr(VI) from Cr(III) in spiked wastewater samples and appeared to be satisfactory.

An extraction mechanism has been proposed for the Cr(VI) extraction case [19], which is based on transfer of the neutral species [HCrO$_4^-$.TBA$^+$] from the lower to the upper phase. Although this proposal is in qualitative agreement with the slope analysis evidencing a dependency close to one H$^+$ per Cr(VI) entity, the involvement of [TBA$^+$] is not demonstrated yet. Two other mechanisms could be envisioned at the moment and would require further studies to be confirmed or ruled out: involvement of the chromium counter-ions, [K$^+$], instead of [TBA$^+$], or anionic exchange involving concomitant transfer (from upper to lower phase) of one [Br$^-$] ion.

As an opening to fascinating technological developments, we refer to the excellent work of Choi and co-workers [66], who investigated extraction of Ru (introduced as "ruthenium red," [Ru(NH$_3$)$_5$Ru-O-Ru(NH$_3$)$_4$-O-Ru(NH$_3$)$_5$]Cl$_6$) in a microfluidic system by the use of the system studied by Akama and collaborators: H$_2$O&[TBA$^+$][Br$^-$]& [NH$_4^+$]$_2$[SO$_4^{2-}$]. Another example of microfluidic extraction technique and results (for biological samples) can be found in [67], and a deeper presentation of the technique itself, together with insights into the theoretical aspects of microfluidic extraction with ABS, is done in [68].

In the microfluidic extraction studies [66], the IL-ABS was prepared under chemical conditions leading to the biphasic state. After equilibration by vigorous shaking, the two phases were separated and fed into the two inlets of the microfluidic device. In such experimental setup, the notions of "upper" and "lower" phases have no meaning anymore as the system is basically 1D and should be replaced by the terms "continuous" and "dispersed" phases. The phase displaying the lower density and the higher viscosity (corresponding to the upper phase in a batch experiment in classical test tubes or beakers) was used as the continuous phase. In the first part of the device, the two phases are under laminar flow and therefore do not mix. Droplets are generated by a pulsed potential difference (150 V, 200 ms). In the tube section where the droplets move in the bulk of the continuous phase, ruthenium red was introduced through the third inlet, as undissolved microparticles in the continuous phase [66]. This ruthenium compound has been chosen because it is highly soluble in the dispersed phase and because of its bright color, which allows an easy detection of mass transfer by the naked eye or any suitable electronic device (digital camera). In such microfluidic devices, kinetics is of tremendous importance: ruthenium diffuses from the inlet to the main flow, then dissolves on the droplet surface and simultaneously transfers from the droplet surface to the droplet bulk until saturation. Extraction is also controlled by the number, size, and guidance of the droplets, all these parameters being monitored through the potential pulse history. Consequently, droplets and/or continuous phase of controlled Ru concentrations can be obtained, and this can be adjusted through time, in view of analytical needs, reaction protocols, etc. Solving the diffusion 1D equations allows recovering the kinetic data quite satisfactorily [66].

9.3.5 IL-ABS with Imidazolium-Based ILs

Three different works which use imidazolium-based ILs for extraction purposes of metallic ions were found in the literature [12, 13, 28]. Bridges and collaborators have studied the extraction of a rather uncommon metallic anion, [TcO$_4^-$], in view of nuclear waste processing from Hanford and Savannah River repositories, by the use of the well-known IL [$C_1C_4im^+$][Cl^-] [28]. The three systems they used, H_2O& [$C_1C_4im^+$][Cl^-]&[K^+]$_3$[PO_4^{3-}], H_2O&[$C_1C_4im^+$][Cl^-]&[K^+]$_2$[HPO_4^{2-}], and H_2O&[$C_1C_4im^+$][Cl^-]&[K^+]$_2$[CO_3^{2-}], are all well-identified IL-ABS [69]. In this study, the amount of IL was varied from 30 to 70 wt% in the initial aqueous phase, while the amount of the other salt was kept equal to 40 wt% in the other initial aqueous phase. Identical volumes of each were mixed to obtain the biphasic state. The Tc(VII) distribution ratio is readily defined as $D_{Tc,u}$. For all three IL-ABS, $D_{Tc,u}$ increases as a function of the tie-line length from ca. 2 to ≈ 700 [28].

Another interesting work concerns the system H_2O&[$C_1C_6im^+$][BF_4^-]&[Na^+][PF_6^-] for Ag(I) extraction (introduced as its nitrate salt) [12]. [$C_1C_6im^+$][BF_4^-] is also a well-known IL and H_2O&[$C_1C_6im^+$][BF_4^-] displays a UCST at 58 °C [40], so the presence of [Na^+][PF_6^-] is not mandatory to observe a mono-/biphasic change. Although they could have used the temperature stimulus, the authors [12] used the [Na^+][PF_6^-] salt to monitor the mono-/biphasic state changes, without any comment on that choice. Furthermore, two additional compounds are present in the studied system, as an acetate buffer, in order to control pH in dedicated experiments, and thio-Michler ketone. This last chemical was not used for extraction enhancement but as a chelating agent for UV-Vis determination of Ag. Therefore, although we denied systems containing molecular compounds to be IL-ABS, we nevertheless consider this system to be perfectly in the scope of this chapter. Finally, as far as [PF_6^-]- or [BF_4^-]-containing systems are concerned, it has been repeatedly acknowledged that such ions suffer hydrolysis when contacted with water, and this point is clearly discussed in the publication presenting the phase diagrams of H_2O&[$C_1C_6im^+$][BF_4^-] [40]. Apart from decomposition and [H^+][F^-] gaseous emission, the production of [F^-] also is detrimental to glassware. Furthermore, on the view point of IL-ABS, hydrolysis of the [PF_6^-] induces large changes in relative volumes of the two phases and changes in the critical temperature. For all these reasons, the author of this chapter strongly recommends, as many others, avoiding such chemicals. Disregarding these experimental and environmental problems, the paper under review presents a rather comprehensive study of factors influencing Ag(I) extraction: pH, IL-ABS composition, addition of a third salt as [Na^+][NO_3^-], temperature, and centrifugation conditions. Under the best operation conditions obtained, Ag(I) extraction reaches 100 % and the system studied is very selective against a broad variety of other elements (Li, Pb, Al, Zn, Mn, Hg, etc.). The method and the system were used for analysis of real samples (photographic wastes and river water) [12].

Recently, the related system H_2O&[$C_1C_4im^+$][Cl^-]&[H^+][NO_3^-]&[K^+]$_2$[HPO_4^{2-}] has also been used for silver extraction [13], but in a very different perspective: in this publication, the aim is the separation of a low amount of ^{109}Cd (carrier-free) arising from the α-irradiation of a silver target, after dissolution of the latter in nitric acid. As previously mentioned, H_2O&[$C_1C_4im^+$][Cl^-]&[K^+]$_2$[HPO_4^{2-}] is an IL-ABS. In view of the applications, the paper focuses on practical details needed to achieve the highest separation factor. The optimized conditions found by the authors are 6 M nitric acid concentration in the IL-ABS, 60% of IL (w/v), and 10 min settling time. With such operating conditions, a separation factor above 100 was obtained, with a recovery of [109]Cd in the range of 90% (Ag(I) transfers to the lower phase, and Cd(II) is found in the upper phase).

9.3.6 New Systems with Imidazolium, Pyrrolidinium, and Pyridinium Cations

In a series of three successive publications, the following ten systems have been investigated for the extraction of the three different precious metallic ions, namely, Au(III) [14], Pt(IV) [26], and Ir(IV) [21]:

- $H_2O\&[C_1C_8im^+][Cl^-]\&[H^+][Cl^-]\&[K^+][AuCl_4^-]$
- $H_2O\&[C_1C_8im^+][Br^-]\&[H^+][Br^-]\&[K^+][AuBr_4^-]$
- $H_2O\&[C_8pyr^+][Br^-]\&[H^+][Br^-]\&[K^+][AuBr_4^-]$
- $H_2O\&[C_1C_8pyrro^+][Br^-]\&[H^+][Br^-]\&[K^+][AuBr_4^-]$
- $H_2O\&[C_1C_8im^+][Cl^-]\&[H^+][Cl^-]\&[H^+]_2[PtCl_6^-]$
- $H_2O\&[C_1C_8pyrro^+][Br^-]\&[H^+][Cl^-]\&[H^+]_2[PtCl_6^-]$
- $H_2O\&[C_{12}(C_1im)_2^{2+}]_2[Br^-]\&[H^+][Cl^-]\&[H^+]_2[PtCl_6^-]$
- $H_2O\&[C_1C_8im^+][Br^-]\&[H^+][Cl^-]\&[K^+]_2[IrCl_6^{2-}]$
- $H_2O\&[C_1C_1C_8pyrid^+][Br^-]\&[H^+][Cl^-]\&[K^+]_2[IrCl_6^{2-}]$
- $H_2O\&[C_1C_1C_8im^+][Br^-]\&[H^+][Cl^-]\&[K^+]_2[IrCl_6^{2-}]$

For all these systems, addition of the IL compound to the aqueous phase containing a mineral acid and the metallic ion leads to a biphasic system, either liquid-liquid or solid-liquid: compounds obtained with $[C_1C_1C_8pyrro^+]$ and Au (III), $[C_1C_1C_8im^+]$ or $[C_{12}(C_1im)_2^{2+}]$ and Pt(IV), $[C_1C_1C_8im^+]$ or $[C_1C_1C_8im^+]$ and Ir(IV) are solids, some of them sticking very well to the walls of the polyethylene test tubes, while all other compounds are liquids (with higher densities than the other phase, therefore becoming the lower phase) under the chemical conditions used. However, as noted by us [14], liquid lower phases could be the result of very hygroscopic solids (that we may also call water-rich phase). Note that none of the water content has been analyzed for any of the obtained compound.

As far as we know from the literature, none of these systems had been shown to present a change from mono- to biphasic state before these three publications [14, 21, 26]. Although no reference has been made in the abovementioned publications to ABS as such, all these systems correspond to IL-ABS under our definition. Interestingly, these systems are one of the very rare examples of IL-ABS containing large amounts (1 M or above) of a mineral acid as one of the key constituents, while most of the known IL-ABS deal with alkaline, neutral, or slightly acidic mixtures [70]. These acidic conditions are clearly an advantage in view of metal extraction, in order to avoid hydrolysis and/or precipitation.

Extraction of the three metals toward the lower phase strongly depends on the global mineral acid concentration and on the exact nature of the IL compound, but for all three metals, chemical conditions leading to very high extraction efficiencies could be found ($E_{x,\ell} > 90\,\%$). The authors [14] note that these experiments correspond to a metathesis procedure leading to the formation of an insoluble IL (i.e., insoluble in an aqueous acidic phase), which is actually a very classical way to synthesize ILs, and thus they proposed a mechanism as

$$Cat^+ + MX_4^- \rightleftarrows CatMX_p \quad (M = Au) \tag{9.6}$$

$$2\,Cat^+ + MX_6^{2-} \rightleftarrows Cat_2MX_p \quad (M = Pt,\,Ir) \tag{9.7}$$

where $X = Cl$ or Br, accordingly. Characterization of the collected fractions (lower phases) by NMR or IR (Pt and Ir cases, collected fractions for Au were too small) confirmed the presence of the IL cation, and UV-Vis spectra confirmed the presence of the halide metallic entity. Equations 9.6 and 9.7 are ascribable to a precipitation mechanism and the authors used this term all along, whatever the exact state (liquid or solid) of the lower phase obtained.[3] The authors [14] consequently developed a simple mathematical treatment, based on the classical definition of a solubility product to determine the K_s values of each compound. Although this formalism well describes the results obtained, a model based on anionic exchange could also be considered for the liquid-liquid biphasic systems as (adaptation to the Ir and Pt cases is straightforward):

$$X^-{}_\ell + AuCl_4^-{}_u \rightleftarrows X^-{}_u + AuCl_4^-{}_\ell \tag{9.8}$$

In the case of the liquid lower phases, we consider both models to be formally identical.

Such experiments lead to a new metal-containing IL quite similar to the $[C_1C_4im^+][FeCl_4^-]$ which has been presented in Sect. 9.3.1. They are also very similar to the synthesis of the liquid compound $[C_1C_8im^+]_2[Ce(NO_3)_6^{2-}]$ by adding $[C_1C_8im^+][Cl^-]$ to a strongly acidic aqueous solution ($[H^+][NO_3^-] > 7$ M) in which $[Ce^{4+}][NO_3^-]_4$ has been dissolved: upon addition of the IL, a red viscous liquid forms at the bottom [15], and the authors propose an anionic exchange similar to Eq. 9.8 to be the rationale for that reaction. Another work in relation with this is the synthesis of imidazolium ILs bearing aminodiacetic moieties as di-*tert*-butyl ester [71]. One of them, put in contact with a water solution containing Cu(II), leads to a cupper complex forming a separate different phase.

---

[3]This corresponds to the counterintuitive (but stimulating) notion of precipitation of a liquid that could possibly be lying as an upper phase. This observation can also be found in the recent paper [53]. Dupont D, Depuydt D, and Binnemans K (2015): Overview of the effect of salts on biphasic ionic liquid/water solvent extraction systems: anion exchange, mutual solubility and thermomorphic properties. J. Phys. Chem. B 119:6747–6757.

9.3.7 Use of Phosphonium-Based ILs

This section presents extraction data for an IL-ABS system for which both T and salt concentration are used in conjunction to induce phase changes. Phosphonium-based ILs are relatively newcomers in the field of metallic ion extraction [72], as compared to imidazolium-based ILs. Under the commercial generic names of Cyphos (Cytec), they are in particular available as chloride, bromide, and dicyanamide salts, thus being fluorine-free ILs. Addition of $[Na^+][Cl^-]$, a cheap and efficient salting-out agent, induces a thermomorphic behavior for $H_2O\&[P_{44414}^+][Cl^-]$. The turnover temperature can vary from 40 °C down to 0 °C as the sodium chloride amount is increased from 2 to 11 wt% [24]. Other examples of mono-/biphasic changes upon composition of the phosphonium-IL/water mixtures, for noncommercial compounds, can be found in [73].

To the best of our knowledge, extraction of metallic ions using such phosphonium-based ILs in IL-ABS was performed only by the Binnemans group so far [18, 61]. Such experiments concern Co, Ni, Cu, and Zn, on the one hand, and Sc on the other hand. Using the system $H_2O\&[P_{44414}^+][Cl^-](40\text{ wt\%})\&[Na^+][Cl^-]$ (5 to 11 wt%), Ni(II) was efficiently separated from Co(II) (1 g.kg^{-1} each), the latter being extracted to the "IL-rich phase" (i.e., most probably the upper phase of the biphasic system, as phosphonium-based ILs are most of the time less dense than aqueous solutions). In this case, the system was cooled below the LCST temperature (monophasic state) and then heated up (biphasic state) but no information on the two operating temperatures could be found. A separation factor of ca. 500 was obtained. This is an interesting example of extraction which benefits both from the addition of a salting-out agent and of the thermal stimulus in order to obtain and drive the biphasic/monophasic changes. This is useful for a complete and fast-mixing, first and second, for phase separation. By contrast, systems as $H_2O\&[P_{444}E_i^+][DEHP^-]$ ($i = 1,2,3$) do not require salt addition to display tractable mono-/biphasic changes upon temperature (T_c from 60 °C to 20 °C depending on the exact IL nature, LCST-type behavior) but do not allow for neither a Co-/Ni-efficient separation nor for separation of Cu (II) and Zn(II), all four elements being extracted with D values in the range 4–25 [18]. Another drawback of these systems is, as stressed above, a large effect of metal concentration onto the turnover temperature, which can be reduced by ca. 20 °C. As discussed by the authors, this effect is interesting on the view point of energy consumption and practical easiness but limits the metal loading to 8000 ppm, in order to prevent a temperature change below 0 °C.

The system $H_2O(500\text{ mg})\&[P_{444}C_1COOH^+][Cl^-](500\text{ mg})\&[Na^+][Cl^-](8\text{ wt} \%)$ was also used for studying Sc(III) extraction (introduced as its chloride salt, at concentration equal to 5 mmol.kg^{-1}) [27]. Again, addition of the metallic ion of interest impacts the binodal curve. Scandium is efficiently extracted from aqueous phases, and the effect of contact time and pH (in the range 0–3.3, in order to avoid Sc hydrolysis) was examined. However, increasing the initial amount of Sc above ca. 10 mmol.kg^{-1} decreases the extraction efficiency, due to saturation of the receiving phase. Insights into the extraction mechanism are given, substantiated by NMR and IR measurements. Note that in an IL phase, the exact state, either associated or not, of any charged species is difficult to ascertain [74], and the same fundamental question arises in "IL-rich phases." We thus suggest that the formation of the neutral species $[P_{444}C_1COO)_3Sc]Cl_3$, as proposed by the authors [27], is as valid as our suggestion of the formation of $[P_{444}C_1COO)_3Sc]^{3+} + [Cl^-]_3$ is, because the differences between the two proposals are merely a question of considering or not a coordination sphere. Stripping experiments have been successfully performed by the use of the classical route of oxalic acid addition, inducing oxalate precipitates.

9.3.8 Stripping and Recycling

The final aim of any extraction process is to concentrate the metal under any suitable form (oxide, salt, neat metal, etc.) in a monophasic phase, while the IL is recovered in view of recycling. Starting from the individual phases of IL-ABS, which both contain large amounts of water and IL, this can be named "metal recovery" or "metal stripping," in reference to traditional extraction process, or "IL recovery." The use of IL-ABS is clearly a disadvantage, because, as pinpointed by the Binnemans group [22], IL-ABS are based on the high solubility of IL into aqueous phases and vice versa. Although, in a successful (i.e., efficient) extraction, the metal has been massively transferred from one phase to the other, by contrast, the IL is distributed between two phases, which implies multiple stripping procedures for its recovery. The recovery of the extractant is an additional problem to cope with, especially considering the large changes in composition of the IL-ABS as a function of all its components, as previously discussed. Furthermore, all other metallic ions present in the starting aqueous phase may interfere with the recovery procedures. These problems are not too often tackled in fundamental studies, and we commend the work and efforts of the Binnemans group [27, 61, 75] on these points. In the following, we simply gather some results on the general problem of metal, extractant, and IL recovery as found in the papers discussed above.

For any recovery procedure requiring acids, the use of $[H^+][Tf_2N^-]$ is limited by the unusual thermomorphic behavior of $H_2O\&[H^+][Tf_2N^-]$, as observed [76] and discussed [77] in previous papers. As for back-extraction of metallic ions, this subject is scrutinized in many papers. Various back-transfer protocols have been tested, depending on metal and samples. Contacting the IL-rich phase with acidic aqueous solution may be efficient [17], but precipitation using oxalic acid is also very common [18, 27, 61, 62]. More specific means, as reduction from Tc(VII) to Tc(V) by addition of $[Sn^{2+}][Cl^-]_2$, have also been tested in the literature but appear to be far from perfect [28]. Most of the time, stripping procedures lead to further losses of the IL components, in a way rather difficult to predict, as this depends, among other parameters, on the nature and amount of acid, but also on the metal concentration, a problem discussed in the studies with the $H_2O\&[Hbet^+][Tf_2N^-]$

system [23]. In some other cases, precipitation of the metallic salt regenerates the IL compound at the same time [27, 62].

In the case an extracting agent has been used, closing the cycle often requests to change its protonation state, which has been modified during the course of the IL-ABS extraction. Washing with water is often used for that purpose [18].

Finally, in view of recovering ionic liquids from aqueous samples, the use of $[Al^{3+}]_2[SO_4^{2-}]_3$ and $[Al^{3+}][K^+][SO_4^{2-}]_2$ has been successfully tested [78]. The system $H_2O\&[Cat^+][Ani^-]\&[Al^{3+}]_2[SO_4^{2-}]_3$ (or $Al^{3+}][K^+][SO_4^{2-}]_2$) is in fact an IL-ABS which does not extract aluminum but generates a biphasic state upon aluminum salt addition, thus cleaning the aqueous phase from its undesired IL content. The ILs under investigation belong to the imidazolium, phosphonium, and pyridinium families. Percentages of IL recovery were all above 90 % [78]. This is an unexpected use of an IL-ABS which is not efficient for Al recovery but is still of interest to a closed cycle.

9.4 Could (Should) All Extraction Systems with H_2O and ILs Be Considered as IL-ABS?

This question may appear rather provocative, of course, but it is our opinion that there is a continuous link between IL-ABS as reviewed above and any extraction system composed of (at least) water and one IL, plus a metallic entity to be extracted. In order to better sustain this paradigm, we will first present, in the two subsections to follow, typical examples of what we consider as two missing links between "real" IL-ABS and any other systems of the type $H_2O\&[Cat^+][Ani^-]$. Then, in the third subsection, few general comments are provided.

9.4.1 Extraction with $[C_1C_nim^+][Tf_2N^-]$ or $[C_1C_4im^+][PF_6^-]$

Systems of the type $H_2O\&[C_1C_nim^+][Tf_2N^-]$ ($n = 2, 3, 4, 5, 6, 7, 8$) also display a UCST behavior [79], but, owing to the physicochemical properties of these ILs, the miscibility curve could not be obtained in the full range of IL contents: in the range of IL mole fraction 0.001–0.7, all systems are biphasic from ca. 17 °C up to 47 °C, which, most of the time, corresponds to a workable T range for laboratory scale experiments. Moreover, as the measured two parts of the miscibility curve are very steep, the biphasic state could be evidenced either for very low ($<10^{-3}$ mole fraction) or rather high (>0.7 mole fraction) amounts of ILs in this accessible T range. To fix ideas, for the rather common $[C_1C_4im^+][Tf_2N^-]$ IL, at T = 25 °C, contacting 5 mL of pure water and that IL leads to a biphasic state for IL volumes from ca. 70 μL to 219 mL. Rather similar values are obtained for the other ILs ($n = 2$–8). Owing to the price of ILs and to the difficulty in handling very low volumes, most workable extraction studies based on $H_2O\&[C_1C_4im^+][Tf_2N^-]$ thus correspond to a very stable biphasic system composed of equal volumes of water and IL. Note that in the publications of concern, acids (nitric, hydrochloric, etc.) and/or acidic extractants could be added to the system to avoid metal hydrolysis and enhance extraction efficiencies, respectively. Those additional compounds may modify the borders of the miscibility curve, as could also large amounts of metal added do, as already stressed in Sect. 9.2, but as a matter of fact monophasic states have not been observed by any of the authors, so we may conclude that acids, extractants, and metallic ions do not dramatically reduce the biphasic envelop of the miscibility curve. Consequently, as far as we know from a broad literature survey (for reviews in the field, see [3, 4]), none of the numerous works performed with the general extraction systems $H_2O\&E\&[C_1C_nim^+][Tf_2N^-]\&[H^+][X^-]$ for metal extraction mention the existence of a UCST behavior, starting from our own publications in the field [80–82]. In our defense, it is most probable that very high temperatures would be needed to obtain a monophasic system, which would limit a lot the advantages of that protocol.

Similar very broad biphasic state is obtained for the system $H_2O\&[C_1C_4im^+][PF_6^-]$ [41], in the T range ≈ 10–87 °C, so the same conclusions as above are derived: apart from a few noticeable exceptions (see below), none of the publications dealing with $H_2O\&[C_1C_4im^+][PF_6^-]$ mention the UCST behavior of the system they are working on. However, considering the instability of $[PF_6^-]$ toward hydrolysis as already discussed, again, we would not recommend this system for metal extraction. Exceptions concern only systems in which the amount of $[C_1C_4im^+][PF_6^-]$ is very low, owing to cost of the compound. Keywords of these publications refer to liquid phase microextraction [83] or micro volume [84]. These two works present extraction of Pb(II) and Ni(II) [83, 84]. In these experiments, volumes are in the range of less than 10 µL for IL versus 3 mL of aqueous phase [83] or 500 µL of IL contacted with 10 mL of aqueous solution [84].

9.4.2 Imidazolium Nonafluorobutanesulfonate ABS

Only two papers [85, 86] take advantage of $[H_2O]\&[C_1C_nim^+][NfO^-]$ ($n = 4, 5, 6$) mixtures for metal extraction (La, Eu, Li, Na, Cs, Ca, and Sr). In one case, only nitric acid was added [86], while a mixture of $[Na^+][ClO_4^-]$ and $[H^+][ClO_4^-]$, together with the acidic extractant $[H^+][TTA^-]$, is used in the other case. To the best of our knowledge, these systems have not proved yet to experience changes from monophasic to biphasic state as a function of T or concentration (fifth criterion in Sect. 9.2). Therefore, strictly speaking, they cannot be considered as real IL-ABS under our definition but we consider them to be emblematic of the general ideas and results when using IL-ABS. In fact, it is our opinion that they are a perfect illustration of the gradation evoked in the introduction of Sect. 9.4.

Jensen and co-workers [85] observed large changes in volumes when equilibrating neat, dry $[C_1C_4im^+][NfO^-]$, and aqueous (perchloric/perchlorate) phases, giving birth to the system $H_2O\&[C_1C_4im^+][NfO^-]\&[H^+][ClO_4^-]\&[Na^+][ClO_4^-]$. As no indication of phase inversion could be found in both papers, we will assume that the lower phase corresponds to what is called an "IL phase" in [85] and [86]. An expansion of ca. 40 % of the volume of the formerly dry IL phase is indicated, accompanied by a substantial decrease of the volume of the other phase [85] but it has to be noted that the starting volumes of both phases are not specified in this paper. However, based on standard procedures, we may assume the preparation protocol to be based on identical volumes of IL and aqueous phase before contact. At equilibrium, the authors note that "the water content of neat IL phase is equal to 20.7 wt% (10.9 mole H_2O per liter)," which corresponds to ca. "6.4 H_2O molecules per $C_1C_4im^+$ NfO^- pair" [85]. It is clear that the so-called neat IL phase obtained by this protocol (with or without $[H^+][TTA^-]$) should be better considered as an aqueous phase containing large amounts of $[C_1C_4im^+]$ and $[NfO^-]$. This perfectly corresponds to the notion of "two immiscible aqueous-rich phases" of Freire and co-workers [35] and is also in line with the results of Kohno et al. [43] detailed in Sect. 9.2. Furthermore, addition of 0.5 M of $[H^+][TTA^-]$ modifies significantly the water amount in the lower phase, with a decrease to 14.2 w%, a phenomenon which resembles previous observations on the impact of Hbet and $[H^+][Cl^-]$ [16, 22]. In the other publication dealing with $[C_1C_4im^+][NfO^-]$ and nitric acid [86], the water content of the $[C_1C_4im^+][NfO^-]$ IL phase is said to be equal to 15.8 wt% (starting volumes of the aqueous and IL phase are identical), a value in line with that obtained by Jensen and co-workers [85] for perchlorate salt/perchloric acid combination. We thus may expect similar volume changes.

The question of water distribution is acknowledged by Jensen et al. [85], who state that "water-saturated $[C_1C_4im^+][NfO^-]$ would seem to have more in common with conventional concentrated salt solutions than (...) molecular organic solvents" so that "in liquid-liquid extraction systems $[C_1C_4im^+][NfO^-]$ appears to resemble an aqueous biphasic extraction system." We fully agree with that statement. However, these authors did not attempt to measure a pH equilibrium value in the lower phase neither did they attempt any experiments by varying the temperature.

9.4.3 Discussion

From the examples discussed above, it is clear that thermomorphic behavior and changes from mono- to biphasic state of systems containing at least water and one IL are a rather common phenomenon, and this book is a clear proof of this statement. By contrast to what could be thought, it is not true that ILs for use in solvent extraction should be immiscible with water: all the examples here provided contradict this statement.

From some striking experiments discussed above [22], it seems that using the thermomorphic behavior of systems in view of metal extraction offers great advantages in terms of efficiency, easiness, and duration and, therefore, cost. Gathering the knowledge of the scientific community dealing with thermomorphic behavior

9 Extraction of Metals with ABS 211

and salt-in/salt-out effects, on the one hand, and that community working on metal extraction by the use of ILs, on the other hand, would be very fruitful, and a first attempt to link the fundamental aspects of the two domains was done previously by us [77]. However, examples given in Sect. 9.4 highlight the regrettable poor connections existing at the moment, and it is one of the aims of this review chapter to bring links between them. We hope that the review of this chapter will bring the two communities closer by showing that "real" ABS have a lot aspects (if not all) in common with "classical" water-IL liquid-liquid extraction systems: water content and transfer of ions from one phase to the other are physicochemical characteristics found in both types of systems. Addition of large amounts of acid and/or metal salts and/or extractant perturbs miscibility curves in a way quite similar in nature (if not in quantitative values) as addition of kosmotropic salts does.

9.5 Conclusions

Despite the obvious considerations on green aspects brought about by the use of large amounts of water and of ILs as nonvolatile and nonflammable salts, one should not overlook some problems to be solved: IL-ABS of the type reviewed here are most probably difficult to handle on an industrial scale because of the large changes in respective volumes of the phases and variations of UCST/LSCT as a function of all component concentrations. Actually, an industrial goal is certainly to feed the system with a highly metal-loaded aqueous phase and to recover a highly metal-loaded lower phase. Another trouble may arise from the distribution of the IL components between the upper and lower phases which render their recovery problematic.

On a more fundamental perspective, elucidation of the extraction mechanism clearly requires a comprehensive analytical determination of ions, molecular compounds, and water fluxes between the two phases. This is a great experimental challenge but it will not be of general help to our understanding of IL-ABS unless tremendous efforts are put also in the search for tractable theoretical approaches, able to describe and predict UCST/LCST and distribution ratios of all components in order to first design and conceive and then apply and use novel systems. Some attempts, which discussion is out of the scope of this review chapter, can be found in [40, 87, 88].

We would like to end this chapter by further opening the door to other systems, again under the idea that the apparent barriers and frontiers existing between them and "real" IL-ABS are merely a question of personal feeling of the reader and are thus not as strict as could be envisioned at first glance. First, in this list of possible extensions, we would like to cite extraction of several lanthanides (La, Ce, Pr, Nd, Sm, Eu, Dy, Er, and Lu) by the use of molten calcium hydrate and $[A336^+][NO_3^-]$ [89]. Such systems are "borderline" in the sense that molten hydrates although containing water, by definition, cannot be considered as aqueous solutions. Nevertheless, distribution ratios in the range 100–700 were obtained. Second, many other thermomorphic systems involving ILs exist. Of special though is the use of carbon dioxide, as a perfect "green solvent" (provided it is captured from the atmosphere and not produced on purpose). Scurto and co-workers demonstrated the feasibility of three-level systems (two meniscuses, for mixtures of IL, water, and CO_2 plus air interface) [90], and this could be put in parallel with metal extraction experiments by the use of supercritical CO_2 and ILs [91]. Another elegant idea, although may be relatively costly, is the replacement of water by a second IL. Successful extraction of Co(II) from biphasic systems composed of two "immiscible" ILs has already been performed [92]. Another wide field that remains to be explored for metal extraction is that of thermomorphic systems between one IL and one organic solvent (no water). Although this could be considered as a step backward on an ecological perspective, some alcohols are not too detrimental to the environment. Numerous examples are found for hexanol [93], butanol [87], propanol and pentanol [94], and other series of alcohols [42], such as octanol [95], for various ILs: $[C_1C_6im^+][BF_4^-]$ [42] but also $[C_1C_4im^+][CF_3SO_2^-]$ [95], $[C_1C_nim^+][PF_6^-]$ [87], or rather unusual ILs [93]. Experiments can already be found for Co(II) in systems composed of either ethanol, propanol, or butanol and nitrile-bearing functionalized ILs, which display UCST values in the range 25–100 °C [96]. Other thermomorphic systems comprise $[C_1C_nim^+][Tf_2N^-]$ ($n=2$, 4, 6, 8, 10) and arenes [97], while $CHCl_3$ or mixtures ($CCl_4 + CHCl_3$) combined with $[C_1C_nim^+][Tf_2N^-]$ ($n=4$, 5) display rather unusual miscibility curves [98]. Going a step further, we think more complex systems are also worth of notice: systems with four phases (three meniscuses at the same time before liquid/air interface, two ILs, and two organic solvents) have been studied by the Seddon group [99], showing that, as already stressed for one-meniscus systems, ions distribute independently from each other [92]. Finally, the circle comes back around with extraction experiments for rare-earth group separation by the use of water, polymer, ammonium sulfate, and an organic solvent, Cyanex 272 [100]: this system mixes "old-fashioned ABS" (water, PEG, and salt) and "newcomers" as ILs and organic solvents. Although extraction ratios are not very high, there are good chances that efficiencies will be enhanced by optimization of the overall system.

As one can see, the scientific playground is *very* broad. Enjoy!

Chemical chart of compounds cited in this chapter:

$[chol^+][Hfac^-]$

9 Extraction of Metals with ABS

[Hbet$^+$][Tf$_2$N$^-$]

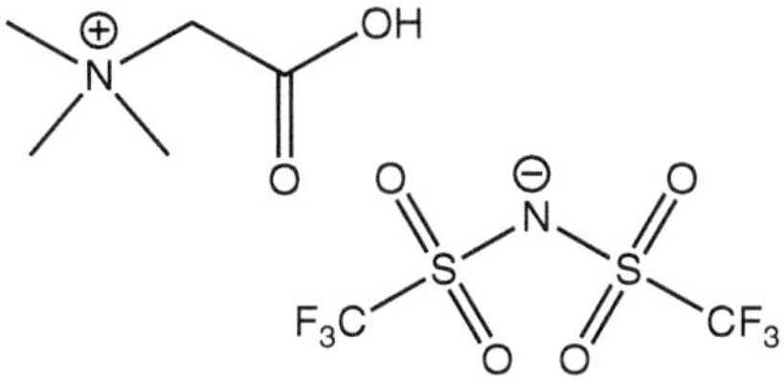

Bet

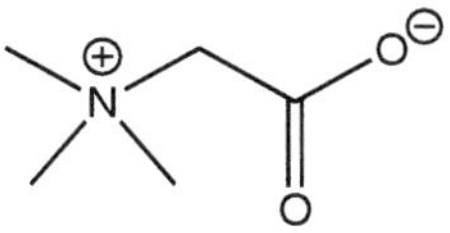

[chol$^+$][Tf$_2$N$^-$]

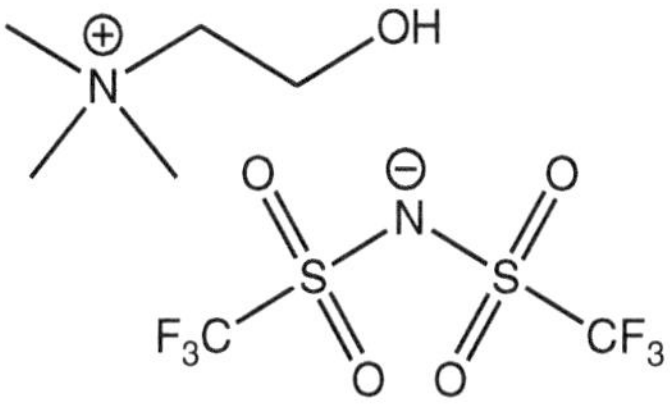

[C$_1$C$_1$C$_8$im$^+$]

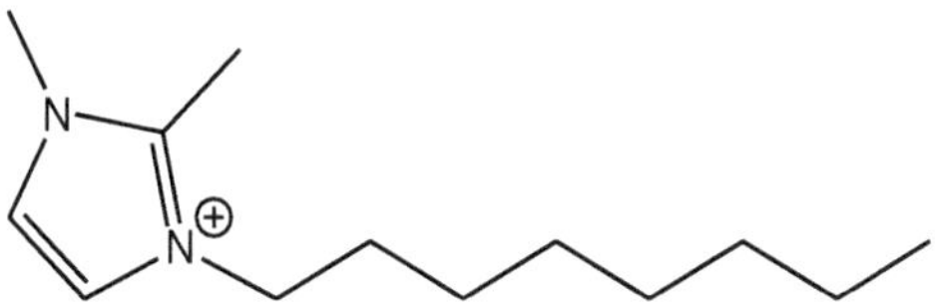

[C$_1$C$_1$C$_8$pirid$^+$]

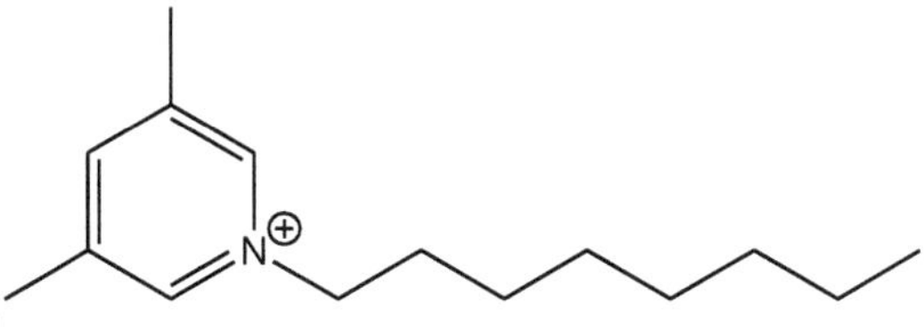

[C$_1$C$_8$pyrro$^+$]

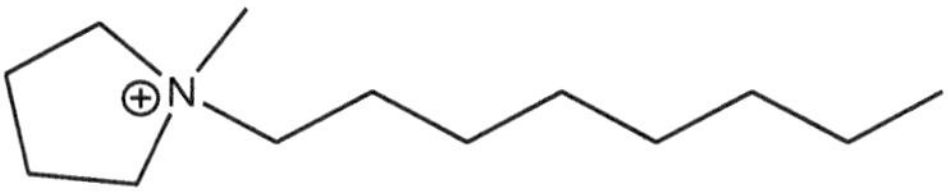

[TMPA$^+$][Tf$_2$N$^-$]

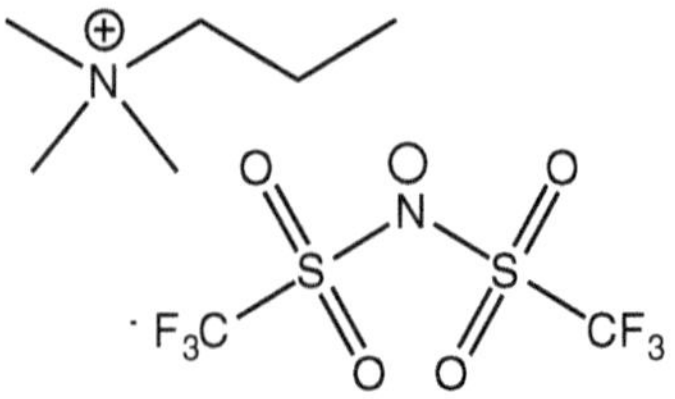

[H$^+$][TTA$^-$]

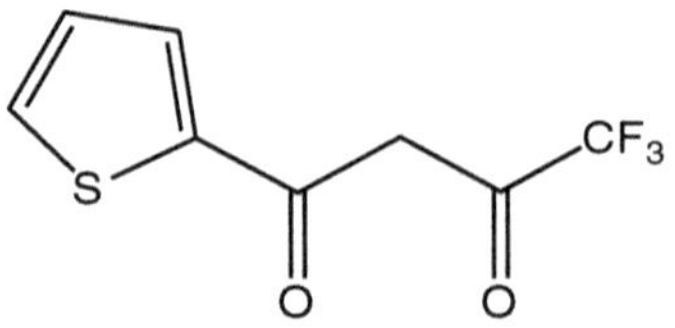

[C$_{12}$(C$_1$im)$_2$$^{2+}$]

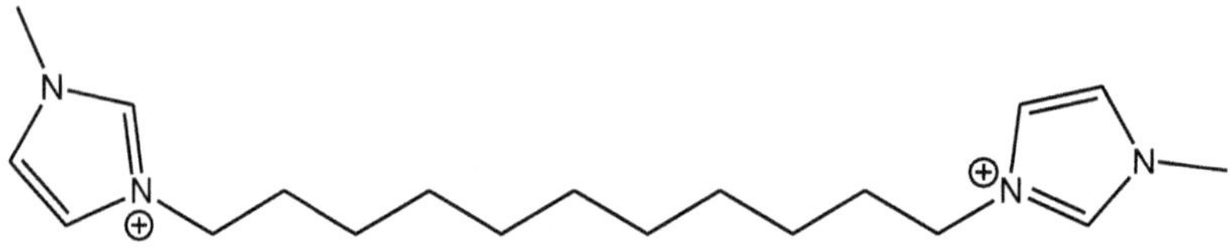

[TBA$^-$]

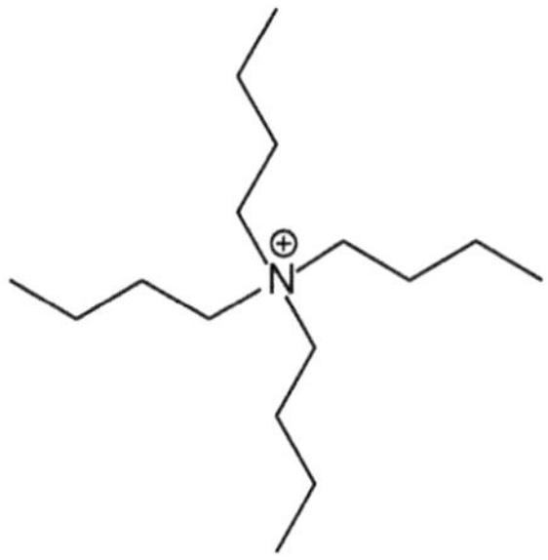

[A336$^+$][Cl$^-$] is a mixture of:

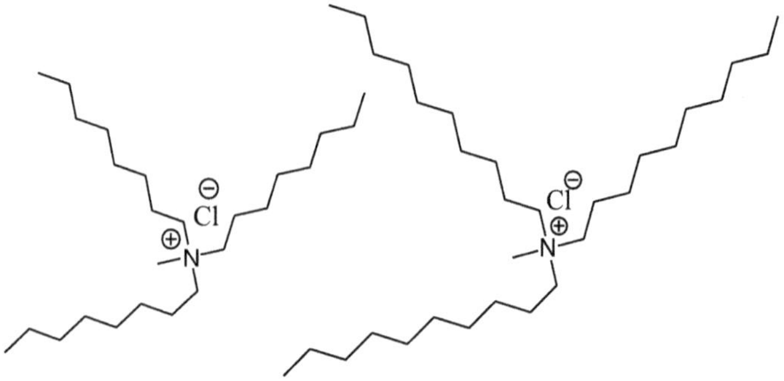

Cyanex 272

[DEHP⁻] and [P$_{444}$E$_n^+$] ($n = 1, 2, 3$)

References

1. Rydberg J (2004) Solvent extraction: principles and practice, 2nd edn. Marcel Dekker, New York
2. Dai S, Yu YH, Barnes CE (1999) Solvent extraction of strontium nitrate by a crown ether using room temperature ionic liquids. J Chem Soc Dalton Trans 8:1201–1202
3. Stojanovic A, Keppler BK (2012) Ionic liquids as extracting agents for heavy metals. Sep Sci Technol 47:189–203
4. Billard I (2013) Ionic liquids: new hopes for efficient lanthanide/actinide extraction and separation? In: Bünzli JCG, Percharsky VK (eds) Handbook on the physics and chemistry of rare earths. Elsevier, Amsterdam
5. Sun X, Luo H, Dai S (2012) Ionic liquids based extraction: a promising strategy for the advanced nuclear fuel cycle. Chem Rev 112:2100–2128
6. Kolarik Z (2013) Ionic liquids: how far do they extend the potential of solvent extraction of f-elements? Solv Ext Ion Exch 31:24–60
7. Messadi A, Mohamadou A, Boudesocque S, Dupont L, Guillon E (2013) Task-specific ionic liquid with coordinating anion for heavy metal ion extraction: cation exchange versus ion-pair extraction. Sep Purif Technol 107:172–178

8. Mohapatra PK, Kandwal P, Iqbal M, Huskens J, Murali MS, Verboom W (2013) A novel CMPO-functionalized task specific ionic liquid: synthesis, extraction and spectroscopic investigations of actinide and lanthanide complexes. Dalton Trans 42:4343–4347
9. Egorov VM, Djigailo DI, Momotenko DS, Cheryshov DV, Torocheshnikova II, Sirnova SV, Pletnev IV (2010) Task-specific ionic liquid trioctylmethylammonium salicylate as extraction solvent for transition metal ions. Talanta 80:1177–1182
10. Visser AE, Swatloski RP, Reichert WM, Mayton R, Sheff S, Wierzbicki A, Davis JH, Rogers RD (2002) Task-specific ionic liquids incorporating novel cations for the coordination and extraction of Hg^{2+} and Cd^{2+}: synthesis, characterization and extraction studies. Environ Sci Technol 36:2523
11. Luo H, Dai S, Bonnesen PV, Buchanan AC (2006) Separation of fission products based on ionic liquids: task specific ionic liquids containing an aza-crown ether fragment. J Alloys Compd 418:195–199
12. Vaezzadeh M, Shemirani F, Majidi B (2012) Determination of silver in real samples using homogeneous liquid-liquid microextraction based on ionic liquid. J Anal Chem 67:28–34
13. Gosh K, Maiti M, Lahiri S, Hussain VA (2014) Ionic liquid-salt based aqueous biphasic system for separation of 109Cd from silver target. J Radioanal Nucl Chem 302:925–930
14. Papaiconomou N, Vite G, Goujon N, Levêque JM, Billard I (2012) Efficient removal of gold complexes from water by precipitation or liquid-liquid extraction using ionic liquids. Green Chem 14:2050–2056
15. Zuo Y, Liu Y, Chen J, Li QD (2008) The separation of cerium(IV) from nitric acid solutions containing thorium(IV) and lanthanides(III) using pure [C8mim]PF6 as extracting phase. Ind Eng Chem Res 47:2349
16. Vander-Hoogerstraete T, Onghena B, Binnemans K (2013) Homogeneous liquid-liquid extraction of metal ions with a functionalized ionic liquid. J Phys Chem Lett 4:1659–1663
17. Blesic M, Gunaratne HQN, Jacquemin J, Nockemann P, Olejarz S, Seddon KR, Strauss CR (2014) Tunable thermomorphism and applications of ionic liquid analogues of Girard's reagents. Green Chem 16:4115–4121
18. Depuydt D, Liu L, Glorieux C, Dehaen W, Binnemans K (2015) Homogeneous liquid-liquid extraction of metal ions with non-fluorinated bis(ethylhexyl)phosphate ionic liquids having a lower critical solution temperature in combination with water. Chem Commun 51:14183–14186
19. Akama Y, Sali A (2002) Extraction mechanism of Cr(IV) on the aqueous two-phase system of tetrabutylammonium bromide and (NH4)2SO4 mixture. Talanta 57:681–686
20. Xie ZL, Taubert A (2011) Thermomorphic behavior of the ionic liquids [C4mim][FeCl4] and [C12mim][FeCl4]. Chem Phys Chem 12:364–368
21. Papaiconomou N, Billard I, Chainet E (2014) Extraction of iridium from aqueous solutions using hydrophilic/hydrophobic ionic liquids. RSC Adv 4:48260–48266
22. Vander-Hoogerstraete T, Onghena B, Binnemans K (2013) Homogeneous liquid-liquid extraction of rare earths with the betaine-betainium(trifluoromethylsulfonyl)imide ionic liquid system. Int J Mol Sci 14:21353–21377
23. Onghena B, Jacobs J, Meervelt LV, Binnemans K (2014) Homogeneous liquid-liquid extraction of neodymium(III) by choline hexafluoroacetylacetonate in the ionic liquid choline bis (trifluoromethylsulfonyl)imide. Dalton Trans 43:11566–11578
24. Onghena B, Opsomer T, Binnemans K (2015) Separation of cobalt and nickel using a thermomorphic ionic-liquid-based aqueous biphasic system. Chem Commun 51:15932–15935
25. Sasaki K, Takao K, Suzuki T, Mori T, Arai T, Ikeda Y (2014) Extraction of Pd(II), Rh(III) and Ru(III) from HNO3 aqueous solution to betainium bis(trifluoromethanesulfonyl)imide ionic liquid. Dalton Trans 43:5648–5661
26. Génand-Pinaz S, Papaiconomou N, Leveque JM (2013) Removal of platinum from water by precipitation or liquid-liquid extraction and separation from gold using ionic liquids. Green Chem 15:2493–2501

27. Depuydt D, Dehaen W, Binnemans K (2015) Solvent extraction of scandium(III) by an aqueous biphasic system with a nonfluorinated functionalized ionic liquid. Ind Eng Chem Res 54:8988–8996
28. Bridges NJ, Rogers RD (2008) Can kosmotropic salt/chaotropic ionic liquid (salt/salt aqueous biphasic systems) be used to remove pertechnetate from complex salt waste? Sep Sci Technol 43:1083–1090
29. Sasaki K, Suzuki T, Mori T, Arai T, Takao K, Ikeda Y (2014) Selective liquid-liquid extraction of uranyl species using task-specific ionic liquid, betainium bis (trifluoromethylsulfonyl)imide. Chem Lett 43:775–777
30. Ohno H (2005) Electrochemical aspects of ionic liquids. Wiley, Hoboken
31. Wasserscheid P, Welton T (2008) Ionic liquids in synthesis. Wiley-VCH, Weinheim
32. Scammells JP, Scott LJ, Singer DR (2005) Ionic liquids: the neglected issues. Aust J Chem 58:155–169
33. Weingartner H (2008) Understanding ionic liquids at the molecular level: facts, problems and controversies. Angew Chem Int Ed 47:654–670
34. Chiappe C (2007) Nanostructural organization of ionic liquids: theoretical and experimental evidences of the presence of well defined local structures in ionic liquids. Monatsh Chem 138:1035
35. Freire MG, Claudio AFM, Araujo JMM, Coutinho JAP, Marrucho IM, Lopes JNC, Rebelo LPN (2012) Aqueous biphasic systems: a boost brought about using ionic liquids. Chem Soc Rev 41:4966–4995
36. Kohno Y, Arai H, Ohno H (2011) Dual stimuli-responsive phase transition of an ionic liquid/water mixture. Chem Commun 47:4772–4774
37. Rebelo LPN, Najdanovic-Visak V, Visak ZP, da Ponte MN, Szydlowski J, Cerdeirina CA, Troncoso J, Romani L, Esperança JMSS, Guedes HJR, de Sousa HC (2004) A detailed thermodynamic analysis of [C4mim][BF4] + water as a case study to model ionic liquid aqueous solutions. Green Chem 6:369–381
38. Suarez PAZ, Einloft S, Dullius JEL, de Souza RF, Dupont J (1998) Synthesis and physical-chemical properties of ionic liquids based on 1-n-butyl-3-methylimidazolium cation. J Chim Phys 95:1626–1639
39. Fukaya Y, Ohno H (2013) Hydrophobic and polar ionic liquids. Phys Chem Chem Phys 15:4066–4072
40. Wagner M, Stanga O, Schroer W (2003) Corresponding states analysis of the critical points in binary solutions of room temperature ionic liquids. Phys Chem Chem Phys 5:3943–3950
41. Najdanovic-Visak V, Esprança JMSS, Rebelo LPN, da Ponte MN, Guedes HJR, Seddon KR, de Sousa HC, Szydlowski J (2003) Pressure, isotope and water co-solvent effects in liquid-liquid equilibria of (ionic liquid + alcohol) systems. J Phys Chem B 107:12797–12807
42. Wagner M, Stanga O, Schroer W (2004) The liquid-liquid coexistence of binary mixtures of the room temperature ionic liquid 1-methyl-3-hexylimidazolium tetrafluoroborate with alcohols. Phys Chem Chem Phys 6:4421–4431
43. Kohno Y, Ohno H (2012) Temperature-responsive ionic liquid/water interfaces: relation between hydrophilicity of ions and dynamic phase change. Phys Chem Chem Phys 14:5063–5070
44. Nockemann P, Binnemans K, Thijs B, Parac-Vogt TN, Merz K, Mudring AV, Menon PC, Rajesh RN, Cordoyiannis G, Thoen J, Leys J, Glorieux C (2009) Temperature-driven mixing demixing behavior of binary mixtures of the ionic liquid choline bis(trifluoromethylsulfonyl) imide and water. J Phys Chem B 113:1429–1437
45. Fukaya Y, Sekikawa K, Murata K, Nakamura N, Ohno H (2007) Miscibility and phase behavior of water-dicarboxylic acid type ionic liquid mixed systems. Chem Commun 2007:3089–3091
46. Wang R, Leng W, Gao Y, Yu L (2014) Microemulsion-like aggregation behaviour of an LCST-type ionic liquid in water. RSC Adv 4:14055–14062

47. Trindade JR, Visak ZP, Blesic M, Marrucho IM, Coutinho JAP, Lopes JNC, Rebelo LPN (2007) Salting-out effects in aqueous ionic liquid solutions: cloud point temperature shifts. J Phys Chem B 111:4737–4741
48. Shahriari S, Tomé LC, Araujo JMM, Rebelo LPN, Coutinho JAP, Marrucho IM, Freire MG (2013) Aqueous biphasic systems: a benign route using cholinium-based ionic liquids. RSC Adv 3:1835–1843
49. Merchuk JC, Andrews BA, Asenjo JA (1998) Aqueous two-phase systems for protein separation studies on phase inversion. J Chromatogr B 711:285–293
50. Kaul A, Pereira RAM, Asenjo JA, Merchuk JC (1995) Kinetics of phase separation for polyethylene glycol-phosphate two phase systems. Biotechnol Bioeng 48:246–256
51. Gutowski KE, Broker GA, Willauer HD, Huddleston JG, Swatloski RP, Holbrey JD, Rogers RD (2003) Controlling the aqueous miscibility of ionic liquids: aqueous biphasic systems of water-miscible ionic liquids and water-structuring salts for recycle, metathesis and separations. J Am Chem Soc 125:6632
52. Neves CMSS, Ventura SPM, Freire MG, Marrucho IM, Coutinho JAP (2009) Evaluation of cation influence on the formation and extraction capability of ionic-liquid-based aqueous biphasic systems. J Phys Chem B 113:5194–5199
53. Dupont D, Depuydt D, Binnemans K (2015) Overview of the effect of salts on biphasic ionic liquid/water solvent extraction systems: anion exchange, mutual solubility and thermomorphic properties. J Phys Chem B 119:6747–6757
54. Fagnant DP, Goff GS, SCott BL, Runde W, Brenecke JF (2013) Switchable phase behavior of [HBet][Tf2N]-H2O upon neodymium loading: implications for lanthanide separations. Inorg Chem 52:549–551
55. Kohno Y, Ohno H (2012) Ionic liquid/water mixtures: from hostility to conciliation. Chem Commun 48:7119–7130
56. Yoshida Y, Saito G (2010) Design of functional ionic liquids using magneto and luminescent active ions. Phys Chem Chem Phys 12:1675
57. Lecocq V, Graille A, Santini C, Baudouin A, Chauvin Y, Basset JM, Arzel L, Bouchu D, Fenet B (2005) Synthesis and characterization of ionic liquids based upon 1-butyl-2,3-dimethylimidazolium chloride/ZnCl2. New J Chem 29:700–706
58. Nockemann P, Thijs B, Pittois S, Thoen J, Glorieux C, Hecke KV, Meervelt LV, Kirchner B, Binnemans K (2006) Task-specific ionic liquid for solubilizing metal oxides. J Phys Chem B 110:20978–20992
59. Brooks NR, Schaltin S, Hecke KV, Meervelt LV, Fransaer J, Binnemans K (2012) Heteroleptic silver-containing ionic liquids. Dalton Trans 41:6902–6905
60. Lee SH, Ha SH, Ha SS, Jin HB, You CY, Koo YM (2007) Magnetic behavior of mixture of magnetic ionic liquid [bmim]FeCl4 and water. J Appl Phys 101:09J102-101-109J102-103
61. Onghena B, Binnemans K (2015) Recovery of scandium(III) from aqueous solutions by solvent extraction with the functionalized ionic liquid betainium bis(trifluoromethylsulfonyl) imide. Ind Eng Chem Res 54:1887–1898
62. Dupont D, Binnemans K (2015) Recycling of rare earths from NdFeB magnets using a combined leaching/extraction system based on the acidity and thermomorphism of the ionic liquid [Hbet][Tf2N]. Green Chem 17:2150–2163
63. Nockemann P, Thijs B, Parac-Vogt TN, Hecke KV, Meervelt LV, Tinant B, Hartenbach I, Schleid T, Ngan VT, Nguyen MT, Binnemans K (2008) Carboxyl-functionalized task-specific ionic liquids for solubilizing metal oxides. Inorg Chem 47:9987–9999
64. Akama Y, Ito M, Tanaka S (2000) Selective separation of cadmium from cobalt copper, iron (III) and zinc by water-based two-phase system of tetrabutylammonium bromide. Talanta 53:645–650
65. Atanassova M, Billard I (2015) Determination of pKaIL values of three chelating extractants in ILs. Consequences on extraction processes of 4f-elements. J Sol Chem 44:606–609
66. Choi YH, Soo Song Y, Kim DH (2010) Droplet-based microextraction in the aqueous two-phase system. J Chromatogr A 1217:3723–3728

67. Zhou Q, Bai H, Xie G, Xiao J (2008) Temperature-controlled ionic liquid dispersive liquid phase micro-extraction. J Chromatogr A 1177:43–49
68. Hardt S, Hahn T (2012) Microfluidics with aqueous two-phase systems. Lab Chip 12:434
69. Bridges NJ, Gutowski KE, Rogers RD (2007) Investigation of aqueous biphasic systems formed from solutions of chaotropic salts with kosmotropic salts (salt-salt ABS). Green Chem 9:177–183
70. Cláudio AFM, Ferreira A, Shahriari S, Freire MG, Coutinho JAP (2011) Critical assessment of the formation of ionic liquid based aqueous two phase systems in acidic media. J Phys Chem B 115:11145–11153
71. Harjani JR, Friscic T, MacGillivray LR, Singer RD (2008) Removal of metal ions from aqueous solutions using chelating task-specific ionic liquids. Dalton Trans 2008:4595–4601
72. Papaiconomou N, Svecova L, Bonnaud C, Cathelin L, Billard I, Chainet E (2015) Possibilities and limitations in separating Pt(IV) from Pd(II) combining imidazolium and phosphonium ionic liquids. Dalton Trans 44:20131–20138
73. Ando T, Kohno Y, Nakamura N, Ohno H (2013) Introduction of hydrophilic groups onto the ortho-position of benzoate anions induced phase separation of the corresponding ionic liquids with water. Chem Commun 49:10248–10250
74. Billard I, Gaillard C (2009) Actinide and lanthanide speciation in imidazolium-based ionic liquids. Radiochim Acta 97:355–359
75. Rout A, Binnemans K (2014) Liquid-liquid extraction of europium(III) and other trivalent rare-earth ions using a non-fluorinated functionalized ionic liquid. Dalton Trans 43:1862–1872
76. Gaillard C, Mazan V, Georg S, Klimchuk O, Sypula M, Billard I, Schurhammer R, Wipff G (2012) Acid extraction to a hydrophobic ionic liquid: the role of added tributylphosphate investigated by experiments and simulations. Phys Chem Chem Phys 14:5187–5199
77. Mazan V, Billard I, Papaiconomou N (2014) Experimental connections between aqueous–aqueous and aqueous ionic liquid biphasic systems. RSC Adv 4:13371–13384
78. Neves CMSS, Freire MG, Coutinho JAP (2012) Improved recovery of ionic liquids from contaminated aqueous streams using aluminium-based salts. RSC Adv 2:10882–10890
79. Freire MG, Carvalho PJ, Gardas RL, Marrucho IM, Santos LMNBF, Coutinho JAP (2008) Mutual solubilities of water and the [Cnmim][Tf2N] hydrophobic ionic liquids. J Phys Chem B 112:1604–1610
80. Billard I, Ouadi A, Gaillard C (2013) Is a universal model to describe liquid-liquid extraction of cations by use of ionic liquid at reach? Dalton Trans 42:6203–6212
81. Bonnaffé-Moity M, Ouadi A, Mazan V, Miroshnichenko S, Ternova D, Georg S, Sypula M, Gaillard C, Billard I (2012) Comparison of uranyl extraction mechanisms in ionic liquid by use of malonamide or malonamide-functionalized ionic liquid. Dalton Trans 41:7526–7536
82. Sypula M, Ouadi A, Gaillard C, Billard I (2013) Kinetics of metal extraction in ionic liquids: Eu3+/HNO3//TODGA/C1C4imTf2N as a case study. RCS Adv 3:10736–10744
83. Abulhassani J, Manzoori JL, Amjadi M (2010) Hollow fiber based liquid phase microextraction using ionic liquid solvent for preconcentration of lead and nickel from environmental and biological samples prior to determination by electrothermal atomic absorption spectrometry. J Hazard Mater 176:481–486
84. Dadfarnia S, Shabani AMH, Bidabadi MS, Jafari AA (2010) A novel ionic liquid/microvolume back extraction procedure combined with flame atomic absorption spectrometry for determination of trace nickel in samples of nutritional interest. J Hazardous Mat 173:534–538
85. Jensen MP, Borkowski M, Lazak I, Beitz JV, Rickert PG, Dietz ML (2012) Anion effects in the extraction of lanthanide 2-thenoyltrifluoroacetone complexes into an ionic liquid. Sep Sci Technol 47:233–243
86. Kozonoi N, Ikeda Y (2007) Extraction mechanism of metal ion from aqueous solution to the hydrophobic ionic liquid 1-butyl3-methylimidazolium nonafluorobutanesulfonate. Monatsh Chem 138:1145

87. Wu CT, Marsh KN, Deev AV, Boxall JA (2003) Liquid-liquid equilibria of room temperature ionic liquids and butan-1-ol. J Chem Eng Data 48:486–491
88. Jin X, Held C, Sadowski G (2013) Modeling imidazolium-based ionic liquids with ePC-SAFT. Fluid Phase Equilib 335:64–73
89. Rout A, Binnemans K (2014) Separation of rare earths from transition metals by liquid-liquid extraction from a molten salt hydrate to an ionic liquid phase. Dalton Trans 43:3186–3195
90. Scurto AM, Aki SNVK, Brennecke JF (2003) Carbon dioxide induced separation of ionic liquids and water. Chem Commun 2003:572–573
91. Mekki S, Wai CM, Billard I, Moutiers G, Yen CH, Wang JS, Ouadi A, Gaillard C, Hesemann P (2005) Cu(II) extraction by supercritical fluid carbon dioxide from a room temperature ionic liquid using fluorinated β-diketones. Green Chem 7:421
92. Wellens S, Thijs B, Möller C, Binnemans K (2013) Separation of cobalt and nickel by solvent extraction with two mutually immiscible ionic liquids. Phys Chem Chem Phys 15:9663–9669
93. Riisager A, Fehrmann R, Berg RW, Rv H, Wasserscheid P (2005) Thermomorphic phase separation in ionic liquid-organic liquid systems: conductivity and spectroscopic characterization. Phys Chem Chem Phys 7:3052–3058
94. Heintz A, Lehmann JK, Wertz C (2003) Thermodynamic properties of mixtures containing ionic liquids. 3. Liquid-liquid equilibria of binary mixtures of 1-ethyl-3-methylimidazolium bis(trifluoromethylsulfonyl)imide with propan-1-ol, butan-1-ol and pentan-1-ol. J Chem Eng Data 48:472–474
95. Crosthwaite JM, Aki SNVK, Maginn EJ, Brennecke JF (2004) Liquid phase behavior of imidazolium-based ionic liquids with alcohols. J Phys Chem B 108:5113–5119
96. Nockemann P, Pellens M, Hecke KV, Meervelt LV, Wouters J, Thijs B, Vanecht E, Parac-Vogt T, Mehdi H, Schaltin S, Fransaer J, Zahn S, Kirchner B, Binnemans K (2010) Cobalt (II) complexes of nitrile-functionalized ionic liquids. Chem Eur J 16:1849–1858
97. Lachwa J, Szydlowski J, Makowska A, Seddon KR, Esperança JMSS, Guedes HJR, Rebelo LPN (2006) Changing from an unusual high temperature demixing to a UCST-type in mixtures of 1-alkyl-3-methylimidazolium bis(trifluoromethylsulfonyl)amide and arenes. Green Chem 8:262–267
98. Lachwa J, Szydlowski J, Najdanovic-Visak V, Rebelo LPN, Seddon KR, da Ponte MN, Esperanca JM, Guedes HJ (2005) Evidence for lower critical solution behavior in ionic liquid solutions. J Am Chem Soc 127:6542–6543
99. Arce A, Earle MJ, Katdare SP, Rodriguez H, Seddon KR (2006) Mutually immiscible ionic liquids. Chem Commun 2006:2548–2550
100. Sui N, Huang K, Zhang C, Wang N, Wang F, Liu H (2013) Light middle and heavy rare earth group separation: a new approach via a liquid-liquid three phase system. Ind Eng Chem Res 52:5997–6008

Chapter 10
Surfactant Self-Assembly Within Ionic-Liquid-Based Aqueous Systems

Kamalakanta Behera, Rewa Rai, Shruti Trivedi, and Siddharth Pandey

Abstract Ionic liquids (ILs) have received increased attention from both academic and industrial research communities all over the world due to their unusual properties and immense application potential in various fields of science and technology. During the past decade, ionic-liquid-based systems have become the subject of considerable interest as a promising media for extraction and purification of several macro-/biomolecules. ILs are attractive designer solvents with tunable physicochemical properties. Using IL-based systems as alternative solvents for forming surfactant self-assemblies has several advantages. For example, the properties of surfactant self-assemblies in these media can be easily modulated by tuning the structure of ILs; ILs can dissolve a large variety of organic and inorganic substances and their properties are designable to satisfy the requirements of various applications. This may enhance the application potential of both ILs and surfactants in many important fields. Consequently, the study on surfactant self-assemblies within IL-based aqueous systems has attracted considerable attention in recent years. This chapter overviews the investigation carried out on the formation of surfactant self-assemblies within IL-based aqueous systems and their applications in various fields.

Keywords Ionic liquid • Surfactant • Micelle • Microemulsion • Self-assembly • Vesicles • Liquid crystals • Phase behavior • Nanomaterial synthesis

10.1 Introduction

Aqueous biphasic systems (ABS) evolved in the past decades as alternatives to traditional liquid–liquid extraction techniques where volatile organic solvents are used [1]. ABS have been extensively used for the separation and purification of drugs and/or biomolecules, such as proteins, nucleic acids, and antibodies, among others [1–3]. Since ABS are mainly composed of water, these are widely recognized as biocompatible media for biologically active substances [1–3]. Conventional

---

K. Behera • R. Rai • S. Trivedi • S. Pandey (✉)
Department of Chemistry, Indian Institute of Technology Delhi, HauzKhas, New Delhi 110016, India
e-mail: sipandey@chemistry.iitd.ac.in

ABS have been largely exploited since the 1980s and mainly consist of two immiscible aqueous-rich phases based on polymer–polymer, polymer–salt, or salt–salt combinations [1]. Although both solutes are water soluble, they separate into two phases above a given concentration: one of the aqueous phases will be enriched in one of the solutes, while the other phase contains the second polymer or salt. Since the last decade, ionic-liquid-(IL)-based ABS have become the subject of considerable interest due to the unusual and interesting properties associated to ILs [4–10]. Rogers and coworkers [4] first reported that the addition of inorganic salts to aqueous solutions of ILs induces the formation of ABS. Further, Abraham et al. [5] showed that IL-based ABS can be easily formed by both hydrophilic and hydrophobic ionic liquids using a salting-out agent. Since then, considerable effort has been made toward the use of ILs as possible phase-forming components to form ABS [6–8]. Although the use of a salting-out agent is a prime requirement for a hydrophilic IL to undergo phase separation in aqueous media, hydrophobic ILs can also form two phases when added to water beyond their solubility limit, even in the absence of any salting-out agent [9, 10]. Both types of IL-based aqueous systems, formed either by hydrophobic or hydrophilic ILs, were foreseen as new media in separation and also as new alternative for carrying out metathesis reactions to form new ILs [4, 6].

ILs are generally salts with their ions poorly coordinated and remain in liquid state at temperatures below 100 °C. These are often termed as designer solvents as their physicochemical properties can be easily tuned just by tuning the structure of the cation and/or the anion. Their main physicochemical properties comprised a wide liquid temperature range, low vapor pressure, high thermal and chemical stabilities, high ionic mobility, good electrical conductivity, and improved solvation ability for a large matrix of compounds [11–14]. Moreover, their negligible vapor pressure and nonflammability at ambient conditions have contributed to their common epithet as "green solvents," and as a result, they have been viewed as alternative replacements for the volatile and hazardous organic solvents largely used in a wide range of applications. Therefore, ILs are applied in many important areas of science and technology, namely, in various synthetic and catalytic reactions, in the conversion of biomass into added-value chemicals, and in extraction and separation processes, among others [11, 14]. Nevertheless, one of the main advantages of using IL in the formation of ABS or other two-phase systems is the ability to tailor their polarities and affinities by choosing suitable combinations of the cation and the anion. This important feature is indeed the major benefit connected to IL-based ABS given the difficulty of overcoming the limited polarity range of polymer-based ABS [6].

Surfactant self-assembled structures known in the form of micelles, microemulsions (water-in-oil and oil-in-water types), vesicles, and lyotropic liquid crystals are a topic of major interest both in academic and in industrial research due to their interesting physicochemical properties and immense application potential as flow field regulators, solubilizing and emulsifying agents, membrane mimetic media, media for chemical analysis and synthesis, and nanoreactors for enzymatic reactions, among many others [15–17]. Due to their widespread application potential, major efforts have been invested in the past decades to investigate the formation, properties, and applications of surfactant self-assembled structures within neat ILs and various aqueous–IL systems [18–21]. The formation of aforementioned surfactant self-assembled nano- or microstructures within aqueous–IL systems may increase the applicability of these biphasic systems in various fields. This may further open up novel avenues for potential applications of ILs. Studies on surfactant self-assembly within IL-based ABS involving a hydrophilic IL have not been reported yet, whereas ample studies have been reported on the formation of self-assembled structures within aqueous two-phase systems formed by water and a hydrophobic IL [19, 20]. Depending upon the nature of the surfactant (ionic or zwitterionic or nonionic) and proportions of the above three components, i.e., IL, surfactant, and water, the surfactant self-assembled structures may be IL-in-water microemulsions (IL-swollen normal micelles where IL resides in the hydrophobic core of the aqueous normal micelles) or water-in-IL microemulsions (water-swollen reverse micelles where IL behaves as the continuous phase and water molecules form the pools of the reverse micelles) or simply IL-modified micelles or vesicles or liquid crystals (where opposite charged ions of IL resides within the micellar palisade layer and surface of the micelles) [18, 20]. In an IL-rich media, water-in-IL microemulsions are formed, while IL-in-water microemulsions are formed in a media containing water as the continuous phase and IL as the dispersed phase [20, 21].

10.2 Formation of Micelles, Vesicles, and Liquid Crystals

The ability of aqueous–IL systems to support surfactant self-assembly into a range of mesophase structures has been established as a widespread phenomenon. Significant amount of work has been reported on the formation of micelles, vesicles, and liquid crystals within aqueous systems involving a hydrophobic IL [19, 22–27]. Pandey et al. [19] reported that the anionic surfactant sodium dodecyl sulfate (SDS) and the zwitterionic surfactant N-dodecyl-N,N-dimethyl-3-ammonio-1-propanesulfonate (SB-12) form IL-modified normal micelles when added to water–IL (water-1-butyl-3-methylimidazolium hexafluorophosphate [bmim][PF_6]) mixtures above their critical micelle concentrations (CMC). The hydrophobic IL [bmim][PF_6] remains insoluble in water when present beyond 2 wt% and forms two phases at temperatures close to room temperature. It was shown that increasing the concentration of the above surfactants increases the miscibility of [bmim][PF_6] in water by forming micelles. Both hydrophobic and electrostatic forces were described to play a crucial role for the above phenomena [19]. It was shown that relatively larger micelles are formed by the anionic SDS, whereas zwitterionic SB-12 forms compact micelles within the aqueous–[bmim][PF_6] system, containing 5 wt% of the IL, as compared to the micelles formed in aqueous medium in the absence of ionic liquid. Addition of SB-12 to aqueous–[bmim][PF_6] results in a change in interaction, as the ions of the IL become involved in electrostatic attractive interactions, thus reducing the electrostatic interactions between the opposite charged ions of the zwitterionic head groups of SB-12 [19]. These electrostatic interactions result in a significant reduction in the polydispersity of the zwitterionic micelles, therefore forming more monodispersed compact micelles in the solution. In case of aqueous anionic SDS, the electrostatic attractions between anionic sulfate ions of SDS head group and the IL cation reduce the electrostatic repulsion, thus increasing the size of SDS micelles [19]. It was also stated that most of the [bmim][PF_6] locates closer to the head groups of both SDS and SB-12 within the micellar pseudo-phase, insignificant [bmim][PF_6] partitions into the micellar hydrophobic region [19]. Further, it was reported that nonionic surfactants, namely, Triton X-100 (TX-100) and Brij-35, form micelles within the same IL–water system with almost similar sizes as compared to those of their micelles formed in water [19]. The nonionic nature of these surfactants allows the IL to partition into the hydrophobic region of the micelles through hydrophobic interactions. Using various methods, such as steady-state molecular fluorescence, electrical conductance, dynamic light scattering (DLS), and transmission electron microscopy (TEM), the researchers [19] have confirmed the formation of IL-modified micelles.

Apart from the use of conventional surfactants, some block copolymers have also been used in the formation of self-assembled structures within aqueous–IL systems [22]. Lodge et al. [22] have investigated the formation of micelles by four poly[(1,2-butadiene)-*block*-ethylene oxide)] (PB–PEO) diblock copolymers within aqueous 1-ethyl-3-methylimidazolium bis(trifluoromethylsulfonyl)imide IL. From the presented DLS results, it was shown that all four copolymers form larger micelles, and, more importantly, the block copolymers show thermoreversible intact micelle shuttling between the ionic liquid and water phases.

The formation of vesicles within hydrophobic IL-based aqueous mixtures has been reported by few research groups [23–25]. Zheng et al. [23] have reported the formation of vesicles by an IL-based cationic surfactant, 1-dodecyl-3-methylimidazolium bromide, within water–IL systems formed by a hydrophobic IL (1-butyl-3-methylimidazolium 2-naphthalenesulfonate [bmim][Nsa]) [23]. Using cryogenic transmission electron microscope (cryo-TEM), the authors demonstrated the formation of vesicles by the cationic surfactant [23]. The formation of vesicles was attributed to various intermolecular interactions taking place within the system, such as hydrophobic interactions between the hydrophobic anion of the IL and the hydrophobic part of the surfactant aggregates, electrostatic attractions of the cation of the surfactant with the IL anion, and π–π stacking interactions [23]. The authors [23] have also shown that the presence of the above hydrophobic IL brings about significant changes in some key physicochemical properties of the systems, such as critical micelle concentration (CMC), surface tension, size, and proton chemical shifts of the above aqueous cationic vesicles. In the same line, Pandey et al. [24] have shown the formation of vesicle-type giant structures by an anionic surfactant, sodium dodecylbenzenesulfonate (SDBS), within the water–[bmim][PF_6] system using DLS and TEM (Fig. 10.1a). In addition to the electrostatic forces between the IL cation and the anionic SDBS head group, π–π stacking between the IL cation and the phenyl moiety of SDBS was proposed to

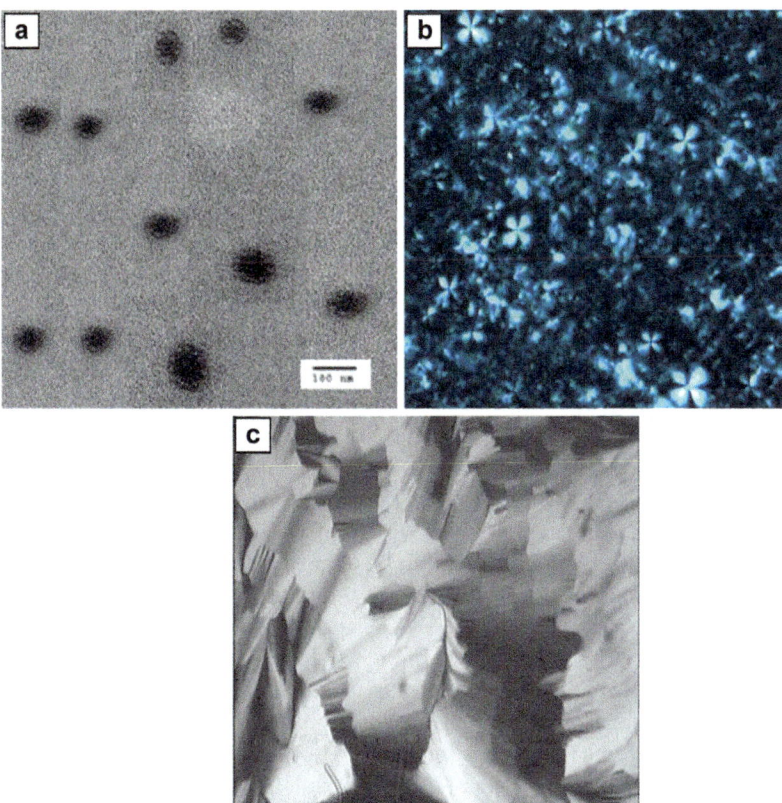

Fig. 10.1 (**a**) TEM images of 300 mM aqueous SDBS in presence of 144 mM [bmim][PF$_6$] (Reprinted with the permission from Ref. [24]. Copyright 2010 American Chemical Society). (**b**) Micrograph of lamellar phase in Brij-30/[bmim]PF$_6$]/H$_2$O system. The weight ratio of Brij-30: [bmim]PF$_6$]H$_2$O = 54:6:40 (Ref. [26] – Reproduced by permission of Springer Science + Business Media, LLC, Copyright @ 2007). (**c**) Textures of hexagonal liquid crystals formed by Brij-97/[bmim][PF$_6$]/water with the composition of 48/12/40 wt % (Reprinted with the permission from Ref. [27]. Copyright 2005 American Chemical Society)

be the major reason for the formation of the gathered large micellar structure. It is noteworthy to mention that the presence of [bmim][PF$_6$] within the above micellar system resulted in drastic changes in the bulk as well as in the micro-fluidity of the micellar solutions. An increase in the bulk dynamic viscosity and increase in the microviscosity of the cybotactic region of the fluorophore 1,3-bis(1-pyrenyl)propane (BPP) were observed in the studied system [24]. Further, it was demonstrated that a block copolymer poly(butadiene-b-ethylene oxide) (PB–PEO) forms vesicles within an aqueous two-phase system formed with the hydrophobic IL 1-ethyl-3-methylimidazolium bis(trifluoromethylsulfonyl)imide [25]. Using cryo-TEM, it was observed that the robust and stable PB–PEO vesicles migrate across the liquid–liquid interface, with their IL interiors intact, and form a stabilized aqueous dispersion of vesicles enclosing microscopic IL pools [25].

Other research groups have reported the formation of liquid crystals within aqueous two-phase systems containing hydrophobic ILs [26–28]. Liquid crystals are liquid anisotropic compounds, whose properties are intermediate between the crystalline solid state and that of the liquid state. Various physical properties, such as the electric permittivity, the refractive index, the magnetic susceptibility, and the mechanical properties of a liquid crystal, depend on the direction in which these quantities are measured [29]. Ge et al. [26] have reported the microstructure of lamellar liquid crystal formed by a nonionic surfactant, Brij-30, within an aqueous biphasic system formed by [bmim][PF_6] (Fig. 10.1b) [26]. The effect of the IL on the lubrication properties of the lamellar mesophase was also investigated to illustrate its relationship with the microstructure obtained. It was reported that the structure strength of the lamellar phase is enhanced with increasing the concentration of Brij-30, thus improving the antiwear capacity of the lamellar phase. The lubrication properties are impaired with increasing water content due to the increased interlayer space and the penetration of water into the amphiphile bilayer [26]. Wang et al. [27] reported the formation of hexagonal liquid crystals by another nonionic surfactant, Brij-97, within aqueous–[bmim][PF_6] mixtures (Fig. 10.1c). Using small-angle X-ray scattering (SAXS) and rheological techniques, the above system was thoroughly investigated, and it was shown that [bmim][PF_6] is dominantly penetrated between the oxyethylene chains of the nonionic Brij-97 molecules [27]. It was further observed that the strength of the network of the hexagonal phase formed in the Brij-97/water/[bmim][PF_6] system is weaker than that of the Brij-97/water/[bmim][BF_4] system [27]. Warr et al. [28] have investigated the self-assembly of a triblock copolymer, $(EO)_{20}(PO)_{70}(EO)_{20}$ (P123), within water–IL systems formed by two hydrophobic IL, namely, [bmim][PF_6] and 1-butyl-3-methylimidazolium tris(pentafluoroethyl)trifluorophosphate ([bmim][FAP]). From SAXS data, it was observed that water-swellable micellar, hexagonal, and lamellar phases of P123 are found in the aqueous–[bmim][PF_6] system, whereas the aqueous–[bmim][FAP] system allows the formation of only a lamellar phase over a narrow composition range [28]. Most importantly, it was shown that the miscibility of [bmim][PF_6] and water was increased by P123 addition, and at sufficiently high P123 concentrations, a single lamellar phase forms in which [bmim][PF_6] and water are miscible in all proportions [28].

10.3 Formation of Microemulsions

Apart from the above self-assembled structures, surfactants may form microemulsions within aqueous systems [17–21, 30–35]. Emulsions are generally thermodynamically unstable mixtures of two immiscible liquids (usually water and an organic solvent) that form two phases: the dispersed phase as liquid droplets and the surrounding continuous phase. When a surfactant of appropriate structure and concentration is added to emulsions, it results in the formation of an optically transparent and thermodynamically stable solution called microemulsion. The surfactant forms a monolayer between the two immiscible liquids and thus enables the formation of either oil-in-water (o/w) or water-in-oil (w/o) microemulsions, depending upon the proportions of the above three components [17]. Sometimes, a cosurfactant is also used in addition to the surfactant in the formation of microemulsions. Microemulsions have been a topic of major research due to their enormous application potential in various fields, such as in synthetic and catalytic reactions, chemical separations and extractions, solubilization, material chemistry, nanotechnology, drug delivery, and biocatalysis, among others [17, 30].

Microemulsions formed within IL-based aqueous two-phase systems are of two types, i.e., IL-in-water or water-in-IL microemulsions, depending upon the proportions of IL, surfactant, and water [18–21]. Water-in-IL microemulsions are formed in an IL-rich media where this fluid acts as the continuous phase and water acts as the dispersed phase, while IL-in-water microemulsions are formed in a media containing water as the continuous phase and the IL as the dispersed phase. In these systems, ILs replace the toxic and environmentally damaging volatile organic solvents (e.g., cyclohexane, benzene, toluene, CCl_4, etc.) broadly used as the oil phase in the formation of microemulsions. Further, a variety of hydrophobic and hydrophilic substances that are insoluble in IL can be solubilized within the IL cores and water pools of microemulsions, respectively, thus broadening the utility of these designer solvents in various important areas of science and technology.

In contrast to the fewer number of reports found on the formation of micelles/vesicles within aqueous–IL systems, considerable efforts have been put forth in recent years to investigate the formation of microemulsions within hydrophobic IL–water systems and their characterization and application in various fields [30–45].

10.3.1 Characterization of Ionic-Liquid-Based Microemulsions

Several reports on the use of the hydrophobic IL [bmim][PF_6] as the oil phase were found for supporting surfactant self-assembly in the formation of IL-based microemulsions [31–38]. Behera et al. [31] presented a clear visual evidence of the formation of water-in-IL ([bmim][PF_6]/TX-100/water) microemulsions by the color change of cobalt chloride ($CoCl_2$) within the ternary system, based on the fact that $CoCl_2$ shows different colors for hexa-coordinated and tetra-coordinated complexes. Prior to the formation of microemulsions, i.e., in the absence as well as in the presence of lower concentrations of water (at $w_0 < 6$, where $w_0 = $[water]/[surfactant]), $CoCl_2$ shows blue color due to the formation of Co(II) tetra-coordinated complexes, whereas by further addition of water to the system, it shows pink color due to the formation of the hexa-coordinated aqua complexes

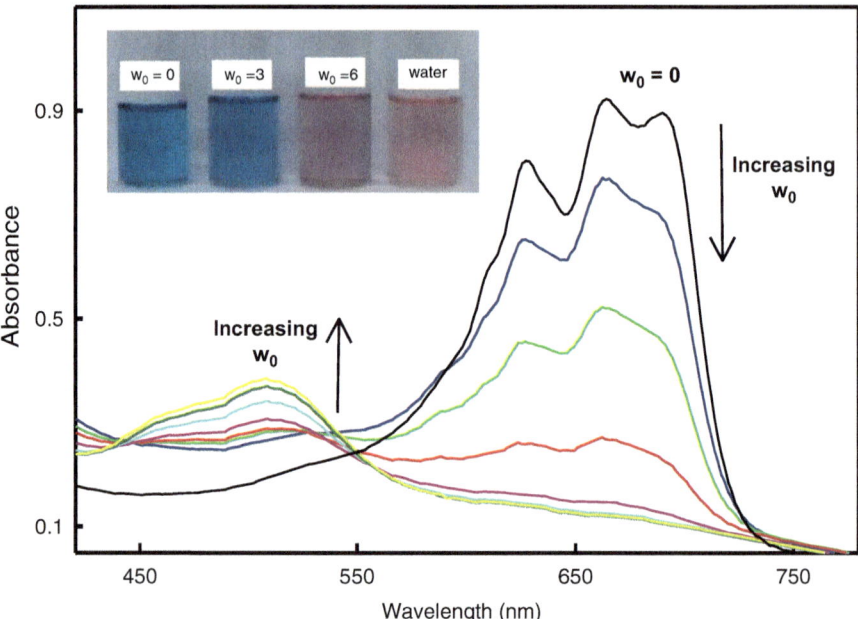

Fig. 10.2 Color change and UV–Vis absorbance spectra of Co(II) within [bmim][PF$_6$]/TX-100 (1.0 M)/water at different w_0 values (w_0 = 0, 2, 3, 4, 5, 6, 8, 10, and 12) (Reproduced from Ref. [31] by permission of John Wiley & Sons Ltd.)

[Co(H$_2$O)$_6$]$^{2+}$ within the water pools of the microemulsions [31]. This effect was further corroborated by the results obtained from the absorbance probe behavior of Co(II) within this system (Fig. 10.2) [31].

It was observed that before the formation of microemulsions, Co(II) shows three peaks around 630 nm, 665 nm, and 690 nm for its tetra-coordinated complexes. On further increasing water content ($w_0 \geq 6$), Co(II) preferentially partitions into the water pools of the water-in-IL microemulsions and is completely solubilized within the water pools, forming the pink hexa-coordinated [Co(H$_2$O)$_6$]$^{2+}$ complex that shows only one peak at 509 nm. Similar molar absorptivities resulting from the linear plots of the absorbance *versus* [Co(II)] for the [bmim][PF$_6$]/TX-100/water system with $w_0 \geq 6$ and for the aqueous medium, respectively, further confirm the complete solubilization of Co(II) within the water pools of the microemulsions. From the DLS results, it was shown that the size of microemulsions increases by increasing w_0, thus confirming that water molecules enter into the pool of the microemulsions formed [31].

Cyclic voltammetry was effectively utilized by Gao et al. [32, 33] to investigate the formation of microemulsions comprising [bmim][PF$_6$] and water stabilized by nonionic surfactants, Tween 20 and TX-100 [32, 33]. Suitable conditions for both the ternary systems to support the formation of water-in-IL and IL-in-water microemulsions were systematically studied using potassium ferrocyanide (K$_4$Fe(CN)$_6$) as an electroactive probe to measure apparent diffusion. From the diffusion

Fig. 10.3 (**a**) The diffusion coefficient of $K_4Fe(CN)_6$ as a function of water content with [bmim][PF_6]-to-TX100 weight ratio I = 0.18 and initial $K_4Fe(CN)_6$ quantity = 0.037 g (Reprinted with permission from Ref. [32]. Copyright 2005 American Chemical Society). (**b**) Electrical conductivity of a microemulsion as a function of water content at 35 °C, I = 0.20. The inset is the water content-dependent electrical conductivity of the water/[bmim][PF_6]/$C_{12}E_{23}$ microemulsions at different I values. (Reproduced from Ref. [34] by permission of the Royal Society of Chemistry)

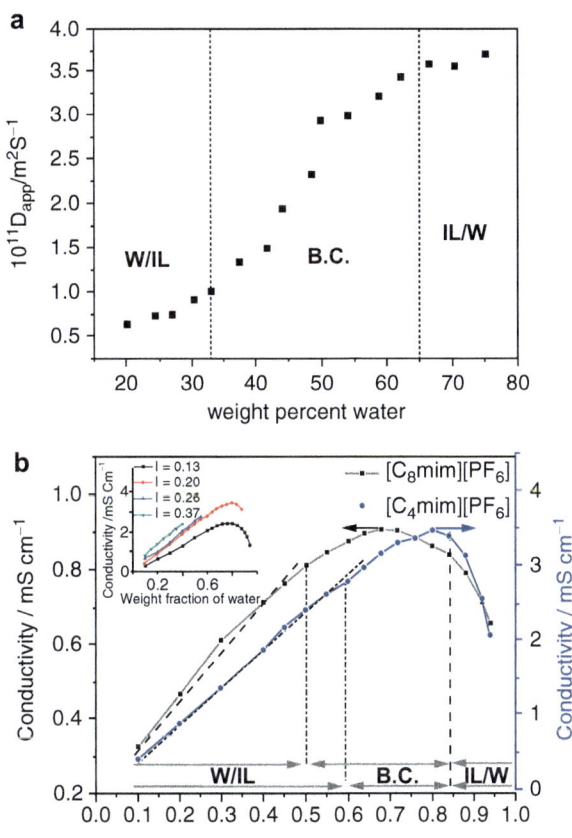

coefficient (D) of [$K_4Fe(CN)_6$] within the above ternary systems, it was observed that both of these systems exhibit large single-phase microemulsion domains with various sub-domains [water-in-IL (W/IL), IL-in-water (IL/W), and irregular bicontinuous (B.C.)] (Fig.10.3a) [32]. The experimental results indicated that the water domains are sufficiently well formed to support the solubility of conventional electrolytes in the water-in-[bmim][PF_6] microemulsions, which was further confirmed from the Ultraviolet-Visible (UV–Vis) absorbance spectral response of $K_3Fe(CN)_6$ within these systems. The formation of microemulsions by another nonionic surfactant polyoxyethylene alkyl ether ($C_{12}E_{23}$) was investigated by Yan et al. [34], within water–[bmim][PF_6] and water–[omim][PF_6] systems, using electrical conductivity (Fig. 10.3b). The authors [34] have presented the influence of the IL cation alkyl chain length on the microstructure of water/IL/$C_{12}E_{23}$ ternary systems. Using FTIR and UV–Vis spectroscopy, the existence of bulk water in both [bmim][PF_6]- and [omim][PF_6]-based $C_{12}E_{23}$-stabilized water-in-IL microemulsions was confirmed. DLS measurements revealed that under the same conditions, the water droplet size dispersed in [bmim][PF_6] is larger than that in [omim][PF_6] [34].

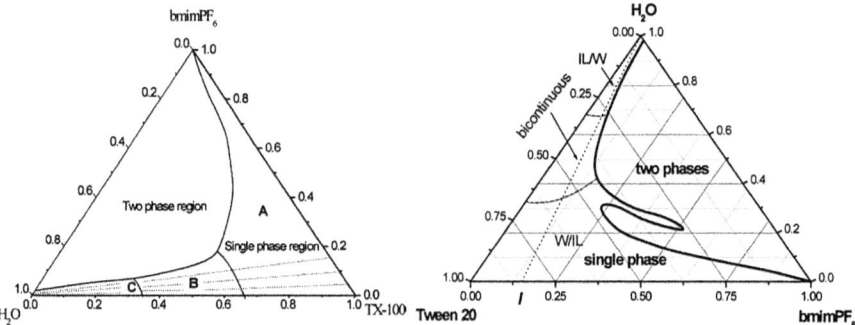

Fig. 10.4 Phase diagrams of the TX-100/[bmim][PF$_6$]/water (Reprinted with permission from Ref. [32]. Copyright 2005 American Chemical Society) and Tween 20/[bmim][PF$_6$]/water (Reproduced from Ref. [33] by permission of the Royal Society of Chemistry) three-component systems, respectively, at 30 °C

The phase behavior of both the TX-100/[bmim][PF$_6$]/water and Tween 20/[bmim][PF$_6$]/water systems was studied thoroughly [32, 33], as shown in Fig. 10.4. The phase diagram was constructed by titration of the "IL + surfactant" mixture with water. For each titration, the [bmim][PF$_6$]-to-surfactant weight ratio (I) was kept fixed. The phase boundaries were determined by observing the transition from turbidity to transparency or from transparency to turbidity.

Pandey et al. [35] have reported the formation of water-in-IL microemulsions within the ternary system containing the nonionic surfactant Brij-35, the hydrophobic IL [bmim][PF$_6$], and water using DLS, FTIR, and SAXS techniques. DLS data confirmed the presence of water-in-[bmim][PF$_6$] microemulsions within the system, and it was observed that size of aggregates increases with an increase in w_0 according to a linear trend [35]. The existence of aggregates in the presence of added water at each w_0 may imply that Brij-35 forms water-in-IL microemulsions within the aqueous–[bmim][PF$_6$] system. To further corroborate this hypothesis, SAXS measurements were performed on samples with no added water, at low w_0, and at high w_0. The existence of a single broad correlation peak followed by q^{-4} decay at higher q values in Brij-35-added [bmim][PF$_6$] solution was an indication of reverse micelle formation, as such a peak is a characteristic of the presence of reverse micelles [35]. Further, it was observed that the scattering intensity of the peak increases as water is added (at $w_0 = 5.2$) to Brij-35–[bmim][PF$_6$], and a careful observation of the SAXS data reveals a slight shift in the peak toward the lower q value, i.e., toward larger diameter sizes, showing the swelling behavior which is a characteristic of water-in-IL microemulsions [35]. FTIR absorbance spectra of water/Brij-35 (0.5 M)/[bmim][PF$_6$] system with different w_0 also afforded insights into the formation of water-in-IL microemulsions [35]. A careful observation of the FTIR absorbance data at $w_0 = 0$ revealed the presence of a broadband between 3350 cm^{-1} and 3570 cm^{-1} along with bands characteristic of [bmim][PF$_6$] (narrow bands at 2924 and 2853 cm^{-1} along with peaks between 3200 and 3000 cm^{-1} are characteristic of the C–H stretching mode of methyl, butyl, and aromatic rings, respectively). The broadband is indicative of "associated" or "bounded" water, corroborating the presence of reverse micelle-type aggregates of Brij-35 within [bmim][PF$_6$] [35]. Again, the origin of this "bounded" water was traced to the residual water present in Brij-35 and/or [bmim][PF$_6$] that may act as glue and thus helps in bringing the surfactant molecules together to form reverse micelles. Interestingly, no sharp peaks were seen in the range 3550 cm^{-1}–3700 cm^{-1} for 0.5 M Brij-35 in the [bmim][PF$_6$] system, strongly hinting at the absence of "free" water within the system; the presence of the "free" water was clearly highlighted by the peaks present in 3550–3700 cm^{-1} range for the ([bmim][PF$_6$] + water) system. It is important to note that, with an increase in w_0, the absorbance of the broadband in the range 3350–3570 cm^{-1} (characterizing the presence of "bound" water) increases along with the appearance of a peak at 3637 cm^{-1} corresponding to "free" water molecules. From the aforementioned observations, it is clear that both "free" and "bounded" water are present within 0.5 M Brij-35 in the water–[bmim][PF$_6$] system; the "free" water is the water that is miscible in neat [bmim][PF$_6$], whereas the "bounded" water is the water that helps to form water-in-[bmim][PF$_6$] microemulsions that swell as the w_0 is increased. In summary, the FTIR absorbance data confirmed the presence of bulk water as well as the swelling of the water-in-[bmim][PF$_6$] microemulsions as w_0 is increased [35].

Apart from the above reports on IL-based microemulsions, many investigations utilized dielectric spectroscopy, small-angle neutron scattering (SANS), transmission electron microscopy (TEM), and various other techniques to characterize the formation of microemulsions within the ternary system containing a nonionic surfactant, a hydrophobic ionic liquid, and water [36–39]. It is noteworthy to mention that microemulsions can also be formed within aqueous–IL systems using ionic and zwitterionic surfactants [40–45]. Goto et al. [40] reported the formation of IL-based microemulsions formed by water, the hydrophobic IL (1-octyl-3-methylimidazolium bis(trifluoromethylsulfonyl)imide ([C$_8$mim][Tf$_2$N]), and the anionic sodium bis(2-ethyl-1-hexyl) sulfosuccinate (AOT) as surfactant and with 1-hexanol as cosurfactant [40]. Rai et al. [41] used a zwitterionic surfactant, N-dodecyl-N,N-dimethyl-3-ammonio-1-propanesulfonate (SB-12), for the formation of water-in-IL microemulsions within aqueous–[bmim][PF$_6$] and aqueous–[emim][Tf$_2$N] systems, respectively, in the presence of ethanol. Further, Safavi et al. [42] reported on the formation of water-in-IL microemulsions by a cationic IL-based surfactant, 1-octyl-3-methylimidazolium chloride ([omim][Cl]), within the aqueous–[bmim][PF$_6$] system. Sun et al. [43] used a mixture of the cationic 1-tetradecyl-3-methylimidazolium bromide ([C$_{14}$mim]Br or TTAB) and the nonionic Triton X-100 for the formation of IL-based microemulsions. Another cationic surfactant, 1-decyl-3-methylimidazolium bromide ([C$_{10}$mim]Br), was used by Mao et al. [44] in the formation of [C$_{10}$mim]Br/[bmim][PF$_6$]/water microemulsions. Xue et al. [45] reported the formation of microemulsions by a mixture of AOT and TX-100 within the aqueous–[bmim][PF$_6$] system.

10.3.2 Excited-State Proton Transfer (ESPT) and Electron Transfer Reactions Within Ionic-Liquid-Based Microemulsions

Excited-state proton transfer (ESPT) in a confined medium helps in understanding many biological processes [46]. Pyranine, with a pK_a ~ 7.4 in the ground state and 0.4 in the first excited state, has been widely used as an ESPT probe [35, 46]. In bulk water, pyranine shows an intense emission peak by its deprotonated format at 520 nm and a weak emission peak of protonated form at 430 nm (Fig. 10.5) [35]. Since ESPT is governed by solvation and dynamics of the ion pair, this is considered to be markedly different in a nanoconfined environment because of the restricted mobility, slower solvation, and proximity of the ion pair [46].

Water-in-IL microemulsions may form highly versatile reaction media for ESPT as they have potential to be used in many applications. The combination of inimitable properties of nano-/micrometer-sized water domains in an IL continuous phase may result in unusual properties and outcomes that cannot be obtained using conventional w/o microemulsions. It has been well established that ESPT of pyranine does not occur in neat ILs and in neat common organic solvents.

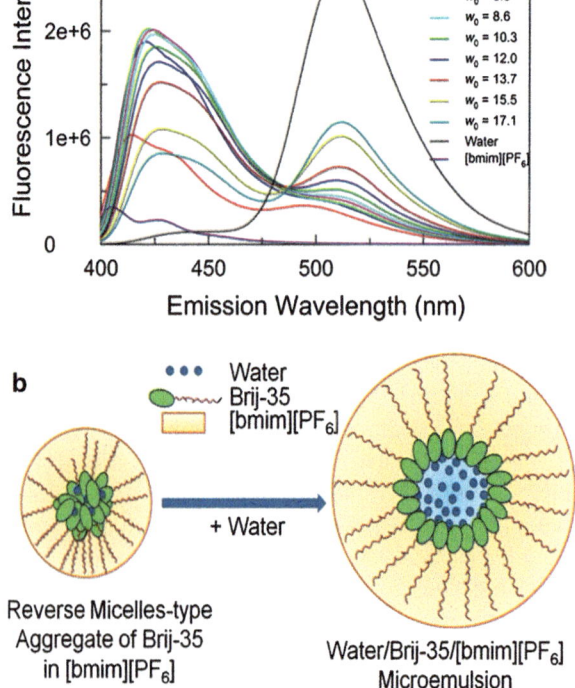

Fig. 10.5 (a) Fluorescence emission spectra of pyranine (10 µM; $\lambda_{ex} = 390$ nm) within water/Brij-35 (0.5 M)/[bmim][PF$_6$] system at different w_0 values, within neat [bmim][PF$_6$], and within water. (b) Cartoon showing formation of water-in-IL microemulsions from Brij-35 reverse micelles (Reprinted with permission from Ref. [35]. Copyright 2014 American Chemical Society)

However, in very recent studies, water-in-IL microemulsions have found application in ESPT of pyranine [35]. The steady-state fluorescence spectra of pyranine in [bmim][PF$_6$] show no ESPT even though water addition is up to its maximum solubility in the IL. However, it is interesting to note that the addition of water in IL in the presence of 0.5 M Brij-35 results in ESPT [35]. These outcomes indicate that the interfacial water of water-in-[bmim][PF$_6$] microemulsions helps in the deprotonation of pyranine and the efficiency of deprotonation gradually increases with the water content. Despite the proton transfer reaction in microemulsions, photoinduced electron transfer reaction between coumarin-153 (C-153) and a perylenetetracarboxylic diimide (PDI) compound, with an attached hydrophilic polyoxyethylene group at the imide nitrogen position, has also been investigated in ternary microemulsions formed by an IL ([bmim][PF$_6$]/TX-100/water) using steady-state electronic absorption and fluorescence spectroscopy [47]. In this study, both PDI and C-153 were found to reside at the interface between the surfactant TX-100 and [bmim][PF$_6$] within the microemulsion. An efficient electron transfer and a less efficient energy transfer from C-153 to PDI were observed [47].

10.3.3 Solvation Dynamics and Rotational Reorientation Within Ionic-Liquid-Based Microemulsions

The estimation of solvation dynamics using ultrafast spectroscopic techniques has been the subject of numerous recent studies [48–53]. Usually, measurements of solvent relaxation of a fluorescence probe that exhibits much smaller dipole moment in ground state in comparison to that in excited state are a useful technique to reveal the dynamics of solvent molecules surrounding the solutes in any solubilizing media. Solvation dynamics in the most fundamental solvent water and in ILs have been addressed by many researchers [48–50]. Both sub-picosecond and nanosecond timescale components have been proposed for the solvation dynamics within ILs. Recently, evidence for the collective motion of both IL cation and anion on sub-picoseconds has been suggested on theoretical and experimental basis [48]. The solvation behavior of ordinary w/o microemulsions was found to strongly depend on water-to-surfactant ratio. Also, it has been found that the water in the core of microemulsions requires a long bimodal relaxation time in comparison to pure water. These observations in neat ILs and microemulsions have motivated researchers to investigate solvation dynamics of IL-based microemulsions [49, 50]. Sarkar et al. [51–53] reported on solvation dynamics of IL-in-water and water-in-IL microemulsions using various forms of coumarin dyes as probes. The authors [51] discussed the behavior of micro-regions of [bmim][PF$_6$]/water microemulsions on the basis of solvation dynamics and rotational relaxation of coumarin-153 (C-153) and coumarin-151 (C-151). Very small variations in the solvent relaxation time constants and almost no change in rotational relaxation time of C-153 were observed with increase in the [bmim][PF$_6$]/TX-100 ratio (R)

[51]. The constant solvation and rotational relaxation times of C-153 showed that the location of C-153 remains the same with an increase in R, and most probably, the location of probe is at the interfacial region of [bmim][PF$_6$] and TX-100 in the microemulsions. On the other hand, in case of C-151, the fast component of the solvation time gradually increases and the slow component gradually decreases with an increase in R. However, the change in solvation time is small in comparison to that of microemulsions containing common polar solvents, such as water, methanol, acetonitrile, etc. The rotational relaxation time of C-151 was found to increase with increase in R. This indicates that the number of C-151 molecules in the core of the microemulsions gradually increases with an increase in the [bmim][PF$_6$] content [51].

The interactions of the IL and water in water/[bmim][PF$_6$] microemulsions have also been studied on the basis of solvation and rotational dynamics of C-153 and C-490 as fluorescence probes [52, 53]. In contrast to that observed for IL-in-water microemulsions, both the rotational relaxation and solvation times of these probes gradually decrease with increasing water content in the microemulsions. However, in comparison to pure water, the reduction of solvation time in water-in-IL microemulsions is very less. The solvation and rotational dynamics of IL-in-water microemulsion were also found to depend on the nature of the surfactant used. In case of IL-in-water microemulsions composed of [bmim][PF$_6$]/Tween 20/water, solvation and rotational relaxation times estimated using C-153 as probe were found to increase with increase in [bmim][PF$_6$]-to-Tween 20 molar ratio. This behavior of IL/Tween 20 microemulsion is different from that observed for ionic liquid/TX-100 microemulsion [52, 53]. Thus, overall observations indicate that the solvation behavior within IL-in-water and water-in-IL microemulsions depends on the type of microemulsions, nature of probe, nature of surfactant, and IL-to-surfactant ratio.

10.4 Applications of Ionic-Liquid-Based Microemulsions

Microemulsions formed within aqueous–IL systems are widely used in various fields, such as in the synthesis of nanomaterials, analytical separations and extractions, biocatalysis, and electropolymerization, among many others [30, 54–73].

10.4.1 Nanomaterial Syntheses

Due to their extraordinary physical and chemical properties, nanoparticles are widely used in catalysis, photonics, optoelectronics, micro-/nanoelectromechanical systems, biological labeling, and information storage, among others [54, 55]. The key factors that condition their properties and determine their applicability are their shape, composition, size, and size distribution. Consequently, the synthesis of nanoparticles of customized shape and size has long been a scientific and technological challenge. Microemulsions have been utilized in synthetic methods to prepare homogeneous and monodispersed small nanoparticles. Water-in-IL microemulsions that contain water droplets of a few nanometers, and act as nanoreactors, were used extensively in various synthesis pathways to form nanoparticles [56–63]. These nanoreactors help in limiting the size and size distribution of the synthesized particles.

Nanoparticles are usually synthesized from reactions initiated by chemical reducing agents, which are quite aggressive. In order to avoid the use of chemical reducing agents, the direct reduction of ions within the water pool of microemulsions has been recently proposed as an alternative method [56]. The high conductivity of ILs over organic solvents helps increase the conductivity of the system allowing significantly higher deposition rates, whereas the higher viscosity of ILs can favor the noncoalescence of the droplets in an aqueous solution. Recently, Serrá et al. [57] used this method to synthesize highly crystalline HCP alloyed CoPt nanoparticles in the 10–120 nm range with a rather narrow size distribution by using nanoreactors of water-in-IL microemulsions containing droplets of aqueous solution [electrolytic solution containing Pt(IV) and Co(II) ions] with the IL [bmim][PF$_6$], as depicted in Fig. 10.6.

Zhang et al. [58] reported the formation of monodispersed Pd nanoparticles of ~3 nm by electrodeposition using water/TX-100/[bmim][PF$_6$] microemulsions. According to the authors [58], the medium containing the Pd nanoparticles has the catalytic application in the well-known Heck reactions of butyl acrylate with iodobenzene. The water-in-IL microemulsions were also used in direct formation of

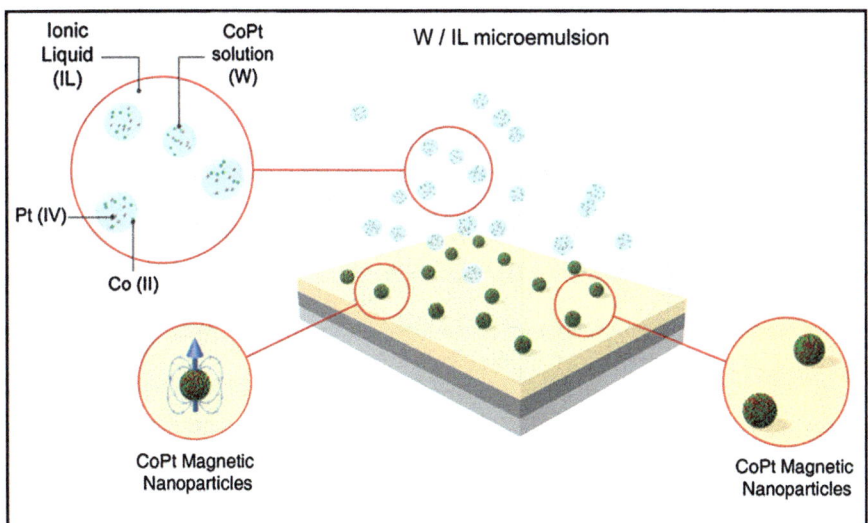

Fig. 10.6 Schematic representation of the electrochemical synthesis of magnetic CoPt nanoparticles in water-in-IL microemulsions (Reprinted with permission from Ref. [57]. Copyright 2014 American Chemical Society)

gold nanoparticles as well as in the formation of cross-linked starch microspheres [58]. The cross-linked starch microspheres that possess good performance in nontoxicity, biodegradability, stability during storage, etc., in comparison to natural starch, are useful in many areas, especially in drug delivery systems.

The electrodeposition method using microemulsions containing water, IL, and surfactant was also used in the growing of nanorods of metals and alloys of extreme porosity [61]. Nanorods prepared by this method present a very good corrosion resistance and stability, with promising applications as electrocatalysers [61]. On the other hand, the synthesis of core–shell structure nanocomposites of organic conjugated polymers and inorganic nanocrystals has attracted great attention because of their potential applications in photonics, photoelectronics, catalysis, and so on. IL-in-water microemulsions have also found application in the synthesis of polyaniline core decorated with TiO_2 (PANI–TiO_2) nanocomposite particles of ~70 nm diameter with strong component interactions [62]. In this method, IL-in-water microemulsions help on the uniform distribution of TiO_2 nanoparticles on the surface of PANI. Cyclic voltammetric studies of the nanocomposites formed using IL-in-water microemulsions revealed their better electrochemical catalytic activity than PANI. Further, Li and coworkers [63] reported the application of water-in-IL (water/TX-100/[bmim][PF_6]) microemulsions for the synthesis of silica microrods with nanosized pores. It was emphasized that [bmim][PF_6] helps in defining the final morphology of silica microrods as the imidazolium species form strong complex with silica surfaces during tetraethyl orthosilicate (TEOS) hydrolysis [63].

10.4.2 Analytical Applications

IL-based microemulsions have further attracted the scientific community due to their application potential in analytical techniques [64–67]. Water-in-IL or IL-in-water microemulsions have been utilized as pseudo-stationary phases (PSP) in microemulsion electrokinetic capillary chromatography (MEEKC). MEEKC, an extended form of capillary electrophoresis (CE), is a separation technique that is based on both electrophoresis and chromatography. This method provides good selectivity and high efficiency for the separation of anionic, cationic, as well as neutral analytes. The potential of IL microemulsion-based MEEKCs is mostly observed in the separation of biomolecules. Li et al. [64] developed a MEEKC method utilizing microemulsions with [bmim][PF_6] as the oil phase and boric acid buffer as the polar phase. This method afforded the simultaneous determination of nucleobases, nucleosides, and nucleotides in biological fluids and herbal medicines, developing the interest for greener analytical methods for various pharmaceutical analyses and disease diagnoses. Li and coauthors [64] further utilized this microemulsion in the determination of three curcuminoids in the rhizome and tuberous root of *Curcuma longa* L. [65]. According to the authors [65], the

MEEKC method of separation is more efficient than CE since the microemulsions significantly protect analytes, even in alkaline conditions. It was finally proposed that microemulsions with [bmim][PF$_6$] have more protective effects on analytes in comparison to organic solvent *n*-octane used as PSP in MEEKC. The efficacy of MEEKC has also been shown in resolution of various isomers [66]. Wang et al. [66] have simultaneously determined the three isomers α-, β-, and γ-asarone in *Acorus tatarinowii* by using MEEKC with [bmim][PF$_6$] as the oil phase.

IL-based microemulsions have also been employed for liquid–liquid extraction of proteins [67]. Proteins can be readily extracted and back-extracted using microemulsions. In addition, the biological activities of proteins remain the same after the extraction process. Wang and coworkers [67] have employed a water/AOT/[bmim][PF$_6$] microemulsion for the selective extraction of hemoglobin. It was reported an extraction of 96 % of hemoglobin from an aqueous solution at pH 6.3 [67]. 73 % of the hemoglobin transferred into the microemulsion system was then back-extracted into an aqueous phase with the addition of 6 M urea as the stripping reagent. The potential of water/AOT/[bmim][PF$_6$] reverse microemulsion system for protein isolation was finally investigated for the extraction of other proteins: cytochrome c, albumin bovine serum, and transferrin [67]. However, except for hemoglobin, the transfer of other proteins from the aqueous phase into the microemulsion was not observed. It was reported that the extraction efficiency of proteins depends on the concentration of surfactant in the microemulsion and the aqueous/microemulsion phase ratio. The authors [67] suggested that the efficiency of extraction of the microemulsion with IL is much greater than the efficiency when microemulsions with ordinary solvents are used. The mechanism proposed for the protein extraction is shown in Fig. 10.7 [67].

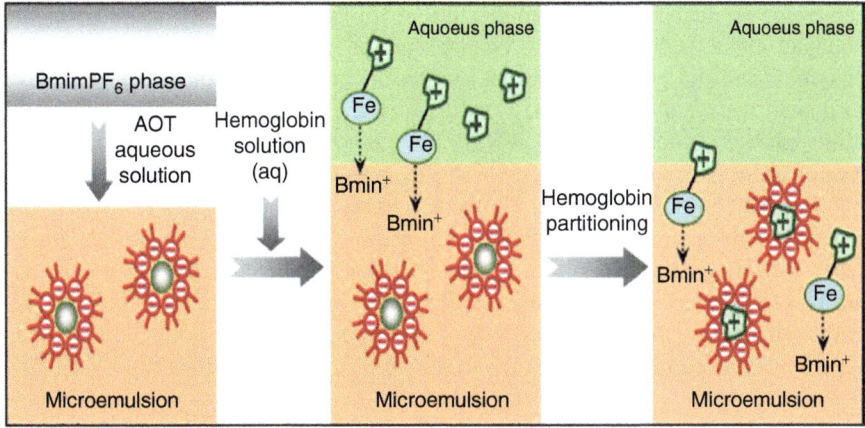

Fig. 10.7 The entrapping process of hemoglobin by the water/AOT/[bmim][PF$_6$] reverse microemulsion system showing two distribution states of hemoglobin (Reprinted from Ref. [67], Copyright 2008, with permission from Elsevier)

10.4.3 Application in Biocatalysis

The main advantage of using microemulsions in the field of biocatalysis is due to the enzymatic conversions of hydrophobic substances that show limited solubility in ordinary solvents. However, the major drawback of using microemulsions with ordinary solvents as reaction media is the poor product recovery and difficulty in recycling enzymes. It is well established that the enzymatic catalytic effect is more significant in some ILs in comparison to ordinary solvent. Moreover, IL media facilitate the product recovery and reuse of the biocatalyst. These solvents also enhance the thermal and the operational stabilities as well as regio- or enantioselectivities of enzymes [68–70]. It has also been observed that the catalytic activity of enzymes is more efficient in hydrophobic ILs in comparison to hydrophilic ones [68, 69]. As a consequence, reverse micelles with IL as a continuous phase have garnered interest of researchers toward investigation of biocatalysis within IL-based microemulsions.

Moniruzzaman and coworkers [70] were the first to propose the use of IL-based microemulsions in enzyme catalysis. They have presented the enzyme activity of *horseradish peroxidase* (HRP) encapsulated in water-in-IL microemulsions composed of water/AOT/1-octyl-3-methylimidazolium bis(trifluoromethylsulfonyl) imide ([omim][Tf$_2$N])/1-hexanol using pyrogallol as substrate. In this system, the catalytic effect of HRP was found to be dependent on w_0, pH, and content of 1-hexanol. The authors [70] found that the enzyme retains almost 70% of its activity, and HRP-catalyzed oxidation of pyrogallol is much more effective in the newly formed microemulsions than that observed in the conventional water/AOT/isooctane microemulsion. According to the authors [70], the high efficiency of HRP within the neoteric water-in-IL microemulsion is due to the partitioning of the substrate, products, and other molecules involved in the reaction between the aqueous phase and the IL, the changes in the microenvironment surrounding enzymes, and the presence of 1-hexanol in the solution. Huang et al. [69] have evaluated the catalytic efficiency of lipase in the lipase-catalyzed hydrolysis of 4-nitrophenyl butyrate (p-NPB) by employing water-in-[bmim][PF$_6$] microemulsions stabilized by both AOT and Triton X-100. In this method, microemulsions have the ability to dissolve both hydrophobic and hydrophilic reagents forming a large lipid–water interface that favors the lipase-catalyzed reactions. Moreover, the composite interfacial membrane consisting of two different types of surfactants facilitates the interfacial regulation of the catalytic performance of lipase. Also, the catalytic activity of lipase from *Candida rugosa* in the IL-based microemulsion has been demonstrated to be much better than that in the water-saturated hydrophobic IL [69]. The authors [69] also studied the transesterification of vinyl acetate with benzyl alcohol catalyzed by lipase hosted in AOT/TX-100 stabilized microemulsions. The respective reaction in microemulsions produced benzyl acetate with high yield, showing IL-based microemulsions to be a promising media for the lipase-catalyzed hydrolytic reaction. Huang and coworkers proposed also other microemulsion formed by 1-tetradecyl-3-methylimidazolium bromide and cationic and neutral surfactants at 35 °C [69]. The authors [69] have proposed that water solubilization capacity of microemulsions increases with increasing TX-100 concentration, enlarging the monophasic region. They [69] used these microemulsions for investigating the catalytic effect of laccase. The interface of the water-in-IL microemulsions was found to have an inhibitory effect on the expression of the laccase activity. Another promising medium composed of water-in-ionic liquid microemulsions formulated with two nonionic surfactants Tween 20 and TX-100 in [bmim][PF$_6$] was studied by Pavlidis and coworkers. The authors studied the catalytic activity of lipases from *C. rugosa*, *C. viscosum*, and *T. lanuginosa* in this microemulsion and reported the higher catalytic activity and operational stability in this novel media. These observations were confirmed by circular dichroism and FTIR spectroscopy indicating the protective environment of water-in-IL microemulsion for the enzyme. In this study, the structural stability of enzyme in water-in-IL microemulsions was compared with that in w/o microemulsions as well as in surfactant-free systems. In all cases, other than water-in-IL microemulsions, decrease in α-helix (up to 32 %) and increase in β-sheet (up to 35 %) content were observed. These observations were found to be more prominent in surfactant-free system. Overall, these observations indicate that ILs can be efficiently used as the reaction media in different biocatalysis processes.

10.4.4 Application in Electropolymerization

Conducting polymers have been extensively studied due to their widespread applications in sensors, supercapacitors, and electrochromic devices [71]. In general, the chemical polymerization or electrochemical polymerization is employed to prepare polymeric materials with various physicochemical properties depending on the characteristics of the monomer. However, electropolymerization has significant advantage in that the growth of polymer can be achieved on an electrode substrate. In this method, electrolyte strongly affects the properties of polymers for the electrochemical process. Traditionally, the electrochemical polymerizations are studied in aqueous solution or organic solvents. However, in these solvents, polymerization is not straightforward because of the low solubility of monomers in water and due to detrimental effect of organic solvents on the environment. ILs have attracted considerable interest in several electropolymerization studies because of their special physicochemical properties. Use of ILs in this field has the advantage that the IL can act both as a growth medium and as an electrolyte leading to significantly altered morphologies and enhanced electrochemical properties. Nevertheless, the electrochemical polymerization in ILs has the disadvantages of higher viscosity, lower conductivity, and weaker solubility for some organic and inorganic compounds that limit the study of this method in ILs. These obstacles have been overcome by using IL-based microemulsions.

Bin Dong and coworkers [72] investigated the electrosynthesis of conducting polymers using microemulsions of water/Tween 20/[bmim][PF_6]. It was described that the high conductivity (k) after the formation of microemulsions with the addition of [bmim][PF_6] in aqueous Tween 20 helps the electropolymerization of 3,4-ethylenedioxythiophene (EDOT). Based on the ionic conductivity results, the authors performed anodic polarization measurements of EDOT in three types of IL microemulsions: IL-in-water, bicontinuous and water-in-IL, and Tween 20 aqueous solution. It was evident that for the IL-in-water microemulsion, the oxidation onset of EDOT was the least, while the current density was the largest. As to water-in-IL, the oxidation onset (1.31 V) was found to be higher than that in Tween 20 aqueous solution (1.14 V). Moreover, no polymer deposition was visually observed on the electrode in water-in-IL. Usually, the lower oxidation potential of the monomer can prevent side reactions leading to the formation of higher-quality polymer films (PEDOT). Therefore, among the three types of IL microemulsions investigated, IL-in-water was found the most suitable for the electropolymerization of EDOT followed by the bicontinuous one. Conducting polymer morphologies were also investigated in this study [71] using scanning electron microscopy (SEM). Two different surface structures, namely, large bundles within the aggregation of small, raised PEDOT grains (ca. 100 nm) and smooth, continuous PEDOT surface, were observed for PEDOT synthesized in the IL-in-water phase. However, PEDOT synthesized in bicontinuous phase was found to exhibit flat, compact, and continuous PEDOT surface (Fig. 10.8) [72]. The differences of PEDOT films synthesized in IL-in-water and bicontinuous microemulsions were reflected in the microdroplet structure of IL-in-water and the IL matrix in the bicontinuous sample. These nanostructures of compact PEDOT films improve electron transfer capability, implying their potential application in ion-selective electrodes, ion-sieving films, and matrices for hosting catalyst particles. Further, authors [73] reported the electrosynthesis of another conducting polymer, poly(3-methoxythiophene) [PMOT], again by direct anodic oxidation of 3-methoxythiophene (MOT) in the water/Tween 20/[bmim][PF_6] system. Electropolymerization of MOT was also found to be the most efficient in IL-in-water subregions of the microemulsions. The PMOT formed has an electrical conductivity of 3.8 S/cm and can be dissolved in many conventional solvents with green light-emitting properties [73]. On the basis of the above observations, it becomes obvious that IL-in-water microemulsions remarkably reduce the amount of expensive IL required and represent a novel microenvironment for electropolymerization. Conducting polymers of various morphologies prepared in this neoteric media would be advantageous in different electrochemical processes.

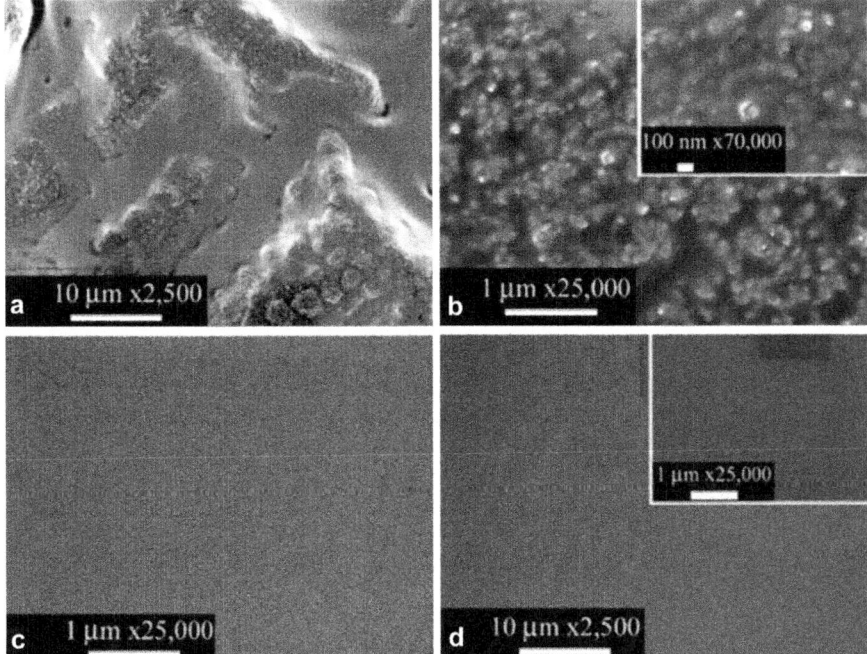

Fig. 10.8 Scanning electron micrographs of PEDOT films on ITO electrode synthesized in ionic-liquid in water (**a**, **b**, and **c**) and bicontinuous (**d**). The insets show the SEM photographs at a higher magnification (Reprinted from Ref. [72], Copyright 2008, with permission from Elsevier)

10.5 Conclusions

Formation of surfactant self-assemblies within aqueous–IL systems is turning out to be of utmost importance due to the immense application potential of both ILs and surfactants in various fields of science and technology. Using aqueous–IL systems as alternatives for conventional solvents for forming surfactant self-assemblies has many advantages. For example, the properties of surfactant self-assemblies in these media can be easily tuned by tuning the chemical structure of ILs. Moreover, ILs can dissolve a variety of organic and inorganic substances. Reports dealing with the formation of a variety of surfactant self-assemblies, such as micelles, vesicles, liquid crystals, and microemulsions, within aqueous–IL systems and their applications in various fields are summarized in this chapter. In most of the reports, the formation of surfactant self-assembled structures within aqueous–IL systems involves a hydrophobic IL without the use of any salting-out agent. No work is reported yet on surfactant self-assembly within IL-based ABS in the presence of a salting-out agent, representing thus an alternative and new path of research. Accordingly, this chapter mainly highlights the formation of the aforementioned surfactant self-assembled structures within aqueous hydrophobic IL-based systems and their applications in various important fields, e.g., nanomaterial synthesis, biocatalysis, analytical applications, etc. It is shown that the formation of the above self-assembled structures within aqueous hydrophobic IL-based systems mainly depends on the nature of the surfactant (ionic or nonionic or zwitterionic) and the proportions of the three components, i.e., IL, water, and surfactant. Several techniques were used to characterize the formation of microemulsions within these systems. A brief report on solvation dynamics and proton transfer reactions within IL-based microemulsions was also presented. Surfactant self-assembly in aqueous–IL systems appears to have enormous potential in a variety of disciplines in science and technology.

Acknowledgments Siddharth Pandey thanks the Department of Science and Technology (DST), Government of India (grant number SB/S1/PC-80/2012), and the Council of Scientific and Industrial Research (CSIR), Government of India [grant no. 01(2767)/13/EMR-II], for generously supporting his work on surfactant self-assembly in aqueous-ionic-liquid systems.

References

1. Albertson PA (1986) Partition of cell particles and macromolecules. Wiley-Interscience, New York
2. Zaslavsky BY (1995) Aqueous two-phase partitioning; physical chemistry and bioanalytical applications. Marcel Dekker, New York
3. Walter H, Brooks DE, Fisher D (1985) Partitioning in aqueous two phase system. Academic, New York
4. Gutowski KE, Broker GA, Willauer HD, Huddleston GJ, Swatloski RP, Holbrey JD, Rogers RD (2003) Controlling the aqueous miscibility of ionic liquids: aqueous biphasic systems of water-miscible ionic liquids and water-structuring salts for recycle, metathesis, and separations. J Am Chem Soc 125:6632–6633
5. Abraham MH, Zissimos AM, Huddleston JG, Swatloski RP, Willauer HD, Rogers RD, Acree WE Jr (2003) Some novel liquid partitioning systems: water-ionic liquids and aqueous biphasic systems. Ind EngChem Res 42:413–418
6. Freire MG, Claudio AFM, Araujo JMM, Coutinho JAP, Marrucho IM, Canongia-Lopes JM, Rebelo LPN (2012) Aqueous biphasic systems: a boost brought about by using ionic liquids. Chem Soc Rev 41:4966–4995
7. Shadeghi R, Golabiazar R, Shekaari H (2010) The salting-out effect and phase separation in aqueous solutions of tri-sodium citrate and 1-butyl-3-methylimidazolium bromide. J Chem Thermodyn 42:441–453
8. Pereira JFB, Lima AS, Freire MG, Coutinho JAP (2010) Ionic liquids as adjuvants for the tailored extraction of biomolecules in aqueous biphasic systems. Green Chem 12:1661–1669
9. Freire MG,Neves CMMSS, Carvalho PJ, Gardas RL, Fernandes AM, Marrucho IM, Santos LMNBF, Coutinho JAP (2007) Mutual solubilities of water and hydrophobic ionic liquids. J Phys Chem B111:13082–13089
10. Freire MG, Santos LMNBF, Fernandes AM, Coutinho JAP, Marrucho AM (2007) An overview of the mutual solubilities of water–imidazolium-based ionic liquids systems. Fluid Phase Equilib 261:449–454
11. Welton T (1999) Room-temperature ionic liquids; solvents for synthesis and catalysis. Chem Rev 99:2071–2083
12. Seddon KR (2003) Ionic liquids: a taste of the future. Nat Mater 2:363–365

13. Wasserscheid P (2006) Chemistry: volatile times for ionic liquids. Nature 439:797
14. Dupont J, de Souza RF, Suarez PAZ (2002) Ionic liquid (molten salt) phase organometallic catalysis. Chem Rev 102:3667–3692
15. Moroi Y (1992) Micelles: theoretical and applied aspects. Springer, New York
16. Jones MJ, Chapman D (1995) Micelles, monolayers, and biomembranes. Wiley-LISS, New York
17. Shah DO (1998) Micelles, microemulsions, and monolayers. CRC Press, Boca Raton
18. Greaves TL, Drummond CJ (2008) Ionic liquids as amphiphile self-assembly media. Chem Soc Rev 37:1709–1726
19. Behera K, Kumar V, Pandey S (2010) Role of the surfactant structure in the behavior of hydrophobic ionic liquids within aqueous micellar solutions. ChemPhysChem 11:1044–1052
20. Qiu Z, Texter J (2008) Ionic liquids in microemulsions. Curr Opin Colloid Interface Sci 13:252–262
21. Hao JC, Zemb T (2007) Self-assembled structures and chemical reactions in room-temperature ionic liquids. Curr Opin Colloid Interface Sci 12:129–137
22. Bai Z, He Y, Lodge TP (2008) Block copolymer micelle shuttles with tunable transfer temperatures between ionic liquids and aqueous solutions. Langmuir 24:5284–5290
23. Gu Y, Shi L, Cheng X, Lu F, Zheng L (2013) Aggregation behavior of 1-dodecyl-3-methylimidazolium bromide in aqueous solution: effect of ionic liquids with aromatic anions. Langmuir 29:6213–6220
24. Rai R, Baker GA, Behera K, Mohanty P, Kurur ND, Pandey S (2010) Ionic liquid-induced unprecedented size enhancement of aggregates within aqueous sodium dodecylbenzene sulfonate. Langmuir 26:17821–17826
25. Bai Z, Lodge TP (2010) Polymersomes with ionic liquid interiors dispersed in water. J Am Chem Soc 132:16265–16270
26. Ge L, Chen L, Guo R (2007) Microstructure and lubrication properties of lamellar liquid crystal in Brij30/[Bmim]PF$_6$/H$_2$O system. Tribol Lett 28:123–130
27. Wang Z, Liu F, Gao Y, Zhuang W, Xu L, Han B, Li G, Zhang G (2005) Hexagonal liquid crystalline phases formed in ternary systems of Brij 97-water-ionic liquids. Langmuir 21:4931–4937
28. Sharma SC, Atkin R, Warr GG (2013) The effect of ionic liquid hydrophobicity and solvent miscibility on pluronic amphiphile self-assembly. J Phys Chem B 117:14568–14575
29. Binnemans K (2005) Ionic liquid crystals. Chem Rev 105:4148–4204
30. Mehta SK, Kaur K (2010) Ionic liquid microemulsions and their technological applications. Ind J Chem 49A:662–684
31. Behera K, Malek NI, Pandey S (2009) Visual evidence for formation of water-in-ionic liquid microemulsions. ChemPhysChem 10:3204–3208
32. Gao Y, Han S, Han B, Li G, Shen D, Li Z, Du J, Hou W, Zhang G (2005) TX-100/water/1-butyl-3-methylimidazolium hexafluorophosphate microemulsions. Langmuir 21:5681–5684
33. Gao Y, Li N, Zheng L, Zhao X, Zhang S, Han B, Hou W, Li G (2006) A cyclic voltammetric technique for the detection of micro-regions of bmimPF$_6$/Tween 20/H$_2$O microemulsions and their performance characterization by UV-Vis spectroscopy. Green Chem 8:43–49
34. Yan K, Sun Y, Huang X (2014) Effect of the alkyl chain length of a hydrophobic ionic liquid (IL) as an oil phase on the phase behavior and the microstructure of H$_2$O/IL/nonionic polyoxyethylene surfactant ternary systems. RSC Adv 4:32363–32370
35. Rai R, Pandey S (2014) Evidence of water-in-ionic liquid microemulsion formation by nonionic surfactant Brij-35. Langmuir 30:10156–10160
36. Lian Y, Zhao K (2011) Study of micelles and microemulsions formed in a hydrophobic ionic liquid by a dielectric spectroscopy method. I. Interaction and percolation. Soft Matter 7:8828–8837
37. Anjum N, Guedeau-Boudeville M-A, Stubenrauch C, Mourchid A (2009) Phase behavior and microstructure of microemulsions containing the hydrophobic ionic liquid 1-butyl-3-methylimidazolium hexafluorophosphate. J Phys Chem B 113:239–244

38. Misono T, Aburai K, Endo T, Sakai K, Abe M, Sakai H (2009) Effect of water on interfacial chemical properties of nonionic surfactants in hydrophobic ionic liquid bmimPF$_6$. J Phys Chem B 113:239–244
39. Kusano T, Fujii K, Hashimoto K, Shibayama M (2014) Water-in-ionic liquid microemulsion formation in solvent mixture of aprotic and protic imidazolium-based ionic liquids. Langmuir 30:11890–11896
40. Moniruzzaman M, Kamiya N, Nakashima K, Goto M (2008) Formation of reverse micelles in a room-temperature ionic liquid. ChemPhysChem 9:689–692
41. Rai R, PandeySh BSN, Vora S, Behera K, Baker GA, Pandey S (2012) Ethanol-assisted, few nanometer, water-in-ionic-liquid reverse micelle formation by a zwitterionic surfactant. Chem Eur J 18:12213–12217
42. Safavi A, Maleki N, Farjami F (2010) Phase behavior and characterization of ionic liquids based microemulsions. Colloids Surf A 355:61–66
43. Sun Y, Yan K, Huang S (2014) Formation, characterization and enzyme activity in water-in-hydrophobic ionic liquid microemulsion stabilized by mixed cationic/nonionic surfactants. Colloids Surf B 122:66–71
44. Mao Q-X, Wang H, Shu Y, Chen X-W, Wang J-H (2014) A dual-ionic liquid microemulsion system for the selective isolation of haemoglobin. RSC Adv 4:8177–8182
45. Xue L, Qiu H, Li Y, Lu L, Huang X, Qu Y (2011) A novel water-in-ionic liquid microemulsion and its interfacial effect on the activity of laccase. Colloids Surf B 82:432–437
46. Mojumdar SS, Mondal T, Das AK, Dey S, Bhattacharyya K (2010) Ultrafast and ultraslow proton transfer of pyranine in an ionic liquid microemulsion. J Chem Phy 132:194505–194513
47. Wu H, Wang H, Xue L, Li X (2011) Photoinduced electron and energy transfer from coumarin 153 to perylenetetracarboxylic diimide in bmimPF$_6$/TX-100/water microemulsions. J Colloid Interface Sci 353:476–481
48. Jarzqba W, Walker GC, Johnson AE, Kahlow MA, Barbara PF (1988) Femtosecond microscopic solvation dynamics of aqueous solutions. J Phys Chem 92:7039–7041
49. Kobrak MN, Znamenskiy V (2004) Solvation dynamics of room-temperature ionic liquids: evidence for collective solvent motion on sub-picosecond timescales. Chem Phys Lett 395:127–132
50. Adhikari A, Sahu K, Dey S, Ghosh S, Mandal U, Bhattacharyya K (2007) Femtosecond solvation dynamics in a neat ionic liquid and ionic liquid microemulsion: excitation wavelength dependence. J Phys Chem B 111:12809–12816
51. Seth D, Chakraborty A, Setua P, Sarkar N (2006) Interaction of ionic liquid with water in ternary microemulsions (Triton X-100/water/1-butyl-3-methylimidazolium hexafluorophosphate) probed by solvent and rotational relaxation of coumarin 153 and coumarin 151. Langmuir 22:7768–7775
52. Seth D, Chakraborty A, Setua P, Sarkar N (2007) Interaction of ionic liquid with water with variation of water content in 1-butyl-3-methylimidazolium hexafluorophosphate ([bmim][PF$_6$])/TX-100/water ternary microemulsions monitored by solvent and rotational relaxation of coumarin 153 and coumarin 490. J Chem Phys 126:224512–224524
53. Seth D, Setua P, Chakraborty A, Sarkar N (2007) Solvent relaxation of a room-temperature ionic liquid [bmim][PF$_6$] confined in a ternary microemulsion. J Chem Sci 119:105–111
54. Astruc D, Lu F, Aranzaes JR (2005) Nanoparticles as recyclable catalysts: the frontier between homogeneous and heterogeneous catalysis. Angew Chem Int Ed 44:7852–7872
55. Wang F, Banerjee D, Liu Y, Chen X, Liu X (2010) Upconversion nanoparticles in biological labeling, imaging, and therapy. Analyst 135:1839–1854
56. Serrà A, Gómez E, Calderó G, Esquena J, Solans C, Vallés E (2013) Microemulsions for obtaining nanostructures by means electrodeposition method. Electrochem Commun 27:14–18
57. Serrá A, Gómez E, López-Barbera JF, Nogués J, Vallés E (2014) Green electrochemical template synthesis of CoPt nanoparticles with tunable size, composition, and magnetism from microemulsions using an ionic liquid (bmimPF$_6$). ACS Nano 8:4630–4639

58. Zhang G, Zhou H, Hu J, Liua M, Kuang Y (2009) Pd nanoparticles catalyzed ligand-free Heck reaction in ionic liquid microemulsion. Green Chem 11:1428–1432
59. Fu C, Zhou H, Xie D, Sun L, Yin Y, Chen J, Kuang Y (2010) Electrodeposition of gold nanoparticles from ionic liquid microemulsion. Colloid Polym Sci 288:1097–1103
60. Serrà A, Gómez E, Vallés E (2014) Electrosynthesis method of CoPt nanoparticles in percolated microemulsions. RSC Adv 4:34281–34287
61. Serrà A, Montiel M, Gómez E, Vallés E (2014) Electrochemical synthesis of mesoporous CoPt nanowires for methanol oxidation. Nanomaterials 4:189–202
62. Guo Y, He D, Xia S, Xie X, Gao X, Zhang Q (2012) Preparation of a novel nanocomposite of polyaniline core decorated with anatase-TiO_2 nanoparticles in ionic liquid/water microemulsion. J Nanomat 2012:1–7
63. Li Z, Zhang J, Du J, Han B, Wang J (2006) Preparation of silica microrods with nano-sized pores in ionic liquid microemulsions. Colloids surf A 286:117–120
64. Li F, Yang F-Q, Xia Z-N (2013) Simultaneous determination of ten nucleosides and related compounds by MEEKC with [BMIM]PF_6 as oil phase. Chromatographia 76:1003–1011
65. Li F, Liu R, Yang F, Xiao W, Chen C, Xia Z (2014) Determination of three curcuminoids in Curcuma longa by microemulsion electrokinetic chromatography with protective effects on the analytes. Anal Methods 6:2566
66. Wang Y, Li F, Yang F-Q, Zuo H-L, Xia Z-N (2010) Simultaneous determination of α-, β- and γ-asarone in Acorus tatarinowii by microemulsion electrokinetic chromatography with [BMIM]PF_6 as oil phase. Talanta 101:510–515
67. Shu Y, Cheng D, Chen X, Wang J (2008) A reverse microemulsion of water/AOT/1-butyl-3-methylimidazolium hexafluorophosphate for selective extraction of hemoglobin. Sep Purif Technol 64:154–159
68. Biasutti MA, Abuin EB, Silber JJ, Correa NM, Lissi EA (2008) Kinetics of reactions catalyzed by enzymes in solutions of surfactants. Adv Colloid Interface Sci 136:1–24
69. Xue L, Zhao Y, Yu L, Sun Y, Yan K, Li Y, Huang X, Qu Y (2013) Choline acetate enhanced the catalytic performance of Candida rugosa lipase in AOT reverse micelles. Colloids Surf B 105:81–86
70. Moniruzzaman M, Kamiya N, Goto M (2009) Biocatalysis in water-in-ionic liquid microemulsions: a case study with horseradish peroxidase. Langmuir 25:977–982
71. Skotheim TA, Elsenbaumer RL, Reynolds JR (1998) Handbook of conducting polymers, 2nd edn. Marcel Dekker, New York/Basel/HongKong
72. Dong B, Zhang S, Zheng L, Xu J (2008) Ionic liquid microemulsions: a new medium for electropolymerization. J Electroanal Chem 619:193–196
73. Dong B, Xu J, Zheng L, Hou J (2009) Electrodeposition of conductive poly (3-methoxythiophene) in ionic liquid microemulsions. J Electranal Chem 628:60–66

Chapter 11
On the Hunt for More Benign and Biocompatible ABS

Jorge F.B. Pereira, Rudolf Deutschmann, and Robin D. Rogers

Abstract The late appearance of more environmentally friendly and biocompatible liquid-liquid extraction systems has triggered a shift in research interest toward aqueous biphasic systems (ABS), including those composed of ionic liquids (ILs). Although most ILs were originally considered as green solvents due to their negligible volatility, the release/escape of these materials into the environment from operations employing IL-water solvent systems rose several concerns. As a consequence, recent studies started to focus on finding more benign and biocompatible ILs for new biocompatible ABS (Bio-ABS) formulations. One of the most thoroughly studied Bio-ABS to date employs cholinium-based ILs combined with various polymers or high-melting inorganic salts; however, other Bio-ABS were also suggested and tested, like those utilizing amino-acid-based ILs as ABS promoters. Although naturally resourced and biodegradable ILs obviously improve the biocompatibility and sustainability of a given ABS, similar attention has to be paid to the selection of the other coexisting phase-forming agents (polymers, biopolymers, sugars, and/or biodegradable high-melting salts) in order to achieve a truly environmentally conscious system. Starting from this basis, in this chapter we (a) review and summarize the available information on Bio-ABS published so far, (b) highlight the most important differences between IL-based Bio-ABS and non-Bio-ABS, (c) show examples of nontoxic and biodegradable phase-forming agents, and (d) evaluate the applicability and future industrial perspectives of IL-based Bio-ABS.

Keywords Aqueous biphasic systems • Ionic liquids • Biocompatible • Biodegradable • Nontoxic • Environmentally friendly • Polymers • Cholinium-amino acid salts

---

J.F.B. Pereira (✉)
Department of Bioprocess and Biotechnology, School of Pharmaceutical Sciences,
UNESP – Univ Estadual Paulista, Araraquara, SP 14800-903, Brazil
e-mail: jfbpereira@fcfar.unesp.br

R. Deutschmann • R.D. Rogers
Department of Chemistry, McGill University, Montreal, QC H3A 0B8, Canada

Abbreviations

AA-IL	Amino acid-based ionic liquid
ABS	Aqueous biphasic systems
AIL	Aprotic ionic liquid
BHb	Bovine hemoglobin
Bio-ABS	Biocompatible ionic liquid-based aqueous biphasic systems
BSA	Bovine serum albumin
Ch-ABS	Cholinium-based aqueous biphasic systems
Ch-IL	Cholinium-based ionic liquid
FLS	Fluorescence spectroscopy
FT-IR	Fourier transform infrared spectroscopy
GB	Good's buffer
IL	Ionic liquid
LCST	Lower critical solution temperature
MW	Molecular weight
OECD	Organization for the Economic Co-operation and Development
OVA	Ovalbumin
PEG	Polyethylene glycol
PEO	Polyethylene oxide
PIL	Protic ionic liquid
PPG	Polypropylene glycol
PVA	Polyvinyl alcohol
TEM	Transmission electron microscopy
TLL	Tie line length
VOC	Volatile organic compounds

11.1 Introduction

The energy crisis in the 1970s has changed the mindset of leaders in politics, technology, and research equally to look for sustainable resources. At the same time, major environmental concerns urged the introduction of stricter legislations about industrial processing technologies worldwide. These factors have opened up the path toward the search for new chemicals, processes, and technologies based on renewable (mostly bio) resources [1]. Solvent-based separation and purification technologies were among the first large-scale operations affected by the new line of environmentally conscious thinking. The high volume of solvents used in these processes was mostly composed of volatile organic compounds (VOCs), and their escape/disposal into the environment represented various risks. They were also shown to be incompatible with many biological products and processes, sometimes even decreasing the quality of the final product [2].

During the race to overcome these environmental issues and satisfy regulations, the end of the last century has seen the rediscovery of aqueous biphasic systems (ABS) as an ideal solution to help design water-rich, more sustainable liquid-liquid extraction processes for the separation and/or extraction of biomolecules [3]. ABS have also shown to be capable of maintaining the native conformation and biological activities/properties of several biomolecules, such as enzymes [4], alkaloids [5], antibiotics [6, 7], and biopharmaceuticals [8], among others [9]. The first studies have focused on the use of traditional ABS based on polymer-polymer and polymer-salt [9] systems, but in 2003, Rogers and coworkers [10] proposed the use of imidazolium-based ILs as ABS promoters, demonstrating their ability to form ABS in the presence of inorganic salts (K_3PO_4). One of the major advantages of ILs over traditional VOC-based solvents is their negligible vapor pressure that could prevent gas phase leaks in a wide range of chemical processes [11]. It was also highlighted, however, that the application of ILs in ABS could be disadvantageous in some cases (costly cation-anion combinations, ILs composed of fluorinated anions exhibiting non-benign character, etc.) [10]. Despite these limitations, the development of the first IL-salt ABS started a "boom" in the study of new IL-based ABS and their possible applications as alternative extraction and purification systems to efficiently separate a wide range of biomolecules [12].

In a recent compilation of the work done so far on ABS composed of ILs, Freire et al. [12] observed that almost all of IL-based ABS are formed with imidazolium-, pyridinium-, piperidinium-, or pyrrolidinium-based ILs. These IL families, however, should not be considered as sustainable and nontoxic compounds, especially those based on imidazolium and pyridinium cations [13–15]. Therefore, there is a strong ongoing motivation to develop and study new ABS incorporating environmentally friendly ILs. Recently, new classes of ILs (cholinium- [16] and amino-acid-based ILs [17]) have been synthesized using bio-precursors through sustainable and easy procedures, exhibiting lower toxicities, higher rate of biodegradation, and higher biocompatibility than conventional ILs [18, 19].

These two classes of ILs were already used and considered as more benign ABS promoters; however, if an IL is more "benign" or more "biocompatible" is an ongoing question. How can one define the benign character of an IL-based ABS? In order to answer this question, we attempt to summarize all recent advances in the field of sustainable and biocompatible IL-based ABS, briefly reviewing the environmental character of some families of ILs and examining if they are capable of promoting an immiscibility region when combined with other compounds (e.g., polymers, salts, or surfactants). However, as the benign behavior of an IL-based ABS is not only dependent on the IL used, it will also be discussed how coexisting compound(s) can be considered in the ABS formulation to increase the "greener" character of ABS. Thus, this chapter will review what has been done to date and discuss what research should be carried out in the future in the liquid-liquid extraction field using ILs.

11.2 Biocompatible and Benign ABS Forming Agents

By definition, an ABS consists of two immiscible water-rich phases formed by the mixing of two or more water-soluble solutes, which, above a certain concentration level, are separated into two coexisting phases [9]. Nowadays, several types of compounds can be used as phase-forming agents, from the more traditional polymers, salts, and surfactants to the more recently reported ILs, peptides, or amino acids. Although there is a wide range of phase-forming agents to choose from, the number of water-soluble solutes that have been used to create ABS in combination with ILs is significantly lower, though it seems that this number will increase in the next few years. Although the recent review by Freire et al. [12] shows that the majority of IL-based ABS are composed of ILs and inorganic salts, other less studied combinations of ILs and other solutes, such as polymers, carbohydrates, amino acids, and surfactants, have also been reported. Herein, in the following subsection, we will go through the concept of benign and biocompatible ILs first and then show other ABS forming agents that have been used (or could be used) in order to increase the biocompatibility of IL-based ABS.

11.2.1 More Benign and Biocompatible ILs

ILs are complex fluids which can express interesting physicochemical properties, such as negligible vapor pressure, nonflammability, large liquid temperature range, high thermal conductivity, and a wide range of electrochemical potential [20, 21]. Moreover, with appropriately chosen synthetic methodologies and the proper manipulation of the ions involved, ILs can be tailor-made for the desired applications, like separation processes, where their polarities [12], safety, and environmentally friendly can be designed to suit certain requirements [22].

According to Seddon's recent claim [23, 24], more than 10^{18} ILs are theoretically possible, although this number includes binary, ternary, and higher mixtures of ILs. With that in mind, it is easy to see that there should be compounds exhibiting complete biocompatibility and benign characteristics on one end of the spectrum and, equally possible, ILs with high toxicity (even higher than in case of many conventional solvents) on the other, among the large array of ILs that have been prepared and characterized to date. So far, the majority of ILs are manufactured synthetically from other chemicals, and this fact questions whether one can consider even the most biocompatible ones as "green" solvents [25]. Usually, the entire life cycle of these chemical compounds is not considered; however, a thorough environmental risk assessment requires one to consider every industrial process involved in the production, application, and ultimate disposal of these ILs, used as phase-forming promoters in different ABS [26]. According to the 12 Green Chemistry principles [27], an IL should have negligible accumulation and persistence in the environment and, for that purpose, its biodegradability is assumed to be a crucial parameter when assessed as "environmentally friendly compound."

A huge number of different ILs have been synthesized and classified according to three generation stages during the last two decades: the first generation of ILs were synthesized in order to obtain specific physicochemical properties; the compounds of the second generation were produced to achieve specific behaviors, suitable to certain final industrial applications; and the third generation of ILs (advanced ILs) is the more recent development and aims to produce ILs with desired biological features [28]. These third-generation potentially environmentally friendly ILs are the central subject of this chapter as they are prepared using more environmentally friendly starting materials, produced by low-energy and easy procedures, composed of ions that are more stable, biodegradable, readily available, and designed to exhibit lower toxicities [29]. ILs of the third generation are mainly composed of anions such as sugars, amino or organic acids, alkylsulfates, or alkylphosphates and cations derived from choline or amino acids.

The late emergence of environmental concerns has pushed a large part of the IL research community toward the search for compound families that are biodegradable and can also be obtained from natural starting materials with well-characterized biodegradability and toxicological properties in order to satisfy the principles of the third IL generation [26]. These new classes of ILs, completely derived from biomaterials, have also been called natural ILs, Bio-ILs [16], or bio-renewable ILs [25]. In addition to being bio-resourced and biodegradable, these ILs have to comply with all the other criteria (mainly toxicity and biocompatibility) themselves as well in order to be considered truly benign and environmentally friendly.

The ecotoxicological features of ILs have been described in numerous studies in the last few years, and a significant number of IL families are already well characterized [30]; however, the concept of biocompatibility is broad and hard to define. By a simplified definition, biocompatibility is a "quality of not having toxic or injurious effects on biological systems" [31]. Thus, if applied to IL research, it should be defined and considered in accordance with the context of its application [32]. Consequently, when it is applied to ILs as ABS promoters, it dictates that only anions and cations that exhibit low toxicities and negligible effects over the target biomolecule to be extracted should be employed. The main criteria that an IL should satisfy in order to be considered as a "green" solvent for separation processes are summarized in Table 11.1.

Several papers highlighted the necessary design and synthetic strategies needed to follow in order to obtain families of ILs that could satisfy the "Green Chemistry" criteria described above [11, 15, 30, 33, 34]. With these strategies in mind, it is possible to form close to ideal Bio-ILs (biodegradable, nontoxic, biocompatible, sustainable) as it was demonstrated through real examples [35–37]. Even though the search for these "Bio-ILs" has been expanding, the number of successful examples obtained from natural sources that fulfills all the "green" requirements is not yet significant. The majority of works executed in this direction so far focused on the synthesis of cholinium- and amino acid-based ILs. A few other classes of

Table 11.1 Summary of the Green Chemistry criteria for an IL to be used in separation processes

Criteria	Standard
Low toxicity	All ILs should exhibit low toxicities against organisms of different trophic levels, using parameters according to the Organization for the Economic Co-operation and Development (OECD) standard tests
Biodegradability	All ILs and derivatives should reach 60 % biodegradation after 28 days of biological treatment to be classified as "readily biodegradable," using parameters defined, for example, according to the closed bottle test (OECD 301D)
Sustainability	All precursor materials used in manufacture should be available from sustainable and renewable feedstocks
Biocompatibility	All compounds should be compatible with the biological compounds to be extracted, maintaining their biological structures and activities

protic ILs have also been suggested as possible environmentally friendly alternative promoters for different ABS formulation; however, generalizations in claiming entire families of ILs as "green" should be avoided since even small changes in the chemical structure of the cation or/and anion can completely modify their character and properties, possibly reversing the environmentally friendly nature of an IL belonging to an assumed bio-family. Moreover, any change that improves one aspect, like biodegradability, might degrade other properties (toxicity and/or biocompatibility) and vice versa.

11.2.1.1 Cholinium-Based ILs

Cholinium-based ILs (Ch-ILs) are derived from choline chloride, N,N,N-trimethylethanolammonium chloride, a naturally occurring essential nutrient frequently used as a food additive [38]. Cholinium and other quaternary ammonium salts play very important roles in living systems, especially in the synthesis of vitamins and enzymes needed in various steps of the carbohydrate metabolism [39, 40]. The biological importance and availability of cholinium chloride moved scientists toward the synthesis of its derivatives, especially those with low melting points [39]. The resulting ILs were termed as "benign" [15], "biocompatible" [18, 41–44], "biodegradable" [18, 19, 22], "environmentally friendly," and/or "nontoxic" [18, 19, 22, 34, 39, 45].

Nockeman et al. [45] were one of the first research groups, in 2007, to examine the toxicity of two Ch-ILs, cholinium saccharinate and cholinium acesulfamate, to the crustacean organism *Daphnia magna*. In that preliminary work, the authors observed that both ILs exhibited very low ecotoxicity, being at least two orders of magnitude less toxic than other traditional families of ILs, such as the imidazolium- and pyridinium-based ones [45]. On the other hand, Pernak et al. [41] synthesized homologous series of Ch-ILs with alkoxy groups containing 2–12 carbons combined with distinct anions and verified that the derivatives were increasingly active against some microorganisms.

The influence of the anion on toxicity was also evaluated by Petkovic et al. [18] using filamentous fungi for a series of nine cholinium-alkanoate ILs. The authors [18] demonstrated that all the alkanoate ILs tested were less toxic than the corresponding sodium salts, and, in addition, it was observed that the cholinium cation can also be digested by the respective fungi strain [18]. Sekar et al. [42] showed that Ch-ILs, like cholinium lactate, cholinium saccharinate, cholinium dihydrogen phosphate, choline lactate, and choline tartrate, can even be employed as co-substrate and carbon source to digest organic dye contaminants with the salt-tolerant bacterium *S. lentus*.

In 2009, Vrikkis et al. [43] performed an extensive study on the biocompatibility/cytotoxicity and protein stabilizing capabilities of choline dihydrogen phosphate and choline saccharinate. The measurement of their effect on metabolic activity of a mouse macrophage cell line using the reduction of resazurin (used as an indicator of activity/viability) has proved both compounds to be relatively benign and generally nontoxic. Lysozyme formulations in 80 wt % choline dihydrogen phosphate were also prepared to investigate the short- and long-term effect of the IL on the thermal stability and activity of the enzyme [43]. The results have showed that the thermal stability and activity of the enzyme were higher than that in an aqueous buffer even after one month of storage [43]. The biocompatibility of cholinium-based salts and ILs was also evaluated for different collagenous biomaterials by Vijayaraghavan and coworkers [44]. It was observed that all cholinium-based compounds tested exhibited good cell viability and adhesion properties fulfilling the requirements of the World Health Organization (WHO) for biomedical applications. These data seem to indicate that ABS prepared with Ch-ILs can be efficient tools to extract, separate, and even store enzymes and proteins [44].

Although, the majority of the literature supports the benign nature of the Ch-ILs, not all cholinium-based salts can be considered harmless as Ventura et al. have pointed out [39]. While eight variants of cholinium-based salts could be classified as "practically harmless" or "harmless" for the bioluminescent marine bacteria *Vibrio fischeri*, in the case of cholinium dihydrogencitrate and cholinium bitartrate, showed moderate toxicities against the same organism [39]. In another work of the same research group, they investigated the ecotoxicological effects of 17 monocationic and eight dicationic cholinium (or cholinium derivative) ILs conjugated with bromide anion on the same microbe (*V. fischeri*). It was observed that the structural features of the cation have a high influence on the harmlessness of the IL, and again, just belonging to the family of cholinium compounds does mean the salts will be nontoxic; long alkyl side or linkage chains, high number of CH_2CH_3OH groups, and multiple covalent bonds can increase toxicity significantly [22]. This evidence reinforces how important the role of structural design is in making "greener" ILs [22], and this is in agreement with the initial assumptions of Wood and Stephens [46], suggesting that "ILs have the potential to make a big difference to the environmental impact of chemical manufacturing processes," but a potential that will be dependent on a judicious selection of the least toxic IL structures.

In order to obtain even more benign IL structures, some researchers [47, 48] have synthesized cholinium-amino acid ILs entirely composed of renewable biomaterials. These new classes of ILs can combine the advantages of two naturally occurring materials, cholinium chloride and amino acids, making them promising candidates as environmentally friendly solvents. The next subsection will provide further details about novel ILs derived from amino acids.

11.2.1.2 Amino-Acid-Based ILs

Amino acids have an enormous structural diversity. They contain both an amino group and a carboxylic acid residue in a single molecule [17, 25, 49]. Additionally, they are one of the most abundant organic compounds in nature and can be produced in large quantities using relatively easy processes. All of these attributes can make amino acids an excellent choice for being used as low-cost feedstock to synthesize new classes of environmentally friendly ILs [17].

Fukomoto et al. [50] prepared the first prototype series of amino acid-based ILs (-AA-ILs) in 2005, combining 1-ethyl-3-methylimidazolium with 20 different amino acids. The preparation of ILs from natural amino acids is not an easy process without significant challenge, since amino acids are zwitterionic species and are stabilized by electrostatic interaction between the carboxylate anion and the protonated amino group [17]. The first successful synthesis of imidazolium-based ILs employing amino acids as the corresponding anion was soon followed by the synthesis of phosphonium- [17] and cholinium-based ILs [46, 51, 52]. Based on a similar approach, a whole new generation of ILs started to emerge, in which the cations are simply derived from natural amino acids or amino acid ester salts [25, 53]. Since the beginning, these ILs were proclaimed as one of the greenest [51] and most biocompatible [25] solvents, mainly due to their natural origin. Nevertheless, in recent years, research has been started to pursue the environmentally friendly (toxicology, biodegradability, etc.) of these new classes of bio-renewable ILs [19, 54, 55].

In order to evaluate the toxic/nontoxic character of AA-ILs, Gouveia et al. [54] prepared 14 ILs using imidazolium, pyridinium, and cholinium cations. Their toxicities were assessed against organisms of various trophic levels, namely, crustacean *Artemia salina*, human cell line HeLa (cervical carcinoma), and Gram-negative and Gram-positive bacteria (*Escherichia coli* and *Bacillus subtilis*, respectively). As expected, the toxicity of each IL appeared to be highly dependent on the chemical nature of both the cation and the anion, as well as the organism tested [54]. Interestingly, however, none of the amino-acid-based ILs exhibited significant toxicity for any of the bacteria strains tested, but it was found to be directly related to the cation used when evaluated against the crustacean and human cell culture. The AA-ILs with imidazolium and pyridinium cations have shown remarkable higher toxicity (10 times higher) than those with the cholinium cation [54]. These data might indicate that ILs produced from bio-materials, such as choline and amino acids, could represent low toxicity to both humans and the environment.

Similarly, Hou et al. [19] assessed the environmental fate of 18 novel ILs prepared with cholinium as cation and amino acids as the anions. The aim of the research was to select and be able to design truly environmentally friendly ILs that are not only produced from bio-resources but are also less toxic and readily biodegradable in the natural environment. For that purpose, the authors [19] examined the toxicity of these novel ILs to enzymes and four strains of representative bacteria and then determined their biodegradability using wastewater microorganisms. Their results showed that the inhibitory effects of cholinium-amino acid ILs were at least an order of magnitude weaker than that of the conventional 1-butyl-3-methylimidazolium tetrafluoroborate, and with the exception of two ILs, these ILs exhibited similar or even lower toxicities than the cholinium chloride salt [19]. According to the biodegradability tests performed by the authors [19], all of the cholinium-amino acid ILs complied with the requirements set by the Organization for the Economic Co-operation and Development (OECD) and thus can be considered as readily biodegradable, since all the compounds tested reached 60 % biodegradation after 28 days of treatment. Hou et al. [19] demonstrated, in general, that the majority of the cholinium *R. satius* was especially sensitive to based ILs are probably "greener," but, as highlighted by the authors as well, other ecotoxicological tests against organisms of different trophic levels are necessary to help design and identify truly "green" chemicals.

Egorova et al. [55] recently proved that the application of amino acids in IL compositions is not always beneficial nor always helps increase the ecological attributes. The authors [55] tested and compared the cytotoxicity of several AA-ILs (incorporation of amino acids either as the cation or the anion) and conventional imidazolium-based ILs on NIH/3 T3 (mouse fibroblasts) and CaCo-2 (colorectal adenocarcinoma) cell cultures [55]. Surprisingly, the presence of an amino acid did not necessarily lower toxicity; especially in the cases when amino acids were employed as the anion, the observed toxicities were comparable with those of the conventional 1-butyl-3-methylimidazolium chloride or 1-butyl-3-methylimidazolium L-lactate [55]. On the other hand, the research group [55] also observed that both the selected conventional and AA-ILs induced the apoptosis of mouse fibroblasts cells. These findings show again the importance of careful design and thorough environmental analysis in the development and application of biomaterials like bio-ILs.

11.2.1.3 Other Classes of Environmentally Friendly ILs

The previous subsections described two of the most important third-generation classes of ILs that are usually claimed as very benign and biocompatible compounds. However, for standard industrial separation or extraction processes, other compounds have also been considered to form environmentally friendly ILs. Brønsted or protic ILs (PILs), as having a proton available for hydrogen bonding, can be also included in this last generation of ILs [56]. These ILs differ from the classic, aprotic ones (AILs), in having both cationic and anionic counterparts formed by low-molecular-weight organic compounds. They usually contain substituted (or polysubstituted) amines as cations and organic acids as anions and were designed exclusively to minimize their environmental impact [56]. PILs have been tested and used for numerous applications, but as the majority of them are fairly new or underdevelopment, ecological assessments of their properties are scarce yet [57].

Peric et al. [56, 57] have been actively studying the ecotoxicological effects and biodegradability of PILs. When they tested the effects of three short aliphatic PILs on soil microbial functions (C and N mineralization) and terrestrial plants (*Lolium perenne*, *Allium cepa*, and *Raphanus sativus*), they found that the PILs were not nontoxic in general for either the plants tested or the soil microbiota and were biodegraded by the soil matrix; plants showed higher sensitivity than microorganisms. *R. satius* was especially sensitive to the 2-hydroxytriethanolamine pentanoate IL which appeared to be somewhat toxic. It was concluded that the more complex molecular structure the ILs has (longer alkyl side chain), the higher rate of inhibition in the organisms is observed [56].

In another study, the same authors [57] examined the toxicity and biodegradability of a wider range of PILs (ten ILs based in three different amines and six organic acids) and AILs (two imidazolium- and two pyridinium-based ILs). The experiments were carried out on various test subjects: three different aquatic organisms (*Vibrio fischeri*, *Pseudokirchneriella subcapitata*, and *Lemna minor*), one enzyme (acetylcholinesterase), and leukemia rat cells (IPC-81). Tests were also performed to assess the potential biodegradability of PILs in water according to the OECD guideline 301. The data showed that the PILs tested were generally nontoxic or at least less toxic than the AILs (various orders of magnitude lower than the aprotic ILs) against most of the organisms studied (aquatic, enzymes, or cells), except for *Lemna minor* that happened to be very sensitive for three of the PILs. The most exciting results obtained by the authors [57], however, were the good rates of biodegradation of the PILs in water, especially if compared with the negligible biodegradation observed for AILs.

Another type of biocompatible ILs was synthesized recently by Taha et al. [36]. The authors developed a series of novel self-buffering and biocompatible ILs using anions derived from biological buffers (designated as "Good's buffers" (GBs)). These GBs are zwitterionic amino acid derivatives, usually acting as strongly hydrated molecules [36]. Twenty GB-based ILs were produced by combining 5 different GBs (N-[tris(hydroxymethyl)methyl]glycine, Tricine; 2-[(2-hydroxy-1,1-bis(hydroxymethyl)ethyl)amino]ethanesulfonic acid, TES; (2-cyclohexylamino)ethanesulfonic acid, CHES; 4-(2-hydroxyethyl)piperazine-1-ethanesulfonic acid, HEPES; 4-morpholineethanesulfonic acid, MES) with four hydroxide bases (1-ethyl-3-methylimidazolium, tetramethylammonium, tetraethylammonium, and tetrabutylammonium). Their capability to extract and stabilize proteins was tested using bovine serum albumin (BSA), and their toxicity was evaluated against *V. fischeri*. All tests provided very promising information; the majority of GB-ILs seemed to bear high self-buffering capacity, showed great potential to extract and stabilize proteins (better than the corresponding buffer solution and conventional ILs), and proved to be nontoxic against *V. fischeri* [36].

All these encouraging results on ecotoxicology, biodegradation, and protein friendliness of PILs and alternative APILs, even despite the scarcity of data available, combined with their low production costs and simple synthesis, make it easy to envision their great potential as alternative and more benign phase-forming agents in the formation of ABS.

11.2.2 Environmentally Friendly Coexisting Phase-Forming Agents

Besides the enormous research effort devoted to the development of more environmentally friendly ILs, the scientific community has also been seeking biocompatible species that can promote the formation of an immiscible region when mixed with aqueous solutions of different ILs. Since the first report of an ABS formed by mixing an aqueous solution of 1-butyl-3-methylimidazolium chloride ([C$_4$mim]Cl) and a concentrated aqueous solution of K$_3$PO$_4$ [10], more than 50 papers have described possible scenarios to form an ABS combining several ILs (mainly conventional ILs) and different inorganic/organic salts [12]. The use of salts with high charge density, normally applied as salting-out agents in the formation of IL-ABS, however, should be avoided and rather replaced with more biodegradable organic salts, carbohydrates, polymers, or amino acids, as pointed out by Freire et al. [12].

A large portion of the literature focuses on IL-based ABS employing inorganic salts with anions such as phosphates, sulfates, or carbonates, compounds that raise environmental concerns when released into aqueous streams [58]. One of the most used inorganic salts, K$_3$PO$_4$, for instance, introduces potassium and phosphate ions into the aqueous system generating a highly alkaline medium that can be harmful to a large number of pH-sensitive biomolecules (such as proteins, enzymes, pharmaceutical active compounds, etc.) and microorganisms, while, on the other hand, its dissolution in the aqueous phase complicates the recycling processes of the ILs [59]. Recent studies, however, presented new formulations of IL-based ABS based on biodegradable and nontoxic organic salts with citrate, tartrate, and acetate anions [58, 60–63]. These systems can increase the benign character and biocompatibility of extraction processes through the use of high-melting salts that are more biodegradable and less toxic than those applied in traditional IL-based ABS formulations. In addition to the organic salts, other benign phase-forming agents, such as amino acids [64, 65], carbohydrates [59, 66–70], and different polymers [3, 12], have also been tested lately to create IL-based ABS.

Amino acids were previously referred as interesting starting materials for the synthesis of new amino acid-based ILs and consequently being applied as more benign phase-forming ILs, but, on the other hand, amino acids can simply be used as phase separation promoters. As described in the earlier sections, amino acids are zwitterionic and, in general, environmentally friendly molecules, and their addition to aqueous solutions of different ILs, above certain concentrations, can lead to the formation of a biphasic region. The application of amino acids in separation processes will produce a more benign extractive environment for biomolecules, as well as help during the recovery of the IL from aqueous effluents [12]. Despite the promising future of these systems, only two publications included ternary phase diagrams for IL-amino acid-water systems [12]. In 2007, Zhang et al. [64] proved that glycine, L-serine, and L-proline are all able to create ABS when combined with water and the hydrophilic IL 1-butyl-3-methylimidazolium tetrafluoroborate. Domínguez-Pérez et al. [65] also demonstrated the formation of ABS combining imidazolium-based ILs with three different amino acids (L-lysine, D,L-lysine HCl, and L-proline). These ABS, however, are contradictory, claiming to be more environmentally friendly due to the use of amino acids as phase-forming agents, while still using ILs based on non-environmentally friendly ions, such as imidazolium or tetrafluoroborate.

Carbohydrates are non-charged, usually nontoxic, and readily biodegradable compounds, almost always obtained from renewable feedstocks, which make them promising candidates to play a leading role in environmentally friendly processes [59]. Carbohydrates are present in different forms in nature: mono-, di-, oligo-, and polysaccharides with sugar alcohols being the most common ones. Regular monosaccharides (simple sugars) are polyhydroxy aldehydes or ketones with –OH groups attached to every available carbon atom, thus capable of forming several hydrogen bonds both as donor or acceptor. Not surprisingly, these molecules have high affinity toward water and can even displace compounds from aqueous solution. The sugar alcohols (low-molecular-weight polyols) are derivatives of simple sugars where the aldehyde/carbonyl group is hydrogenated, resulting in an even higher water affinity [59]. In the few publications to date, the following monosaccharides, disaccharides, and sugar alcohols were tested to create a biphasic region by mixing them with aqueous solutions of ILs: sucrose [59, 66–69], glucose [59, 67, 69], fructose [59, 67, 70], xylose [59, 67, 69], maltose [59, 68, 69], mannose [59], arabinose [59], maltitol [59], sorbitol [59], and xylitol [59]. It is important to draw attention to the fact that the majority of the ILs tested were of the most conventional and non-benign classes of ILs.

Organic salts, amino acids, and simple sugars can all be considered as environmentally friendly phase-forming agents. However, the most biocompatible, benign, and biodegradable compounds with the best applicability can probably be found among oxygen-rich (potential for forming multiple hydrogen bonds) hydrophilic polymers, like polyethylene glycol (PEG), polypropylene glycol (PPG), polyvinyl alcohol (PVA), polyethylene oxide (PEO), and polysaccharides. They were firstly used in the formulation of ABS in the mid-1950, when Albertsson demonstrated the formation of two immiscible water-rich phases combining two hydrophilic polymers, dextran and PEG [71]. Studies afterward tried to identify other suitable ABS forming candidates among various polymers [72, 73], salts [9, 71], surfactants [74], carbohydrates [75], and ILs [3, 76].

The majority of the abovementioned hydrophilic polymers are less toxic (or nontoxic) and more biocompatible than most of the other materials of choice; some can even be obtained from natural resources. One of the most promising examples is dextran, a naturally occurring complex, branched glucose-based polysaccharide. It is water-soluble, completely biodegradable and biocompatible, and can be produced from renewable sources as it is synthesized from sucrose by certain lactic acid bacteria, such as *Leuconostoc mesenteroides*, *Lactobacillus brevis*, and *Streptococcus mutans* [77, 78]. Despite these advantages the high cost associated with its production represents a major obstacle in the way of widespread industrial application [79].

The most extensively used oxygen-rich polymers are PEGs and PPGs. They are chemically similar compounds; their lower molecular weight variants are soluble in water and capable of creating two immiscible water-rich phases with almost all the ABS phase-forming agents reported so far. PEGs have a long history of employment in industrial processes, from traditional chemical procedures to the newest technologies in life sciences [80]. Their potential biodegradability, high water solubility, low vapor pressure, low melting point, and low production cost make these materials favorable for many applications [74]; they can have important roles in the formulation of polymer surfactants or polymer electrolytes [81], in drug delivery processes [82], and can act as solvents for several separation or purification approaches [71] or as protein stabilizing and partitioning agents [83]. The popularity of PEG in biotechnological applications and the requirement for more sustainable ILs have brought the idea to employ it in IL-based ABS [3, 84–87]. PPGs, although less frequently used, are thermosensitive, biodegradable, and nontoxic materials having similar properties to those of PEGs and can be easily recovered from an ABS by convenient heating [84]. These examples might well illustrate the future potential of oxygen-rich polymeric materials in the formulation of IL-based ABS.

11.3 Biocompatible Ionic-Liquid-Based Aqueous Biphasic Systems

Although authors frequently termed many of IL-based ABS as "biocompatible," "benign," "biodegradable," or "greener," only a handful of ILs and phase-forming agents exhibit true environmentally friendly characteristics. According to the previously described considerations, it is evident that environmentally friendly and biocompatible IL-ABS will only result from a proper combination of already benign (nontoxic and biodegradable) ILs and phase-forming compounds preferably obtained from renewable resources.

This section will review the recently published IL-based ABS that are the closest ones to an ideal, truly benign, and biocompatible system. First the IL-ABS, employing frequently studied Ch-ILs, will be discussed; then, the systems formulated with AA-ILs will be summarized; and finally other potentially highly benign and biocompatible IL-based ABS will be evaluated. Moreover, the main physicochemical properties and some interesting thermodynamic and/or modeling properties will also be presented for each type of IL-based ABS.

11.3.1 Cholinium-Based ABS

Since 2003, a large number of IL-based ABS were published, but the first work about the use of Ch-ILs (or salts) in the formulation of ABS was only reported recently in 2012, where Wang and coworkers [35] demonstrated the possibility to form ABS combining different Ch-ILs and PPG-400. Other researchers then confirmed the formation of ABS using IL-based ABS (Ch-ABS) as demixing promoters [2, 63, 88–90]. Table 11.2 shows the types of Ch-ABS assessed to date with the temperature at which stable ABS were observed.

In contrast with more conventional IL-based ABS, the majority of Ch-ABS were obtained with oxygen-rich polymers (PPG, PEG, PEG-PPG copolymers) instead of inorganic/organic salts. There were only two exceptions where liquid-liquid demixing was achieved with either aqueous solutions of K_3PO_4 according to Shahriari et al. [89] or when two cholinium-based salts (benzyldimethyl (2-hydroxyethyl)ammonium chloride, [BCh]Cl, and cholinium salicylate, [Ch][Sal]) were mixed with aqueous solutions of potassium citrate buffer ($C_6H_5K_3O_7/C_6H_8O_7$), potassium phosphate buffer (KH_2PO_4/K_2HPO_4), and potassium carbonate (K_2CO_3) in the results provided by Sintra et al. [63].

In the case of [Ch]Cl-based ABS, formation was observed with practically all of the polymers, regardless of the molecular weight, as well as with K_3PO_4. This behavior is worth searching for as having the potential for giving more flexibility to the design of future ABS. With this in mind, the systems with [Ch][OAc], [Ch][Gly], [Ch][Lac], [Ch][Glu], and [Ch][Suc] look promising; however, they are not thoroughly tested yet. [Ch][Sal] and [BCh][Cl] were shown to be able to form ABS with various salts and buffers, but were not assessed with PEG- or PPG-based systems.

Regarding the temperature effect on phase diagrams, the majority of the Ch-ABS were determined at 15 °C or 25 °C. A more complete evaluation of the temperature influence was only carried out for some particular examples of Ch-ABS, namely, [Ch]Cl + PPG-400 + H_2O, [Ch][Gly] + PPG-400 + H_2O, [Ch][Lac] + PPG-400 + H_2O, [Ch][Pro] + PPG-400 + H_2O, [Ch]Cl + PEG-600 + H_2O, and [Ch][DHph] + PEG-600 + H_2O. Although detailed information regarding the effect of temperature on ABS is scarce, some examples suggest that, depending on the IL used, elevated temperatures mostly enhance the immiscibility region [88, 90].

Shahriari et al. [89] used a series of Ch-ILs created with anions derived from organic acids or HCl to promote ABS in the presence of K_3PO_4. The trend in capability of forming ABS was observed in the following order:

Table 11.2 List of Ch-ABS studied to date. Each Ch-IL (or salt) is correlated with the phase forming agents studied[a]

	Coexisting phase forming agents								
	Polymers						Copolymer	Salts	Other
	Polypropylene glycol		Polyethylene glycol						
Ionic liquids	PPG-400	PPG-1000	PEG-400	PEG-600	PEG-1000	PEG-4000	EO$_{10}$PO$_{90}$	K$_3$PO$_4$	
Cholinium acetate [Ch][OAc]	At 25 °C [35]	–	At 25 °C [90]	At 25 °C [90]	At 25 °C [90]	–	–	At 25 °C [89]	–
Cholinium benzoate [Ch][Benz]	At 25 °C [35]	–	–	–	–	–	–	–	–
Cholinium bicarbonate [Ch][Bic]	–	–	At 25 °C [90]	At 25 °C [90]	At 25 °C [90]	–	–	–	–
Cholinium bitartrate [Ch][Bit]	–	–	At 25 °C [90]	At 25 °C [90]	At 25 °C [90]	–	–	–	–
Cholinium butyrate (or butanoate) [Ch][But]	At 25 °C [35]	–	N/A [90]	N/A [90]	N/A [90]	–	–	–	–
Cholinium chloride [Ch]Cl	At 5 °C [88], At 15 °C [88], At 25 °C [88], At 35 °C [88], At 45 °C [88]	At 15 °C [88]	At 25 °C [90]	At 25 °C [90], At 50 °C [90]	At 25 °C [90]	–	At 15 °C [88]	At 25 °C [89]	–
Cholinium dihydrogencitrate [Ch][DHcit]	–	–	At 25 °C [90]	At 25 °C [90]	At 25 °C [90]	At 25 °C [2]	–	N/A [2]	–

(continued)

Table 11.2 (continued)

	Coexisting phase forming agents								
	Polymers							Salts	
	Polypropylene glycol		Polyethylene glycol				Copolymer		
Ionic liquids	PPG-400	PPG-1000	PEG-400	PEG-600	PEG-1000	PEG-4000	$EO_{10}PO_{90}$	K_3PO_4	Other
Cholinium dihydrogenphosphate [Ch][DHph]	–	–	At 25 °C [90]	At 25 °C [90] / At 50 °C [90]	At 25 °C [90]	–	–	–	–
Cholinium formate [Ch][For]	At 25 °C [35]	–	–	–	–	–	–	–	–
Cholinium fumarate [Ch][Fum]	–	–	–	At 25 °C [2]	–	At 25 °C [2]	–	N/A [2]	–
Cholinium glutarate [Ch][Glu]	–	–	–	At 25 °C [2]	–	At 25 °C [2]	–	At 25 °C [89]	–
Cholinium glycolate [Ch][Gly]	At 15 °C [35] / At 25 °C [35] / At 35 °C [35]	At 15 °C [88]	At 25 °C [90]	At 25 °C [90]	At 25 °C [90]	–	At 15 °C [88]	–	–
Cholinium lactate [Ch][Lac]	At 15 °C [35] / At 25 °C [35] / At 35 °C [35]	At 15 °C [88]	N/A [90]	At 25 °C [90]	At 25 °C [90]	–	At 15 °C [88]	–	–

Cholinium levulinate [Ch][Lev]	—	—	—	—	—	—	N/A [2]	—	—	At 25 °C [89]	—
Cholinium L-malate [Ch][L-Ma]	—	—	—	—	At 25 °C [2]	—	At 25 °C [2]	—	N/A [2]	—	—
Cholinium malonate [Ch][Mal]	—	—	—	—	At 25 °C [2]	—	At 25 °C [2]	—	N/A [2]	—	—
Cholinium oxalate [Ch][Ox]	—	—	—	—	At 25 °C [2]	—	—	—	N/A [2]	—	—
Cholinium propionate (or propanoate) [Ch][Pro]	At 15 °C [35] At 25 °C [35] At 35 °C [35]	—	—	At 15 °C [88]	N/A [90]	—	N/A [90]	—	At 15 °C [88]	—	—
Cholinium salicylate [Ch][Sal]	—	—	—	—	—	—	—	—	—	At 25 °C [89]	At 25 °C with: $C_6H_5K_3O_7$/$C_6H_8O_7$ KH_2PO_4/K_2HPO_4 K_2CO_3 [63]
Cholinium succinate [Ch][Suc]	—	—	—	—	At 25 °C [2]	—	—	—	At 25 °C [2]	At 25 °C [89]	—
Di-cholinium oxalate [Ch]$_2$[Ox]	At 25 °C [35]	—	—	—	—	—	—	—	—	—	—
Tri-cholinium citrate [Ch]$_3$[Cit]	At 25 °C [35]	—	—	—	—	—	—	—	—	—	—

(continued)

Table 11.2 (continued)

Ionic liquids	Coexisting phase forming agents								
	Polymers							Salts	
	Polypropylene glycol		Polyethylene glycol				Copolymer		
	PPG-400	PPG-1000	PEG-400	PEG-600	PEG-1000	PEG-4000	$EO_{10}PO_{90}$	K_3PO_4	Other
Benzyldimethyl(2-hydroxyethyl)ammonium chloride [BCh]Cl	–	–	–	–	–	–	–	At 25 °C [89]	At 25 °C with: $C_6H_5K_3O_7$/ $C_6H_8O_7$ KH_2PO_4/ K_2HPO_4 K_2CO_3 [63]

[a]N/A – ABS formation was not observed

[Ch][Sal] > [BCh]Cl > [Ch][Lev] > [Ch][Glu] ≈ [Ch][Suc] ≈ [Ch][Ac] > [Ch]Cl [89]. This is in a good agreement with the findings for more conventional ILs families, such as imidazolium-based ILs, where the more hydrophobic ILs are more easily salted out by K_3PO_4 [91, 92]. The phase formation is driven by the competition of the IL and salt species for water molecules, and the parameters of the immiscible region were determined by the relative hydrophilic/hydrophobic character of the IL compared to the high charge density salt [89]. Accordingly, ILs with anions having more aliphatic or aromatic moieties form ABS more readily (salted out), while the ones with more –OH groups less likely induce phase separation [89].

Sintra et al. [63] studied the formation of ABS by mixing members of several IL families, including the cholinium-based [Ch][Sal] and [BCh]Cl, with appropriate aqueous solutions of $C_6H_5K_3O_7/C_6H_8O_7$, KH_2PO_4/K_2HPO_4, and K_2CO_3. Although the authors [63] showed the expected two-phase regions at 25 °C, their aptitude to form ABS was found to be lower than that of the conventional ILs [63]. When the authors compared the effect of each salt/buffer solution on the phase-forming ability of [Ch][Sal], the following tendency was found: KH_2PO_4/K_2HPO_4 > $C_6H_5K_3O_7/C_6H_8O_7$ > K_2CO_3. Since this was somewhat different from the expectations, where the formation of an ABS is directly related with the affinity of the salt component toward being hydrated, it was deduced that the dominating interactions in these systems are more complex and a more comprehensive model is needed for the future to explain the liquid-liquid demixing [63].

Ch-ILs are capable of forming ABS in the presence of inorganic/organic salts, but, as observed by Shahriari et al. [89], even when strong salting-out species, such as K_3PO_4, are applied, the phase separation potential of the given system could be lower than in the case of other, more conventional IL families. The presence of –OH groups and/or aliphatic/aromatic regions in the ILs has a large influence on their ability to form two phases with salt solutions as pointed out earlier. However, this attribute can be exploited to use Ch-ILs as tunable phase promoting agents. This idea was first presented by the Wang group [35] in 2012, when they intended to design novel, more environmentally friendly ABS systems for protein extraction by combining Ch-ILs with the less hydrophilic PPG-400 polymer instead of an inorganic salt. In this work, the Ch-ILs acted as the phase-forming/salting-out agent. During the study of nine Ch-ILs conjugated with several organic acid anions, the phase-forming potential of the ILs was found to decrease in the following order: $[Ch]_3[Cit]$ > $[Ch]_2[Ox]$ > [Ch][Gly] > [Ch][Pro] ≈ [Ch][Lac] ≈ [Ch][OAc] > [Ch][For] > [Ch][But] [35]. The hydration capacity of the IL anion, and so its capability to salt out PPG-400, appeared to be the main driving force behind the ABS formation and thus establishing the observed trend. Consequently, the ILs with citrate and oxalate anions, the ones with the strongest salting-out character, appeared to be the most potent to induce the formation of an immiscibility region with aqueous solutions of PPG-400 [35]. Additionally, the authors [35] obtained the phase diagrams for three specific systems, [Ch][Pro]/PPG-400/H_2O, [Ch][Gly]/PPG-400/H_2O, and [Ch][Lac]/PPG-400/H_2O, at 15, 25, and 35 °C in order to assess the influence of temperature on the phase behavior. The results were in agreement with previous studies on conventional ABS composed of PPG and inorganic salts [93–95]; an increase of temperature enhances the liquid-liquid demixing region. As highlighted by the authors [35], these findings are in accordance with the salting-out mechanism of the IL over PPG-400, since PPG becomes more and more hydrophobic at elevated temperatures and, consequently, water will be drawn from the polymer-rich phase to the IL-rich phase [35].

Later, the same group of researchers [88] performed a more detailed study in order to have a better understanding of the phase equilibrium of ABS formed by Ch-ILs and PPG [88]. Phase diagrams of four different Ch-ILs in combination with PPGs of different molecular weight (MW 400 and 1000 g·mol^{-1}) or a PEG-PPG copolymer ($EO_{10}PO_{90}$) were measured; this way, the effects of (1) the type of anion present in the IL, (2) the hydrophilic/hydrophobic nature of the polymers, and (3) the temperature could be analyzed. The chemical makeup of the anion had an effect on phase separation similar to those previously mentioned; more hydrophilic anions had a more pronounced ability to induce ABS ([Ch][Lac] ≈ [Ch][Gly] > [Ch][Pro] > [Ch]Cl) in the presence of PPG-1000 [88]. The change in the MW of the polymer also strongly influenced the phase equilibrium as the ones with longer chains ($EO_{10}PO_{90}$ and PPG-1000) were salted out more easily with even less IL due to their more hydrophobic character [88]. The effect of temperature variation on the Ch-ABS closely followed the trend observed in their earlier work with PPG-400 [35], where an increase in temperature enhanced the immiscibility region [88].

Both studies performed by Wang and coworkers [35, 88] clearly demonstrated how effectively Ch-ILs can salt out PPG polymers and form ABS. However, in order to develop liquid-liquid extraction systems with coexisting aqueous-rich phases, it would be desirable to use more hydrophilic polymers. Accordingly, at the beginning of this decade, Coutinho and coworkers [3, 87, 90] started to study how the formation of ABS can be initiated using different combinations of ILs and polyethylene glycol polymers. Contrary to the simple salting-out effect of cholinium-based ILs in Ch-IL/PPG ABS reported by the Wang group [35, 88], the studies on IL-PEG ABS using conventional ILs (such as imidazolium, piperidinium, etc.) suggested that the molecular phenomenon behind the formation of an immiscible regime is more complex and intricate [3]. In order to gain deeper understanding of the IL/PEG ABS, recently, phase diagrams for several ternary systems composed of water, Ch-ILs (or salts), and PEGs (PEG-400, PEG-600, and PEG-1000) at different temperatures were determined [90]. The cholinium-based compounds were distinguished according to their melting points (mp): five salts with high mp (in the range of 103–302 °C), two ILs with mp below 100 °C but above room temperature, and three that are liquid at room temperature [90]. Interestingly, this broad study showed that two of the liquid salts, [Ch][Pro] and [Ch][But], were unable to promote phase demixing when combined with any of the PEGs tested (PEG-400, PEG-600, and PEG-1000), while [Ch][Lac] only forms ABS when mixed with the higher MW PEGs (600 and 1000 g·mol^{-1}) as highlighted in Table 11.2 [90]. It was observed that all the other cholinium salts could form ABS with any of the PEGs tested and, in general, exhibited the following trend in phase separation potential: [Ch][DHph] > [Ch][Bit] > [Ch][Bic] > [Ch][DHcit] ≈ [Ch][OAc] ≈ [Ch][Lac] ≈ [Ch][Gly] > [Ch]Cl [90].

Based on the equilibrium phase diagrams, as well as on the solubility of each cholinium-based compound in PEG or water, the water activities of the species studied, and the binary and ternary excess enthalpies, the ability to promote an ABS seems to be quite different for higher melting cholinium-based salts than for the lower melting or liquid Ch-ILs. It was demonstrated that in case of the cholinium-based salts with high mp, the formation of an ABS is controlled by the solvation of the salt anion (its salting-out aptitude over PEG), similarly to the phase demixing of the conventional PEG/inorganic salt or IL/inorganic salt ABS [90]. On the other hand, aqueous solutions containing Ch-ILs with mp lower than 100 °C and PEG go through phase separation driven by a more complex mechanism, which is related to the balance of competition between all the components in equilibrium to form hydrogen bonds [90]. It became evident that hydrogen bonding interactions between the IL anion and the terminal –OH groups of the polymer are more predominant, allowing a fine adjustment of the biphasic systems [90].

Later on, Mourão et al. [2] investigated the formation of ABS employing hydrophilic Ch-ILs (or salts) in combination with PEG polymers. The authors measured the ternary phase diagrams at 25 °C for seven Ch-ILs with anions derived from naturally occurring organic acids and thus evaluated the influence of different substituent groups and alkyl chains with various lengths [2]. After the evaluation of the influence that different alkyl chains embedded into cholinium IL dicarboxylate anions can have on ABS formation, the following order was observed: [Ch][Glu] > [Ch][Suc] > [Ch][Ox] > [Ch][Mal] [2]. In general, the longer the alkyl chain of the IL anion, the larger is the biphasic region. Initially, the authors [2] assumed this trend as consequence of the salting out of the more hydrophobic ILs by the more hydrophilic PEG; however, the results of water activities of these Ch-ILs urged the authors to revise this conclusion, since the most hydrophilic [Ch][Glu] was the IL that showed the highest capability to form ABS [2]. The effect of the substituent groups attached to the anion was probed with the [Ch][Fum] and [Ch][L-Ma] systems, and their phase diagrams were compared with those obtained for [Ch][Bit] and [Ch][Suc]. The data provided the following ABS forming tendency: [Ch][Fum] > [Ch][Bit] > [Ch][L-Ma] > [Ch][Suc] [2]. [Ch][Bit] and [Ch][L-Ma] are two high-melting salts; however, in contrast to previous findings observed by Pereira et al. [90], no differences were observed in their binodal curves. It is important to note that Mourão et al. [2] presented the binodal curves in molality units (mol/kg), while in Pereira's work [90] they were expressed as initial, total weight fraction of PEG (wt%) instead of initial, total molality of salt (mole of salt per kg of H_2O). We have been trying to alert the scientific community in the field, that the binodal curves should be displayed in units that do not hide trends due to the high MW of polymers when compared with the MW of ILs or water. Thus, although the authors observed similar shapes in the binodal curves, some intricate phenomena could have stayed hidden.

The authors [2] compared the results from the phase equilibria to the ones on water activities for those cholinium-based compounds, but could not observe any relationship. Thus, the authors [2] assumed the predominance of the PEG-600-IL interactions over the water-IL and water-PEG-600 interactions as a possible explanation [2]. In agreement with our previous work [90], Mourão et al. [2] affirmed that the mechanism behind phase splitting in systems containing very hydrophilic Ch-ILs and hydrophilic PEG polymers is not always obvious; ABS formation can be a result of a delicate balance between all possible interactions (PEG-water, PEG-ion, water-ion). Additionally, the authors also evaluated how MW increase of PEG chains affects the phase equilibria by testing all of the seven Ch-ILs (or salts) in combination with PEG-4000. Behavior and trends, similar to those for PEG-600, were observed [2].

Phase separation was also investigated by Mourão et al. for Ch-ILs/K_3PO_4 systems as well, to gain more mechanistic insight [2]. As presented in Table 11.2, five of the Ch-ILs tested were not able to induce a two-phase system when mixed with the strong salting-out K_3PO_4; the two exceptions, [Ch][Glu] and [Ch][Suc] [2], were in agreement with the earlier description of these systems [89]. On the other hand, the majority of Ch-ILs studied did promote two immiscible phases in the presence of aqueous PEG solutions [89]. Similarly to our earlier remarks [90], however, it was noted by the authors that the majority of the cholinium monocarboxylate ILs did not undergo phase separation with PEGs (especially with the ones having low MW), while the IL anions derived from di- and tricarboxylic acids exhibited an aptitude for biphase formation [2].

Table 11.3 summarizes the main conclusions drawn from the published phase diagrams of Ch-IL/K_3PO_4, Ch-IL/PPG, and Ch-IL/PEG systems based on the

Table 11.3 Summary of Ch-ABS and the suspected mechanisms behind phase separation

Ch-ABS type	Ability to induce ABS formation	Mechanism
K_3PO_4 [89]	[Ch][Sal] > [BCh]Cl > [Ch][Lev] > [Ch][Glu] ≈ [Ch][Suc] ≈ [Ch][Ac] > [Ch]Cl	Salting-out aptitude of K_3PO_4 over Ch-ILs
PPG-400 [35]	[Ch]$_3$[Cit] > [Ch]$_2$[Ox] > [Ch][Gly] > [Ch][Pro] ≈ [Ch][Lac] ≈ [Ch][OAc] > [Ch][For] > [Ch][But]	Salting-out aptitude of Ch-ILs over PPG-400
PPG-1000 [88]	[Ch][Lac] ≈ [Ch][Gly] > [Ch][Pro] > [Ch]Cl	Salting-out aptitude of Ch-ILs over PPG-400
PEG-600 [90]	[Ch][DHph] > [Ch][Bit] > [Ch][Bic] > [Ch][DHcit] ≈ [Ch][OAc] ≈ [Ch][Lac] ≈ [Ch][Gly] > [Ch]Cl	*Ch-salts:* solvation of salt anion (salting-out aptitude) *Ch-ILs:* complex hydrogen bonding competition between all the components of the system
PEG-600 [2]	[Ch][Glu] > [Ch][Suc] > [Ch][Ox] > [Ch][Mal] – effect of alkyl chain length [Ch][Fum] > [Ch][Bit] > [Ch][L-Ma] > [Ch][Suc] – effect of substituent groups	Complex interactions between PEG-600 and ILs

discussions presented in this section and in previously published research. Although the number of phase diagrams obtained for cholinium-based compounds is not yet significant, tendencies in the mechanism of phase separation can be observed in Table 11.3. When Ch-ILs are mixed with K_3PO_4, the preferential solvation of the inorganic salt appears to be the driving force toward ABS formation. On the other end of the spectrum, the mixture of Ch-IL and PPG, the most hydrophobic of the tested phase inducers, will separate due to the higher affinity of the IL ions toward water molecules. However, when the systems incorporating PEG polymers are studied, the mechanism controlling the phase splitting at the molecular level becomes far more complex since all the dissolved components can contribute significantly to the solvation processes. The most accepted mechanism describing the behavior of these systems is based on a balance of competition for hydrogen bonding between the ternary phase components.

Despite the complexity of the ABS composed of Ch-ILs and the abovementioned polymers, these systems are among the most interesting and promising ones to be utilized as a sustainable alternative in several industrial processes, mainly due to their environmentally friendly characteristics, including high biodegradability, low toxicity, high biocompatibility, and low cost.

11.3.2 Amino-Acid-Based ABS

Amino acids are one of the most abundant groups of natural compounds. Their applicability in the synthesis of new amino acid-based ABS, both as ILs and/or phase-forming agents, has been meticulously studied by several research groups (see Sects. 11.2.1 and 11.2.2). The introduction of amino acids in these complex materials is hoped to increase the sustainability and environmentally friendly of the related synthesis and separation processes.

The first successful syntheses of ILs from 20 natural amino acids were published in 2005 by Fukumoto et al. [50]. Shortly after that, as noted in Sect. 11.2.1.2, several other classes of AA-ILs were obtained, some of them with high biodegradability and biocompatibility. Additionally, the majority of these IL families are easy to obtain in large quantities at low cost, which also makes them promising candidates for separation and extraction processes [96, 97]. Since their first report about amino acid ILs [50], Ohno group [96, 97] has focused their studies on the ability of these ILs to promote two immiscible phases when mixed with water above certain concentrations and at an appropriate temperature. They found [96, 97] that mixtures of water and ILs composed of phosphonium or ammonium cations and various anions derived from amino acids showed phase separation with a lower critical solution temperature (LCST), suggesting that these IL-water mixtures can later be applied as functional fluids for reversible phase transitions (i.e., can be reversed by temperature change or bubbling gases) and used as a rapid separation method for extraction of water-stable proteins. However, these LCST-type IL-water systems, despite the large number of systems published by the

Ohno group [96, 97] and their foreseeable future applications, cannot be considered as true ABS; ABS are composed of two non-mixing yet water-rich phases, while one of the phases of these IL-water mixtures contains mainly hydrophobic ILs with only traces of water. Nevertheless, attempts to obtain true ABS with AA-ILs have been carried out by other researchers.

In 2011, Wu et al. [98] published the first ternary phase diagrams for mixtures of AA-ILs, K_3PO_4, and water. The authors studied the phase equilibria of the obtained ABS in order to gain more insight onto the impact of the structural diversity of the IL components on the relative hydrophobicity and immiscibility of the phases. The extractive capabilities of these systems for proteins were also examined by measuring the partition coefficients as a function of pH and hydrophobicity of the IL for cytochrome-c as a model protein. In their first study, they prepared a series of ILs with $[C_4mim]^+$ as cation and derivatives of four amino acids, L-serine ($[Ser]^-$), glycine ($[Gly]^-$), L-alanine ($[Ala]^-$), and L-Leucine ($[Leu]^-$), as the corresponding anion. The solubility curves of these AA-IL/K_3PO_4 ABS at 25 °C showed that the capability to form two distinct phases decreased in the following order: $[C_4mim]$[Leu] > $[C_4mim]$[Ala] > $[C_4mim]$[Gly] > $[C_4mim]$[Ser], a trend resulted from the hydrophobic nature of the amino acid anions [98]. This suggests that the mechanism driving the phase separation is similar to the one hypothesized for Ch-ILs/K_3PO_4 ABS [89]; the liquid-liquid demixing is a consequence of the competition between the IL ions and the strong salting-out inorganic salt for hydration, where the more hydrophobic IL is easier to be salted out by K_3PO_4 to form another phase [98].

In order to clarify the relationship between relative hydrophobicity of the phases in the AA-ILs ABS and the ability to induce phase separation, the authors [98] calculated the amount of Gibbs free energy required to transfer a nonpolar –CH_2 group from the bottom salt-rich phase to the top AA-IL-rich phase ($\Delta G_T(CH_2)$). The $\Delta G_T(CH_2)$ values measured by Wu et al. [98] are negative, indicating the favorable transfer of CH_2 from the bottom phase to the top phase and consequently the more hydrophobic nature of the AA-ILs-rich phase [98]. The observed relative hydrophobicity values showed the same trend among the systems tested as the one obtained while analyzing the phase separation efficiency [98]. Interestingly, it was also found that the relative hydrophobicities of these AA-IL/salt ABS are equivalent to hydrophobicities of polymer/salt ABS and higher than those obtained for polymer/polymer ABS.

Despite their capability to induce liquid-liquid demixing, this series of AA-ILs exhibits some non-benign and non-environmentally friendly characteristics, due to the $[C_4mim]^+$ cation employed. In order to overcome these environmental issues, the same research group proposed alternative solutions in which amino acids are combined with ammonium or phosphonium cations, thus being less toxic for aquatic environments than the imidazolium- (or pyridinium)-based ILs [99]. They synthesized five glycine-based ILs, four of them were prepared with ammonium-based cations (tetramethylammonium glycine, $[N_{1111}]$[Gly]; tetraethylammonium glycine, $[N_{2222}]$[Gly]; tetra-n-butyl-ammonium glycine, $[N_{4444}]$[Gly]; tetra-n-pentylammonium, $[N_{5555}]$[Gly]) and one with tetra-n-butylphosphonium cation

([P$_{4444}$][Gly]). In this manner, they were able to gain information about how the structure of the IL cation affects phase formation, phase hydrophobicity, and the phase equilibrium after mixing these AA-ILs with aqueous solutions of K$_2$HPO$_4$ at 25 °C. The results indicated that, with the exception of the [N$_{1111}$][Gly]/K$_2$HPO$_4$ system, immiscible phases were successfully promoted and the phase-forming ability decreased in the following order: [N$_{5555}$][Gly] > [P$_{4444}$][Gly] > [N$_{4444}$][Gly] > [N$_{2222}$][Gly] [99]. As expected from previous observations on other IL/salt ABS, the formation of ABS was controlled by the relative hydration of the ions in equilibrium, especially that of K$_2$HPO$_4$ being the provider of the most competitive ions, thus salting out the IL into the other phase. The analysis of the relative hydrophobicity between the coexisting phases showed the same trend found to illustrate the ability to induce phase separation; the system containing the [N$_{5555}$][Gly] IL exhibited the highest hydrophobicity while the one with [N$_{2222}$][Gly] the lowest [99].

Although the synthesis of AA-ILs opened a wide window of new applications, their use in the formation of ABS is quite limited yet. The two works considered in this section proved that this new class of ILs should be taken into account as an interesting alternative to the conventional ILs that are already heavily studied for different ABS-based separation processes. We would like to emphasize again, however, how important it is to consider the toxicity, sustainability, and environmentally friendly of both the IL cations and anions during their design and application in ABS. A good example might be the class of new cholinium-amino-acid ILs, members of which were recently synthesized and characterized [19].

11.3.3 Other Biocompatible IL-ABS

In two previous subsections, we provided examples of biocompatible ABS formulations based on cholinium- and various amino acid-based ILs. Besides these, other new classes have been recently discovered and reported using benign, coexisting phase-forming agents like biodegradable organic salts, carbohydrates, and bio-friendly polymers. We will briefly go through these classes of ABS discussing their phase equilibria and the possible mechanisms behind their formation.

One of the classes of new, more biocompatible ABS are based on ILs generated with biological buffers (Good's buffers (GBs)) as described in Sect. 11.2.1.3. [36]. Taha et al. [36] synthesized a series of 20 new ILs conjugating five GBs (Tricine, TES, CHES, HEPES, and MES) with four hydroxide bases (1-ethyl-3-methylimidazolium, [C$_2$mim]$^+$; tetramethylammonium, [N$_{1111}$]$^+$; tetraethylammonium, [N$_{2222}$]$^+$; tetrabutylammonium, [N$_{4444}$]$^+$) and then analyzed their environmental impact and biocompatibility [36]. As a next step, they tried to apply some of the GB-ILs in ABS formulations by mixing the obtained ILs with aqueous solutions of inorganic/organic salts. They were able to achieve phase separation with both Na$_2$SO$_4$ and potassium citrate (K$_3$C$_6$H$_5$O$_7$) for every

[N₄₄₄₄]⁺-based GB-ILs they tested [36]. The binodal curves of the ternary phase diagrams clearly displayed that the ILs with more hydrophobic buffer anions, such as CHES or MES, are more easily salted out than the GB-ILs having more hydrophilic anions (HEPES, TES, or Tricine) [36]. These observations are in parallel with the results of earlier IL/salt ABS experiments, thus the mechanism controlling the phase equilibria at a molecular level is most probably the same competition between the ions in equilibrium for hydration. The differences in the phase diagrams of systems with Na_2SO_4 and $K_3C_6H_5O_7$ were also pointing in the same direction; the inorganic salt with high charge density is a stronger salting-out agent and consequently induced two-phase formation much easier than the weakly salting-out organic salt [36]. Even though the ABS composed of GB-ILs and potassium citrate exhibited lower capability to promote liquid-liquid demixing, the more environmentally friendly nature of the organic salt employed, compared to Na_2SO_4, still makes their application preferable.

Taha et al. [36] also compared the phase equilibria of these new GB-ILs-based ABS with the more conventional ABS composed of [N₄₄₄₄]Cl IL and potassium citrate, a system previously examined by Passos et al. [58]. All the [N₄₄₄₄][GB] ILs exhibited higher ability to form ABS than [N₄₄₄₄]Cl, since these GB-ILs are easier to salt out by $K_3C_6H_5O_7$ than the ILs comprising Cl⁻. Thus, these biological buffer ILs/organic salt systems can be effective and bio-friendly phase formers with an additional benefit of delicate pH control in the separation media, a useful process parameter when proteins or biopharmaceuticals have to be handled in a more biocompatible manner. However, biocompatibility can be enhanced even in the cases of more conventional ABS, like the ones tested by Passos et al. [58] containing [N₄₄₄₄]Cl and [P₄₄₄₄]Cl ILs, by the environmentally conscious selection of the coexisting phase-forming agents.

Efforts have been made in order to improve the "greener" character of frequently used ABS through the application of less toxic and more environmentally friendly phase-forming agents. As recently reviewed by Freire et al. [12], many of the approaches, however, were initiated from a false starting point by adding biocompatible phase formers into systems composed of highly toxic IL families. Two studies, for example, confirmed the formation of ABS for the combination of ILs and amino acids, but unfortunately the ILs tested (imidazolium- and tetrafluoroborate-based ILs) were members of nonbiodegradable families and considered to be highly toxic to aquatic life forms. Similar works were also published describing the combination of carbohydrates or biocompatible polymers and not very environmentally friendly ILs, with the exception of Ch-ILs, to form aqueous biphasic systems.

Another halfway approach was recently published by Wang and coworkers [37, 100]. They suggested the use of guanidinium-based ILs, as more biocompatible phase-forming ABS promoters, supposedly suitable for protein extractions. Their shorter synthesis time and better designability could be advantageous over traditional ILs; however, as we highlighted in Sect. 11.2, more information is needed about their biodegradability and aquatic toxicity. The authors analyzed the phase equilibria of 1,1,3,3-tetramethylguanidinium acrylate [TMG][Acr] systems in the presence of inorganic salts, concluding salting-out strength in the following order: $K_3PO_4 > K_2HPO_4 > KH_2PO_4 > K_2CO_3$ [100]. They also evaluated the phase-forming ability of nine guanidinium-based ILs when mixed with aqueous solutions of K_2HPO_4 and observed a trend that follows: acetate > sorbate > itaconate > acrylate > methacrylate > lactate > cinnamate > maleate [100]. Based on this trend, Wang et al. [100] proposed that the IL viscosity, which depends on the IL anion, may play a key role in the phase separation mechanism. The same group then went further down this road by synthesizing another series of guanidinium-based ILs and testing their ABS forming efficiency with aqueous solutions of K_2HPO_4 [37]. The ternary phase diagrams showed differently shaped curves for the guanidinium-based ILs, which could be related to many complex factors, such as hydration capacity (IL components with higher charge density are strongly hydrated and consequently decrease the number of water molecules available for hydration of the salt ions) or longer alkyl chain length that leads to easier liquid-liquid demixing [37].

Sintra et al. [63] obtained phase diagrams for systems formed by cholinium-, phosphonium-, or ammonium-based ILs (or salts) and aqueous solutions of biodegradable salts, in particular, potassium citrate buffer. It was concluded that, in general, phase separation was controlled by the pH and the length of the alkyl chains appended to the IL ions, that is, the hydrophobic/hydrophilic nature of the IL, and their susceptibility toward being salted out [63].

Herein we summarized the major recent advancements in creating biocompatible ABS, highlighting the most important classes of ILs able to create them and showing the main effects governing the liquid-liquid demixing. However, as we have seen, judging results and chemical products can be troublesome when even the concepts of "biocompatibility" and "environmentally friendly" are too ambiguous themselves.

11.4 Applications of Biocompatible Aqueous Biphasic Systems

The major applications of IL-based ABS, proven and/or predicted, could be the extraction and/or purification of a wide range of solutes from simple alcohols to complex proteins and enzymes and the recovery of the ILs themselves from aqueous media [12]. Bio-ABS should be able to fulfill the same or similar roles; however, in most cases, experimental results describing phase equilibria and, in some reports, intended applications tested on model compounds are the only data yet available.

Proposed applications of biocompatible ABS are summarized in Table 11.4. The ABS considered were prepared using new biocompatible ILs families, cholinium-, amino-acid-, and guanidinium-based ILs; guanidinium-based ILs were included as

Table 11.4 Applications of biocompatible ABS

Type of Bio-ABS	Ternary system	Application
Ch-ILs-based ABS	Salt + K_3PO_4 + H_2O	Separation of antibiotics: *tetracycline, tetracycline-HCl, ciprofloxacin-HCl* [89]
		Recovery of *saponins* and *polyphenols* from two matrixes (tea and mate) [103]
	Salt + K_2HPO_4 + H_2O	Extraction of proteins: *bovine serum albumin (BSA)* [101]
	Salt + PEG-600 + H_2O	Extraction of antibiotics from fermentation broth: *tetracycline* [102]
		Extraction/separation of antioxidants: *tert-butylhydroquinone (TBHQ)* [2]
	Salt + PPG-400 + H_2O	Partitioning of proteins: *lysozyme, papain, trypsin, BSA* [35]
AA-ILs-based ABS	IL + K_3PO_4 + H_2O	Partitioning of proteins: *cytochrome-c* [98]
	IL + K_2HPO_4 + H_2O	Partitioning of proteins: *cytochrome-c* [99]
Guanidinium-ILs-based ABS	IL + K_2HPO_4 + H_2O	Extraction of proteins: *lysozyme, BSA, ovalbumin (OVA), bovine hemoglobin (BHb)* [100]
		Purification of proteins: *lysozyme, trypsin, OVA, BSA* [37]
GB-ILs-based ABS	IL + $K_3C_6H_5O_7$ + H_2O	Extraction of proteins: *BSA* [36]

benign classes, since these ABS are suggested to be able to extract proteins with low degradation rates [37]. Other ABS that use conventional ILs with more benign coexisting phase-forming agents (polymers, organic salts, carbohydrates, etc.) were recently reviewed by Freire et al. [12].

As it can be seen from Table 11.4, the majority of Bio-ABS, 7 out of the 11 systems shown, are intended for extraction and/or separation of various proteins, while other possible extractive applications were also suggested (extraction of antibiotics, recovery of saponins and polyphenols, and separation of antioxidants). The next two subsections will discuss these applications in the same order, starting with the extraction of proteins.

11.4.1 Protein Extraction and Purification

The list of biotechnological applications of proteins and enzymes (therapeutic biopharmaceuticals, diagnostic biomarkers, etc.) is continuously growing. The traditional methods to extract and purify proteins are too complex, thus too expensive, time-consuming, and hard to scale-up. Moreover, most organic solvents usually applied in these separation processes are harmful to these sensitive and extremely unstable molecules and can cause their denaturation and consequently the loss of their biological activity. Thus, there is an urgent need for more benign processes, and the use of benign and biocompatible ILs in stable, water-rich, liquid-liquid extraction systems seems to be the ideal approach for a more "natural" step in protein purification.

In the seven studies summarized in Table 11.4, the applicability of Bio-ABS for protein extraction was frequently proved with model proteins/enzymes (cytochrome-c, BSA, OVA, BHb, lysozyme, papain, trypsin). Most of these studies were tested to examine the partitioning mechanisms, the interaction phenomena governing protein migration between the phases, and the effect of the ABS on protein stability. The high purification efficiency of Ch-ILs-based ABS was observed even in the earliest work of the Wang group in 2012 [35]. After one single-step extraction of four model proteins, yields in the range of 86.4–99.9 % were achieved with a trend in extraction efficiency decreasing in the following order: lysozyme > papain > trypsin > BSA [35]. The authors [35] justified that the size of the protein has a major effect on its extractability; the transfer of a protein molecule into the IL-rich phase depends on the interactions of the phase-forming components that have to be broken in order to create a cavity where the protein will be entrapped. Thus, larger proteins require larger cavities that need more energy to break solvent interactions [35].

The influence of two other parameters, tie line length (TLL) and pH, was also examined [35]. It was found that the IL will be more concentrated in the bottom, IL-rich phase with increased TLL, and consequently the energy required to disrupt the IL-water network will be higher, decreasing the partitioning coefficient of the protein [35]. The effect of pH change, however, could not be explored reliably as it was observed that, when the pH was adjusted between pH 5 and 11, more than 86 % of the protein molecules were already transferred into the IL-rich phase and thus not sensitive to any change in the pH administered in the other phase [35].

This extensive study [35] also incorporated the evaluation of the change in enzymatic activity accountable to Ch-ILs/PPG-400 ABS and the recyclability of the PPG polymer. According to their results [35], the addition of some ILs (especially [Ch][Lac]) to water mixtures can apparently enhance the activity and thermal stability of certain enzymes, trypsin in particular. On the other hand, the PPG-400 phase-forming polymer could easily be recovered and recycled by convenient heating [35]. All these interesting data obtained by the Wang group [35] are fundamental to show how to increase the more benign and biocompatible character of ABS, combining cholinium-based ILs and polymers, which suggests that their application can be a viable alternative to the common protein separation processes.

Following this work, Huang et al. [101] compared the extraction efficiency of a cholinium-like IL-based ABS to that of imidazolium-based IL-ABS, all formed with aqueous solutions of K_2HPO_4, using BSA as a model protein. The authors [101] found the cholinium-like IL-ABS to be the most efficient extractive system, and, after optimizing the most crucial parameters (temperature, extraction time, IL-salt content), yields of 92–100 % were achieved in single-step extractions [101]. They also determined that intermolecular processes, micelle formation, and salting-out effects were the major driving forces in the protein transfer process, while aggregation phenomena were responsible for protein separation [101]. Analytical results (including UV-Vis, FLS, FT-IR, and TEM) also proved that the conformation of the model protein BSA was not altered by this IL extraction process, indicating that this system has a potential future in biotechnological separations [101].

Wu et al. [98, 99] reported similar observations with AA-ILs. These ILs were synthesized with imidazolium cations [98] or ammonium or phosphonium cations [99] and then ABS were prepared with K_3PO_4 [98] or K_2HPO_4 [99] solutions, and their ability to extract cytochrome-c was tested. The determination of cytochrome-c partition coefficients and the polarity of the corresponding ionic-liquid-rich phase suggested that hydrophobic interactions between the IL and cytochrome-c are mainly responsible for achieving the higher partition coefficients [98, 99]. The authors also elucidated, however, that protein separation can be effected by other types of interactions as well, like specific electrostatic interactions between the ions of the AA-ILs and the protein or salting out of the protein by the inorganic salt [98, 99].

The occurrence of liquid-liquid demixing in aqueous mixtures containing 1,1,3,3-tetramethylguanidinium-based ILs was also proved [37, 100]. Zeng et al. [100] examined the extractive potential of the obtained ABS with three model proteins, namely, BSA, BHb, and OVA. The authors [100] maximized the extraction efficiency for every system using single factor experiments by optimizing time, temperature, and the concentration of the components. However, the efficiency and the mechanism of protein partitioning were observed to be dependent on the type of the protein and the IL as well. In the case of BSA, yields up to 99.62 % were achieved, while no change in the conformation of the protein was observed. The authors also demonstrated that the mechanism of protein partitioning was mainly controlled by the aggregation and embrace phenomena, which depends on the interactions of the phase-forming components that have to be broken in order to create a cavity where the protein will be entrapped [100].

Similar experiments with another set of model proteins (lysozyme, trypsin, OVA, and BSA) and guanidinium-IL-based ABS also proved that protein separation using these systems can be more effective than with conventional IL-ABS [37]. It was further concluded that the trend in the observed extraction efficiencies (in decreasing order: lysozyme > trypsin > OVA > BSA) was governed by the size effect of proteins, that is, larger proteins require a larger inclusion cavity to fit in, necessitating the separation of more interacting solvent components [37], a mechanism similar to the one discussed earlier with regard to Ch-ILs-based ABS [35]. These systems were also found to be very effective in protein separation without causing any change in the structural features and biological activity of the model subjects [37].

Taha et al. [36] performed similar experimental studies with GB-ILs. The authors [36] explored if these ILs can be employed as self-buffering compounds, protein stabilizers, and, ultimately, extracting agents in ABS formulations. GB-ILs with buffering capacity at pH 7 were combined with a biodegradable and nontoxic organic salt, potassium citrate, and then the extraction of the model protein BSA was examined. In every experiment, extraction efficiencies close to 100 % were observed, with no apparent signs of BSA loss in the mass balances, either by precipitation or denaturation [36].

Thus, as we saw in the examples mentioned in this subsection, these ABS, formulated with more benign and/or biocompatible ILs, have the ability to extract proteins with high efficiency without altering their conformation or biological activity. It has to be highlighted for future ABS design, however, that energy is required to transport the protein into the IL-rich phase, and this energy is apparently proportional to the size of the protein molecule and probably to the strength of the solvent-solvent interactions.

11.4.2 Approaches for Extraction of Other Biomolecules

The examples in Table 11.4 show that the extractive potential of Bio-ABS can probably be extended toward other solutes, such as antibiotics, antioxidants, saponins, and polyphenols. Antibiotics are high-added value pharmaceutical agents that can be naturally produced by some microorganisms, however, in relatively low concentrations. The separation of antibiotics is usually a complex and consequently costly process (processes), including extraction with organic solvents, precipitation or crystallization techniques, and/or even more expensive methods, such as ultrafiltration or chromatography. The extensive production and usage of pharmaceuticals, including antibiotics, on the other hand, have raised several concerns about their potential adverse effects on human health and the environment. Accidental release of pharmaceuticals/antibiotics, even by passing through the human body with no chemical change into the environment, combined with their usually low degradation rate, hence their accumulation and migration among water systems, have already generated severe issues [12]. To address these issues, the design and widespread application of more benign and environmentally friendly, and hopefully more economic, separation/concentration processes are necessary. ABS incorporating the new and biocompatible IL families are certainly viable alternatives for traditional methods and hold great future potential.

It was proposed in 2013 that ABS composed of cholinium salts and PEG-600 can be cheaper, more sustainable, and biocompatible methods for the extraction of tetracycline (used as a model) from the fermented broth of *Streptomyces aureofaciens* [102]. ABS, composed of aqueous solutions of PEG-600 and five different cholinium-based salts ([Ch]Cl, [Ch][OAc], [Ch][DHcit], [ChDHph], and [Ch][Bic]), were tested and optimized first with commercially available tetracycline, in order to gain information about the partitioning coefficients of tetracycline. The composition of the IL and the pH had the major influence on the partitioning coefficient of tetracycline; the highest values of partitioning were obtained for the most alkaline ABS, the [Ch][OAc]- and [Ch][Bic]-based ones, while the ABS composed of PEG-600 and [Ch]Cl were not able to induce a preferential partitioning [102]. The most fascinating observation was that, depending on the cholinium salt employed, tetracycline can be concentrated either in the polymer-rich or in the salt-rich phase, suggesting further options for future tailored applications [102]. When the two systems showing the highest and lowest partitioning, [Ch][Bic] and [Ch]Cl, respectively, were used to extract tetracycline directly from fermented broth, extraction efficiencies over 80 % were obtained [102]. These promising results further support the idea of applying these biocompatible Ch-ILs-based ABS as an alternative technique in a pre-purification stage of downstream processes designed to recover antibiotics from complex and natural matrices.

Shahriari et al. [89] also examined the applicability of ABS composed of Ch-ILs and K_3PO_4 for the separation of antibiotics using two model compounds, tetracycline and ciprofloxacin, and their hydrochloride forms [89]. In general, the observed partitioning coefficients decreased with the following rank: [Ch][OAc] > [Ch]Cl > [Ch][Lev] >> [Ch][Suc] > [Ch][Glu]. On the other hand, the IL-rich phases appeared to be more successful in extracting the hydrochloride forms in the following order: tetracycline· HCl > ciprofloxacin·HCl > tetracycline [89]. The mechanisms behind the migration of these antibiotics seem to be far more complex than the phenomenon observed for proteins, where the dominant factor was the size of the protein molecule. The authors supposed, however, that partitioning of these antibiotics may be influenced by specific ionization effects and hydrogen bonding interactions (since the portioning was favorable toward the IL-rich phase) and that the hydrochloride forms have higher affinity for water [89]. Although the extraction mechanisms are not well understood yet, this study provided another example of the potential applicability of the Ch-ILs-based ABS as a real alternative to the traditional extraction methods.

Attempts to extract other biomolecules with the aid of Ch-ILs-based ABS were published as well. The antioxidant *tert*-butylhydroquinone (TBQH) was extracted using four different IL/PEG-600 ABS, where the ILs of choice were [Ch][Mal], [Ch][Suc], [Ch][Fum], and [Ch][L-Ma] [2]. The obtained partition coefficients for the PEG-rich phase varied between 3.55 and 11.79, while the extraction efficiencies ranged from 83 to 93 % [2]. Interestingly, the results are in accordance to those measured when the partitioning of tetracycline in certain ABS, composed of cholinium-based salts and PEG-600, also favored the polymer-rich phase [102].

Ribeiro et al. [103] examined if [Ch]Cl/K_3PO_4 ABS were suitable to extract saponins and polyphenols from complex biological extracts of tea and mate. The high partitioning coefficients obtained suggest that the application of Ch-ILs-based ABS can be a viable method for the extraction of these metabolites even from complex natural matrices [103].

The examples in this section clearly illustrated that Bio-ABS have the potential to be used as biocompatible and stable methods for purification and separation of various biomolecules with high efficiency, even at the industrial level. Proper selection of phase-forming agents and optimized conditions can further improve selective partitioning of the desired compounds.

11.5 Conclusions and Future Prospects

Environmental concerns and the growing market for bio-products with added value have initiated new research activities to find and design more benign and biocompatible methods and processes to produce them. ABS incorporating ILs may one day replace conventional organic solvent-based extraction processes assuming that their environmental and economic impacts are reduced below a required level. This chapter tried to collect and explain the outcomes of the emerging research focusing on biocompatible ILs and ABS thereof. Various aspects and combinations of the three components required to form IL-based ABS, cation and anion of the IL complemented with a phase-forming agent, were discussed. The most promising ABS variants were those composed of cholinium- and/or amino-acid-based ILs, especially the ones that employ a cholinium cation and an amino acid anion. Both compound classes are abundant in nature, thus highly biocompatible, and therefore can be obtained from renewable natural sources, vehemently sought after and highly appreciated properties. The study of these ILs, however, is far from being complete leaving plenty of room available for further research.

Besides the research on more variations of Ch- and/or AA-ILs, it is also important to identify and thoroughly test possible candidates for the role of benign and biocompatible phase-forming agents. They can most likely be found among (bio)polymers, (bio)surfactants, and biodegradable salts. The cholinium-based ILs/PEG ABS can be considered as early examples since they exhibit strong potential toward the environmentally friendly extraction of various bio-products at low cost.

Although the research of these environmentally friendly IL-ABS has already produced promising results, the mechanisms governing the liquid-liquid demixing at the molecular level or the phenomena behind the migration of biomolecules between the phases are still not fully understood. The theoretical understanding of these processes can lead to further developments and, combined with necessary improvements in the modeling and simulation of the scale-up processes, will be highly beneficial at a future industrial level. The full picture obliges us, however, to continuously study ways of recovering phase-forming agents, as well as ILs, in order to minimize the environmental impact of the entire industrial production system. In summary, it is evident, even from the few studies reporting the use of Bio-ABS so far, that these systems can become an effective, environmentally friendly sustainable, and biocompatible alternative to the more traditional extraction systems, including IL-ABS, in the near future to obtain high-value proteins, such as biopharmaceuticals.

References

1. Ragauskas AJ, Williams CK, Davison BH, Britovsek G, Cairney J, Eckert CA, Frederick WJ Jr, Hallet JP, Leak DJ, Liotta CL, Mielenz JR, Murphy R, Templer R, Tschaplinski T (2006) The path forward for biofuels and biomaterials. Science 311:484–489

2. Mourao T, Tomé LC, Florindo C, Rebelo LPN, Marrucho IM (2014) Understanding the role of cholinium carboxylate ionic liquids in PEG-based aqueous biphasic systems. ACS Sustain Chem Eng 2:2426–2434
3. Freire MG, Pereira JFB, Francisco M, Rodríguez H, Rebelo LPN, Rogers RD, Coutinho JAP (2012) Insight into the interactions that control the phase behavior of new aqueous biphasic systems composed of polyethylene glycol polymers and ionic liquids. Chem Eur J 18:1831–1839
4. Chen XC, Xu GM, Li X, Li Z, Ying H (2008) Purification of an alpha-amylase inhibitor in a polyethylene glycol fructose-1,6-bisphosphate trisodium salt aqueous two-phase system. Process Biochem 43:765–768
5. Li SH, He CY, Gao F, Li D, Liu H, Li K, Liu F (2007) Extraction and determination of morphine in compound liquorice using an aqueous two-phase system of poly(ethylene glycol)/K_2HPO_4 coupled with HPLC. Talanta 71:784–789
6. Pereira JFB, Santos VC, Johansson HO, Teixeira JAC, Pessoa A Jr (2012) A stable liquid-liquid extraction system for clavulanic acid using polymer-based aqueous two-phase systems. Sep Purif Technol 98:441–450
7. Bora MM, Borthakur S, Rao PC, Dutta NN (2005) Aqueous two-phase partitioning of cephalosporin antibiotics: effect of solute chemical nature. Sep Purif Technol 45:153–156
8. Rosa PAJ, Ferreira AM, Azevedo MR, Aires-Barros MR (2010) Aqueous two-phase systems: a viable platform in the manuscript of biopharmaceuticals. J Chromatogr A 1217:2296–2305
9. Albertsson PA (1986) Partition of cell particles and macromolecules. Wiley, New York
10. Gutowsky KE, Broker GA, Willauer HD, Huddleston JG, Swatloski RP, Holbrey JD, Rogers RD (2003) Controlling the aqueous miscibility of ionic liquids: aqueous biphasic systems of water-miscible ionic liquids and water-structuring salts for recycle metathesis, and separations. J Am Chem Soc 67:6632–6633
11. Siedlecka EM, Czerwicka M, Neumann J, Stepnowski P, Fernández JF, Thöming J (2011) Ionic liquids: methods of degradation and recovery, ionic liquids: theory, properties, new approaches, Prof. Kokorin A (ed) ISBN: 978-953-307-349-1, InTech, doi:10.5772/15463
12. Freire MG, Cláudio ADM, Araujo JMM, Coutinho JAP, Marrucho IM, Canongia CJN, Rebelo LPN (2012) Aqueous biphasic systems: a boost brought about by using ionic liquids. Chem Soc Rev 41:4966–4995
13. Docherty KM, Kulpa JCF (2005) Toxicity and antimicrobial activity of imidazolium and pyridinium ionic liquids. Green Chem 7:185–189
14. Ventura SPM, Gonçalves AMM, Sintra T, Pereira JL, Gonçalves F, Coutinho JAP (2012) Designing ionic liquids the chemical structure role in the toxicity. Ecotoxicology 22:1–12
15. Petkovic M, Seddon KR, Rebelo LPN, Pereira CS (2011) Ionic liquids: a pathway to environmental acceptability. Chem Soc Rev 40:1383–1403
16. Fukaya Y, Iizuka Y, Sekikawa K, Ohno H (2007) Bio ionic liquids: room temperature ionic liquids composed wholly of biomaterials. Green Chem 9:1155–1157
17. Ohno H, Fukomoto K (2007) Amino acid ionic liquids. Acc Chem Res 40:1122–1129
18. Petkovic M, Ferguson JL, Gunaratne HQN, Ferreira R, Leitão MC, Seddon KR, Rebelo LPN, Pereira CS (2010) Novel biocompatible cholinium-based ionic liquids-toxicity and biodegradability. Green Chem 12:643–649
19. Hou XD, Liu QP, Smith TJ, Zong MH (2013) Evaluation of toxicity and biodegradability of cholinium amino acids ionic liquids. PLoS One 8:e59145
20. Wilkes JS (2002) A short history of ionic liquids—from molten salts to neoteric solvents. Green Chem 4:73–80
21. Earle MJ, Seddon KR (2000) Ionic liquids. Green solvents for the future. Pure Appl Chem 72:1391–1398
22. Silva FA, Siopa F, Figueiredo BFHT, Gonçalves AMM, Pereira JL, Gonçalves F, Coutinho JAP, Afonso CAM, Ventura SPM (2014) Sustainable design for environment-friendly mono and dicationic cholinium-based ionic liquids. Ecotoxical Environ Saf 108:302–310
23. Holbrey JD, Seddon KR (1999) Clean products and processes, vol 1. Springer, New York

24. Plechkova N, Seddon KR (2008) Applications of ionic liquids in the chemical industry. Chem Soc Rev 37:123–150
25. Tao G, He L, Sun N, Kou Y (2005) New generation ionic liquids: cations derived from amino acids. Green Chem 28:3562–3564
26. Yu Y, Lu X, Zhou Q, Dong K, Yao H, Zhang S (2008) Biodegradable naphthenic acid ionic liquids: synthesis, characterization, and quantitative structure–biodegradation relationship. Chem Eur J 14:11174–11182
27. Anastas PT, Warner JC (1998) Green chemistry: theory and practice. Oxford University Press, New York
28. Hough WL, Smiglak M, Rodríguez H, Swatloski RP, Spear SK, Daly DT, Pernak J, Grisel JE, Carliss RD, Soutullo MD, Davis JH Jr, Rogers RD (2007) The third evolution of ionic liquids active pharmaceuticals ingredients. New J Chem 31:1429–1436
29. Tavares APM, Rodríguez O, Macedo EA (2013) New generations of ionic liquids applied to enzymatic biocatalysis, Prof. Kadowaka J (ed) ISBN: 978-953-51-0937-2, InTech, doi:10.5772/51897
30. Egorova KS, Ananikov P (2014) Toxicity of ionic liquids: eco(cyto)activity as complicated, but unavoidable parameter for task-specific optimization. ChemSusChem 7:336–360
31. Dorland WAN (2011) Dorland's illustrated medical dictionary, 32nd edn. Saunders, Philadelphia
32. Pereira C, Ferreira R, Garcia H, Petkovic M (2014) Ionic liquids, biocompatible in encyclopedia of applied electrochemistry. Springer, New York
33. Coleman D, Gathergood N (2010) Biodegradation studies of ionic liquids. Chem Soc Rev 39:600–637
34. Gorke J, Srienc F, Kazlauskas R (2010) Toward advanced ionic liquids. Polar enzyme-friendly solvents for biocatalysis. Biotechnol Bioprocess Eng 15:40–53
35. Li Z, Liu X, Pei Y, Wang J, He M (2012) Design of environmentally friendly ionic liquid aqueous two-phase systems for the efficient and high activity extraction of proteins. Green Chem 14:2941–2950
36. Taha M, Silva FA, Quental MV, Ventura SPM, Freire MG, Coutinho JAP (2014) Good's buffers as a basis for developing self-buffering and biocompatible ionic liquids for biological research. Green Chem 16:3149–3159
37. Ding X, Wang Y, Zeng Q, Chen J, Huang Y, Xu K (2014) Design of functional guanidinium ionic liquid aqueous two-phase systems for the efficient purification of protein. Anal Chim Acta 815:22–32
38. Zeisel SH, da Costa KA (2009) Choline: an essential nutrient for public health. Nutr Rev 67:615–623
39. Ventura SPM, Silva FA, Gonçalves AMM, Pereira JL, Gonçalves F, Coutinho JAP (2014) Ecotoxicity analysis of cholinium-based ionic liquids to *Vibrio fischeri* marine bacteria. Ecotoxical Environ Saf 102:48–54
40. Meck WH, Williams CL (1999) Choline supplementation during prenatal development reduces proactive interference in spatial memory. Dev Brain Res 118:51–59
41. Pernak J, Syguda A, Mirska I, Pernak A, Nawrot J, Pradzynska A, Griffin ST, Rogers RD (2007) Choline-derivative-based ionic liquids. Chem Eur J 13:6817–6827
42. Sekar S, Surianarayanan M, Ranganathan V, MacFarlane DR, Mandal AB (2012) Choline-based ionic liquids-enhanced biodegradation of azo dyes. Environ Sci Technol 46:4902–4908
43. Vrikkis RM, Fraser KJ, Fujita K, MacFarlane DR, Elliott GD (2009) Biocompatible ionic liquids: a new approach for stabilizing proteins in liquid formulation. J Biomech Eng 131:074514–074518
44. Vijayaraghavan R, Thompson BC, MacFarlane DR, Kumar R, Surianarayanan M, Aishwarya S, Sehgal PK (2010) Biocompatibility of choline salts as crosslinking agents for collagen based biomaterials. Chem Commun 46:294–296

45. Nockeman P, Thijs B, Driesen K, Janssen CR, Van Hecke K, Van Meeervelt L, Kossmann S, Kirchner B, Binnemans K (2007) Choline saccharinate and choline acesulfamate: ionic liquids with low toxicities. J Phys Chem B 111:5254–5263
46. Wood N, Stephens G (2010) Accelerating the discovery of biocompatible ionic liquids. Phys Chem Chem Phys 12:1670–1674
47. Liu QP, Hou XD, Li N, Zong MH (2012) Ionic liquids from renewable biomaterials: synthesis, characterization and application in the pretreatment of biomass. Green Chem 14:304–307
48. Tao D, Cheng Z, Chen FF, Li ZM, Hu N, Chen XS (2013) Synthesis and thermophysical properties of biocompatible cholinium-based amino acid ionic liquids. J Chem Eng Data 58:1542–1548
49. Plaquevent JC, Levillain J, Guillen F, Malhiac C, Gaumont AC (2008) Ionic liquids: new targets and media for α-amino acid and peptide chemistry. Chem Rev 108:5035–5060
50. Fukumoto K, Yoshizawa OH (2005) Room temperature ionic liquids from 20 natural amino acids. J Am Chem Soc 127:2398–2399
51. Hu S, Jiang T, Zhang Z, Zhu A, Han B, Song J, Xie Y, Li W (2007) Functional ionic liquid from biorenewable materials: synthesis and application as a catalyst in direct aldol reactions. Tetrahedron Lett 48:5613–5617
52. Moriel P, García-Suárez EJ, Martínez M, García AB, Montes-Morán MA, Calvino-Casilda V, Bañares MA (2010) Synthesis, characterization, and catalytic activity of ionic liquids based on biosources. Tetrahedron Lett 51:4877–4881
53. Tao G, He L, Liu W, Xu L, Xiong W, Wang T, Kou Y (2006) Preparation, characterization and application of amino acid-based green ionic liquids. Green Chem 8:639–646
54. Gouveia W, Jorge TF, Martins S, Meireles M, Carolino M, Cruz C, Almeida TV, Araújo ME (2014) Toxicity of ionic liquids prepared from biomaterials. Chemosphere 104:51–56
55. Egorova KS, Seitkalieva MM, Posvytenko AVP (2014) Unexpected increase of toxicity of amino acid-containing ionic liquids. Toxicol Res. doi:10.1039/C4TX00079J
56. Peric B, Martí E, Sierra J, Cruañas R, Iglesias M, Garau MA (2011) Terrestrial ecotoxicity of short aliphatic protic ionic liquids. Environ Toxicol Chem 30:2802–2809
57. Peric B, Sierra J, Martí E, Cruañas R, Garau MA, Arning J, Bottin-Weber U, Stolte S (2013) (Eco)toxicity and biodegradability of selected protic and aprotic ionic liquids. J Hazard Mater 261:99–105
58. Passos H, Ferreira AR, Cláudio AFM, Coutinho JAP, Freire MG (2012) Characterization of aqueous biphasic systems composed of ionic liquids and citrate-based biodegradable salt. Biochem Eng J 67:68–76
59. Freire MG, Louros CLS, Rebelo LPN, Coutinho JAP (2011) Aqueous biphasic systems composed of a water-stable ionic liquid + carbohydrates and their applications. Green Chem 13:1536–1545
60. Han J, Yu W, Wang X, Xie X, Yan Y, Yin G, Guan W (2010) Liquid-liquid equilibria of ionic liquid 1-butyl-3-methylimidazolium tetrafluoroborate and sodium citrate/tartrate/acetate aqueous two-phase systems. J Chem Thermodyn 42:39–47
61. Zafarani-Moattar MT, Hamzehahzadeh S (2010) Phase diagrams for the aqueous two-phase ternary system containing the ionic liquid 1-butyl-3methylimidazolium bromide and tri-potassium citrate. J Chem Eng Data 54:833–841
62. Zafarani-Moattar MT, Hamzehahzadeh S (2010) Salting-out effect, preferential exclusion, and phase separation in aqueous solutions of chaotropic water-miscible ionic liquids and kosmotropic salts: effects of temperature, anions, and cations. J Chem Eng Data 55:1598–1610
63. Sintra T, Cruz R, Ventura SPM, Coutinho JAP (2014) Phase diagrams of ionic liquids-based aqueous biphasic systems as a platform for extraction processes. J Chem Thermodyn 77:206–213

64. Zhang J, Zhang Y, Chen Y, Zhang S (2007) Mutual coexistence curve measurement of aqueous biphasic systems composed of [bmim][BF$_4$] and glycine, L-serine, and L-Proline respectively. J Chem Eng Data 52:2488–2490
65. Domínguez-Pérez M, Tomé LIN, Freire MG, Marrucho IM, Cabeza O, Coutinho JAP (2010) (Extraction of biomolecules using) aqueous biphasic systems formed by ionic liquids and amino acids. Sep Purif Technol 72:85–91
66. Wu B, Zhang YM, Wang HP (2008) Aqueous biphasic systems of hydrophilic ionic liquids + sucrose for separation. J Chem Eng Data 53:983–985
67. Wu B, Zhang Y, Wang H (2008) Phase behavior for ternary systems composed of ionic liquid + saccharides + water. J Phys Chem B 112:6426–6429
68. Chen Y, Meng Y, Zhang S, Zhang Y, Liu X, Yang J (2010) Liquid-liquid equilibria of aqueous biphasic systems composed of 1-butyl-3-methyl imidazolium tetrafluoroborate + sucrose/maltose + water. J Chem Eng Data 55:3612–3616
69. Chen Y, Wang Y, Cheng Q, Liu X, Zhang S (2009) Carbohydrates-tailored phase tunable systems composed of ionic liquids and water. J Chem Thermodyn 43:1153–1158
70. Zhang Y, Zhang S, Chen Y, Zhang J (2007) Aqueous biphasic systems composed of ionic liquid and fructose. Fluid Phase Equilib 257:173–176
71. Albertsson PA (1958) Particle fractionation in liquid two-phase systems: the composition of some phase systems and the behaviour of some model particles in them application to the isolation of cell walls from microorganisms. Biochim Biophys Acta 27:378–395
72. Zaslavsky BY (1995) Aqueous two-phase partitioning, physical chemistry and bioanalytical applications. Academic, New York
73. Diamond AD, Hsu JT (1992) Aqueous two-phase systems for biomolecule separation. In: Advances in biochemical engineering biotechnology. Springer, Berlin
74. Sivars U, Bergfeldt K, Piculell L, Tjerneld F (1996) Protein partitioning in weakly charged polymer-surfactant aqueous two-phase systems. J Chromatogr B Biomed Sci Appl 680:46–53
75. Larsson M, Mattiasson B (1988) Characterization of aqueous two-phase systems based on polydisperse phase forming polymers: enzymatic hydrolysis of starch in a PEG-starch aqueous two-phase system. Biotechnol Bioeng 31:979–983
76. Pereira JFB, Lima AS, Freire MG, Coutinho JAP (2010) Ionic liquids as adjuvants for the tailored extraction of biomolecules in aqueous biphasic systems. Green Chem 12:1661–1669
77. Sidebotham RL (1974) Dextrans. Adv Carbohydr Chem Biochem 30:371–444
78. Monsan P, Bozonnet S, Albenne C, Joucla G, Willemot RM, Remaud-Siméon M (2001) Homopolysaccharides from lactic acid bacteria. Int Dairy J 11:675–685
79. Banik RM, Santhiagu A, Kanari B, Sabarinath C, Upadhyay SN (2003) Technological aspects of extractive fermentation using aqueous two-phase systems. World J Microbiol Biotechnol 19:337–348
80. Powell GM (1980) Handbook of water soluble gums and resins. McGraw-Hill, New York
81. Kato Y, Hasumi K, Yokoyama S, Yabe T, Ikuta H, Uchimoto Y, Wakihara M (2002) Polymer electrolyte plasticized with PEG-borate ester having high ionic conductivity and thermal stability. Solid State Ionics 150:355–361
82. Kumar BS, Saraswathi R, Kumar KV, Jha SK, Venkates DP, Dhanaraj SA (2014) Development and characterization of lecithin stabilized glibenclamide nanocrystals for enhanced solubility and drug delivery. Drug Deliv 21:173–184
83. Rawat S, Raman Suri C, Sahoo DK (2010) Molecular mechanism of polyethylene glycol mediated stabilization of protein. Biochem Biophys Res Commun 392:561–566
84. Wu C, Wang J, Pei Y, Wang H, Zhiyong L (2010) Salting-out effect of ionic liquids on poly (propylene glycol) (PPG): formation of PPG + ionic liquid aqueous two-phase systems. J Chem Eng Data 55:5004–5008
85. Li X, Hou M, Zhang Z, Han B, Yang G, Wang Z, Zou L (2008) Absorption of CO_2 by ionic liquid/polyethylene glycol mixture and thermodynamic parameters. Green Chem 10:879–884
86. Rodríguez H, Rogers RD (2010) Liquid mixtures of ionic liquids and polymer as solvent systems. Fluid Phase Equilib 924:7–14

87. Pereira JFB, Rebelo LPN, Rogers RD, Coutinho JAP, Freire MG (2013) Combining ionic liquids and polyethylene glycols to boost the hydrophobic-hydrophilic range of aqueous biphasic systems. Phys Chem Chem Phys 15:19580–19583
88. Liu X, Li Z, Pei Y, Wang H, Wang J (2013) (Liquid + liquid) equilibria for (cholinium-based ionic liquids + polymers) aqueous two-phase systems. J Chem Thermodyn 60:1–8
89. Shahriari S, Tomé LC, Araújo JMM, Rebelo LPN, Coutinho JAP, Marrucho IM, Freire MG (2013) Aqueous biphasic systems: a benign route using cholinium-based ionic liquids. RSC Adv 3:1835–1843
90. Pereira JFB, Kurnia KA, Cojocaru AO, Gurau G, Rebelo LPN, Rogers RD, Freire MG, Coutinho JAP (2014) Molecular interactions in aqueous biphasic systems composed of polyethylene glycol and crystalline *vs.* liquid cholinium-based salts. Phys Chem Chem Phys 16:5723–5731
91. Ventura SPM, Neves CMSS, Freire MG, Marrucho IM, Oliveira J, Coutinho JAP (2009) Evaluation of anion influence on the formation and extraction capacity of ionic-liquid-based aqueous biphasic systems. J Phys Chem B 113:9304–9310
92. Mourão T, Cláudio AFM, Boal-Palheiros I, Freire MG, Coutinho JAP (2012) Evaluation of the impact of phosphate salts on the formation of ionic-liquid-based aqueous biphasic systems. J Chem Thermodyn 54:398–405
93. Zafarani-Moattar MT, Samadi F, Sadeghi R (2004) Volumetric and ultrasonic studies of the system (water + polypropylene glycol 400) at temperatures from (283.15 to 313.15) K. J Chem Thermodyn 36:871–875
94. Zafarani-Moattar MT, Emanian S, Hamzehzadeh S (2008) Effect of temperature on the phase equilibrium of the aqueous two-phase poly(propylene glycol) + tripotassium citrate system. J Chem Eng Data 53:456–461
95. Sadeghi R, Jamehbozorg B (2009) The salting-out effect and phase separation in aqueous solutions of sodium phosphate salts and poly(propylene glycol). Fluid Phase Equilib 280:68–75
96. Fukumoto K, Hiroyuki O (2007) LCST-type phase changes of a mixture of water and ionic liquids derived from amino acids. Angew Chem Int Ed 46:1852–1855
97. Kohno Y, Saita K, Nakamura N, Ohno H (2011) Extraction of proteins with temperature sensitive and reversible phase change of ionic liquid/water mixture. Polym Chem 2:862–867
98. Wu C, Wang J, Wang H, Pei Y, Li Z (2011) Effect of anionic structure on the phase formation and hydrophobicity of amino acid ionic liquids aqueous two-phase systems. J Chromatogr A 1218:8587–8593
99. Wu C, Wang J, Li Z, Jing J, Wang H (2013) Relative hydrophobicity between the phases and partition of cytochrome-c in glycine ionic liquids aqueous two-phase systems. J Chromatogr A 1305:1–6
100. Zeng Q, Wang Y, Li N, Huang X, Ding X, Lin X, Huang S, Liu X (2013) Extraction of proteins with ionic liquid aqueous two-phase systems based on guanidine ionic liquid. Talanta 116:409–416
101. Huang S, Wang Y, Zhou Y, Li L, Zeng Q, Ding X (2013) Choline-like ionic liquid-based aqueous two-phase extraction of selected proteins. Anal Methods 5:3395–3402
102. Pereira JFB, Vicente F, Santos-Ebinuma VC, Araújo JM, Pessoa A, Freire MG, Coutinho JAP (2013) Extraction of tetracycline from fermentation broth using aqueous two-phase systems composed of polyethylene glycol and cholinium-based salts. Process Biochem 48:716–722
103. Ribeiro BD, Coelho MAC, Rebelo LPN, Marrucho IM (2013) Ionic liquids as additives for extraction of saponins and polyphenols from mate (*Ilex paraguariensis*) and tea (*Camellia sinensis*). Ind Eng Chem Res 52:12146–12153

Chapter 12
Toward the Recovery and Reuse of the ABS Phase-Forming Components

Sónia P.M. Ventura and João A.P. Coutinho

Abstract Ionic liquids (ILs) have attracted significant interest as solvents in the extraction and purification of diverse biomolecules. Despite the so many different applications of ILs as solvents or as phase forming components of aqueous biphasic systems (ABS), little is known about the economic impact and scale-up of these processes. In fact, for any process to be of industrial relevance while using ILs as solvents, it is crucial to study their effective recovery, removal, and recyclability. In this sense, this chapter intends to summarize the approaches and strategies of recycling and reuse of ILs as pure compounds, in aqueous solution, and as IL-rich phases.

Keywords Aqueous biphasic systems • Ionic liquids • Recycle • Reuse • Economic impact • Sustainability

12.1 Introduction

Ionic liquids (ILs) have been used in a wide range of applications in both organic environments and aqueous media. The crescent interest in these compounds is closely related with the recurrent identification of ILs as "tunable compounds" and "green" or even "biocompatible solvents" and their unique properties such as their high solvation ability for a wide range of organic and inorganic solutes, negligible vapor pressure, wide electrochemical window, and good chemical and thermal stabilities [1, 2]. In spite of the many applications reported using ILs, among which their use in the extraction/purification of biomolecules [3], the number of studies addressing the scale-up of extraction processes, including the recycling and reuse of ILs, is scarce, and those describing real industrial extraction applications are indeed inexistent, mainly because ILs continue to be considered fancy and expensive solvent chemicals, despite some efforts showing that this is not always correct [4].

The study of aqueous biphasic systems (ABS) based on ILs and salts is a decade old [5], and it has been extended to ABS combining ILs and other phase-forming

---

S.P.M. Ventura (✉) • J.A.P. Coutinho
CICECO, Departamento de Química, Universidade de Aveiro, 3810-193 Aveiro, Portugal
e-mail: spventura@ua.pt

components, namely, polymers, carbohydrates and amino acids [6], and, more recently, organic solvents [7] and surfactants [8]. Different authors have looked for the fundamentals and applications of this type of liquid–liquid extraction process [6], which promoted, in the last decade, a significant increase on their number of applications as recently reviewed [6]. ABS based on ILs can be used for multiple applications, as pointed by Rogers and co-workers on their seminal article [5], but up to present, they have been restricted to two major applications [6], namely, their use as liquid–liquid extraction/purification processes and as a route for the recovery or concentration of hydrophilic ILs from aqueous effluents or aqueous solutions. Actually, most works in literature deal with the use of IL-based ABS as separation and purification technologies, in which the extraction of alkaloids [9–11], pharmaceuticals [12–16], metals [17], amino acids [18, 19], proteins and enzymes [20–24], dyes [8, 25] and colorants [26], and aromatic and phenolic compounds [27, 28] was investigated. Several of these works highlight the potential of these processes to be scaled up; however, no relevant studies were yet developed on this direction, being this normally attributed to economic factors (high price of ILs), but also to the environmental footprint (toxicity [29–40], persistence [29–31, 41], and bioaccumulation [29–31, 37, 42], among others). If in one hand, the industry is still worried about the high costs of ILs, when compared with the prices of the conventional solvents, on the other hand, the number of processes capable to regenerate and recover ILs, allowing their recycling into the extraction system, is still scarce.

Despite the crescent number of publications focusing on IL-based ABS at laboratory scale (more than 380 publications reported in the *ISI Web of Knowledge* by March 2016), the scale-up of these technologies lagged behind. To minimize the environmental and economic impact of IL-based ABS, various conditions need to be investigated, but one aspect stands out: the necessity to recover and reuse the phase-forming components, principally the IL, which represents the most expensive solute. The number of methodologies published regarding the recycling of ILs from aqueous solutions, although scarce, is quite relevant, as recently described in different publications [43–45]. In this chapter, different techniques will be described and analyzed as depicted in Fig. 12.1. In this figure, the main steps on the development of an integrated process of extraction and purification, coupled to the IL recovery, and considering these systems' industrial application, are described. It should be noted that most studies addressing the use of ILs in the extraction and purification fields are focused in the two first steps and practically none in the final task.

12.2 State of the Art on the IL Recyclability

Despite the fact that the number of articles on IL-based ABS, considering not only their fundamental aspects, but also their application as extractive technologies is nowadays very large, the number of works addressing the recycle, recovery, and reuse of ILs, applied to ABS extractions, is still limited.

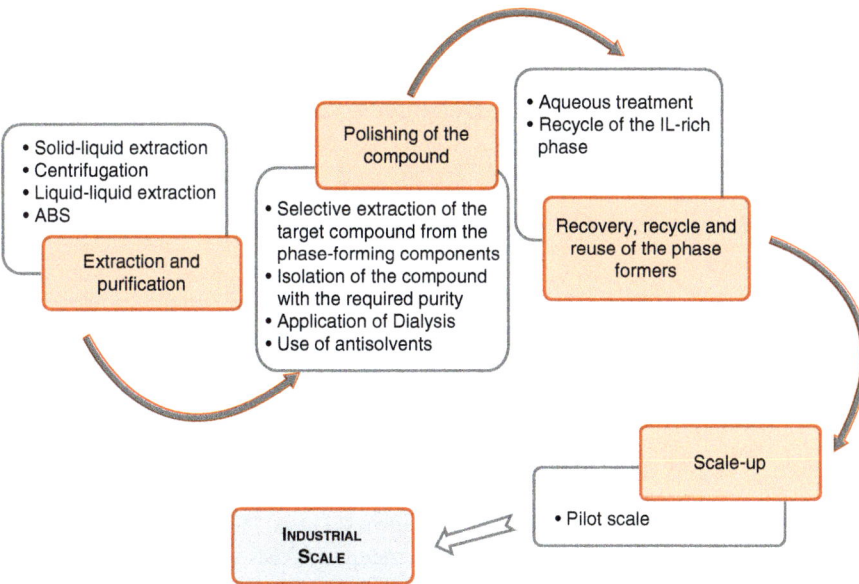

Fig. 12.1 Schematic representation of the principal steps in the development of a scaled-up process for the extraction and purification of added-value compounds

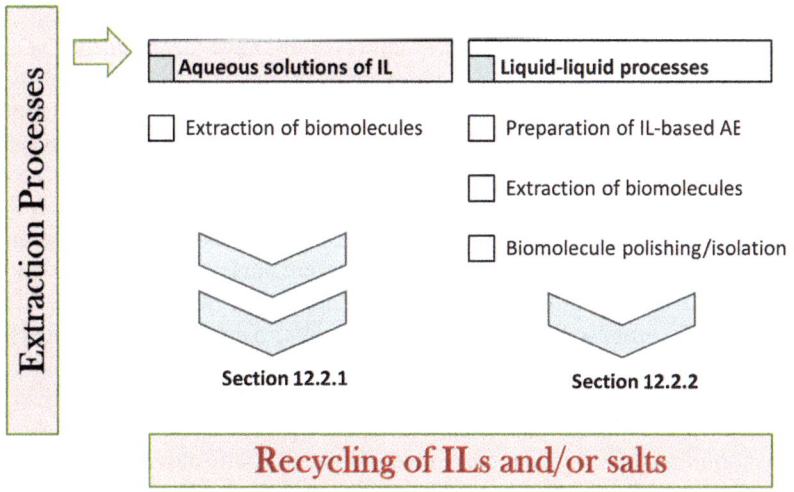

Fig. 12.2 Extraction processes considered in this chapter

The results summarized in this chapter are divided into two major sections. The first section (Sect. 12.2.1) presents the description of works dealing with the recovery, regeneration, removal, and reuse of ILs from aqueous solutions (or aqueous effluents), while the second section (Sect. 12.2.2) describes the studies regarding the recovery and reuse of the phase-forming components from IL-ABS, whose studies are depicted schematically in Fig. 12.2. The first section is relevant in the context of IL-ABS because the techniques reported for the recovery of ILs in process streams or effluents can be extended to other processes and can provide new insights and directions on potential methodologies to be applied in the recyclability (recovery, removal, and reuse) of ILs used in ABS formation.

12.2.1 Recovery of ILs from Aqueous Solutions

This section deals with the recycling of pure ILs or IL aqueous solutions, since this information can be helpful in the implementation of IL-based ABS for extraction purposes.

12.2.1.1 The Use of IL-Based ABS to Recover ILs

In the first work reporting IL-based ABS, Gutowski et al. [5] proposed the concept of recycling ILs from aqueous solutions by means of ABS formation. This allows the creation of an IL-rich phase from a diluted aqueous solution. Meanwhile, some authors have been working on the IL recovery from aqueous solutions along these lines [46–49]. Deng et al. [46] investigated the recovery of 1-allyl-3-methylimidazolium chloride, [aC$_1$im]Cl, using three inorganic salts, namely, tripotassium phosphate, K_3PO_4; dipotassium phosphate, K_2HPO_4; and dipotassium carbonate, K_2CO_3. The authors [46] have determined the ABS phase diagrams of [aC$_1$im]Cl with those three inorganic salts, followed by the proper characterization of the binodal curves and tie-lines and additional recovery tests. The results reported showed that it is possible to increase the recovery efficiency of ILs by increasing the "salting-out" strength, following the order $K_3PO_4 > K_2HPO_4 > K_2CO_3$, representative of the Hofmeister series [50]. Besides the ionic strength of the salt, higher concentrations also lead to an increase on the recovery efficiency of the IL. The best results were achieved using 46.48 wt% of K_2HPO_4, and the maximum recovery efficiency found was 97 % [46]. Li and coauthors [47] proposed the use of sodium salts, namely, sodium phosphate, Na_3PO_4; sodium carbonate, Na_2CO_3; sodium sulfate, Na_2SO_4; sodium phosphate monobasic, NaH_2PO_4; and sodium chloride, NaCl, to promote the recovery of 1-butyl-3-methylimidazolium tetrafluoroborate – [C$_4$C$_1$im][BF$_4$] – from water or aqueous solutions. The binodal curves were experimentally described for different temperatures and the respective tie-lines calculated. The study of the IL recovery was performed considering the addition of a known mass fraction of salt to an IL aqueous solution of known concentration until the formation of the ABS (the system was placed for 5 min in a centrifuge at 2000 rpm) [47]. The sample was placed under controlled temperature until phase separation, and the concentration of the IL in the top phase was quantified by UV–vis spectroscopy at 211 nm (since an imidazolium-based IL was used). The percentage recovery efficiency (*100R*) was then calculated (for more details, see [47]). The results obtained suggest that the recovery efficiency of [C$_4$C$_1$im][BF$_4$] increases with the mass fraction of each salt (due to a "salting-out" effect). The authors [47] demonstrated that different salts have distinct effects on the recovery efficiency of this specific IL, following the order Na$_3$PO$_4$ > Na$_2$CO$_3$ > Na$_2$SO$_4$ > NaH$_2$PO$_4$ > NaCl. The best results regarding the IL recycling were achieved with Na$_2$CO$_3$, with a maximum recovery of 98.77 % [47].

Wu and collaborators [48] studied the removal of ILs from an aqueous solution by means of ABS formation using carbohydrates. The authors used xylose, sucrose, fructose, and mannose to promote the ABS formation with [C$_4$C$_1$im][BF$_4$] in which the upper phase is rich in IL and the bottom phase is rich in carbohydrate (Fig. 12.3a). The ABS was prepared by the addition of specific compositions of carbohydrate to the IL in water. The top phase, richer in IL, was removed, and a known volume of ethyl acetate was added to the top phase, inducing the formation of a second phase rich in ethyl acetate (Fig. 12.3b). The authors [48] clarified that, when ethyl acetate is added to the IL-rich top phase, a third phase is originated in which the bottom phase is sugar rich; the intermediate phase is rich in IL and does not contain ethyl acetate due to their mutual immiscibility; and the top phase is concentrated in ethyl acetate (Fig. 12.3b).

Although it was found to be easy to separate this particular IL from water, after the process of regeneration/purification, the recovered IL (water content < 1 %) was sugar-free (tests were performed to prove the absence of carbohydrates in the IL-rich phase). However, the recoveries of IL measured were not so satisfactory in the authors' perspective (74 % for sucrose, 72 % for xylose, 64 % for fructose, and 61 % for glucose) [48]. In the same year, the same group of authors published an additional work focusing on the separation of ILs by using IL + sugars ABS [49]. In this work, the same idea was applied to other ILs, particularly 1-allyl-3-

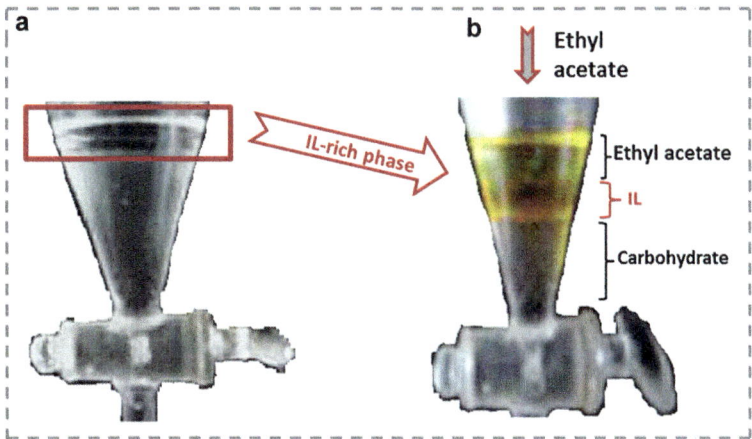

Fig. 12.3 IL-carbohydrate-based ABS: (**a**) [C$_4$C$_1$im][BF$_4$] + sucrose + water and (**b**) top phase of the system [C$_4$C$_1$im][BF$_4$] + sucrose + water + ethyl acetate (Adapted with permission from Ref. [48]. Copyright 2008 American Chemical Society)

methylimidazolium chloride, [aC$_1$im]Cl; 1-allyl-3-methylimidazolium bromide, [aC$_1$im]Br; and 1-butyl-3-methylimidazolium tetrafluoroborate, [C$_4$C$_1$mim][BF$_4$]. The formation of ABS was again studied using sucrose. This study reported some effects of the IL structure on the phase equilibrium, being described the favorable effect promoted by the increase of the ILs' alkyl side chain length [49]. Regarding the anion effect, the authors concluded that the phase separation varies with the halogenated anions [49], being this behavior related with the decrease of the ILs' hydrophobicity as follows: [C$_4$C$_1$im][BF$_4$] > [aC$_1$im]Br > [aC$_1$im]Cl. The same methodology previously described [48] was again applied with success [49], being the results regarding the recovery of the ILs around 74 % for [C$_4$C$_1$im][BF$_4$], 65 % for [aC$_1$im]Br, and 63 % for [aC$_1$im]Cl. One year later, Wu and collaborators provided a mini-review [51] on the recycling of ILs focusing on the methods needed for the removal of ILs from different "working environments".

In 2010, Wu et al. [52] studied the "salting-out" effect of different ILs on the formation of ABS with poly(propylene glycol). They determined the respective phase diagrams, with the binodal curves for PPG400 + [aC$_1$im]Cl or [C$_4$C$_1$im][CH$_3$COO] or [C$_4$C$_1$im]Cl, and the respective tie-lines and tie-line lengths. The authors [52] described the use of these ABSs in the development of new methodologies to promote the recycling and/or enrichment of ILs from aqueous solutions; however, no data on this matter were provided [52].

Neves et al. [53] proposed the use of the aluminum salts, Al$_2$(SO$_4$)$_3$ and AlK(SO$_4$)$_2$.12 H$_2$O, to concentrate and then eliminate imidazolium-, pyridinium-, and phosphonium-based ILs from aqueous solutions. The authors [53] justified the choice of these salts by their high "salting-out" aptitude to promote the formation of ABS [50] and by the fact that these salts are normally used in water treatment processes [54]. The phase diagrams with the different ILs conjugated with the two salts were determined, and some additional properties of the coexisting phases were addressed, namely, density, viscosity, conductivity, and pH values. The study of the ILs' reuse in various (removal/recovery) cycles was also performed to test the suitability of the proposed approach [53]. The main results showed the great ability of these salts to remove and recover ILs from aqueous solutions, being the recovery efficiency results achieved in the range of 96–100 %. The residual concentrations of ILs in the aqueous solution were between 0.01 and 6 wt%. Considering the recovery efficiencies obtained, four cycles of ILs' removal/recovery were tested with the IL tri(isobutyl)methylphosphonium tosylate, [P$_{i(444)1}$][C$_7$H$_7$SO$_3$]. The results obtained showed that for each cycle, the recovery was maintained at around 100 %, proving the recyclability of the IL as shown in Fig. 12.4.

The results obtained by Neves et al. [53] suggest that, as initially proposed by Gutowski et al. [6], ABS can indeed be used to reconcentrate and recover ILs from aqueous solutions (e.g., process streams or aqueous effluents). Given the mechanism of phase formation in IL-ABS, strongest salting-out-inducing salts should be used (phosphates, sulfates, and carbonates are thus recommended), and high concentrations also favor the IL recovery.

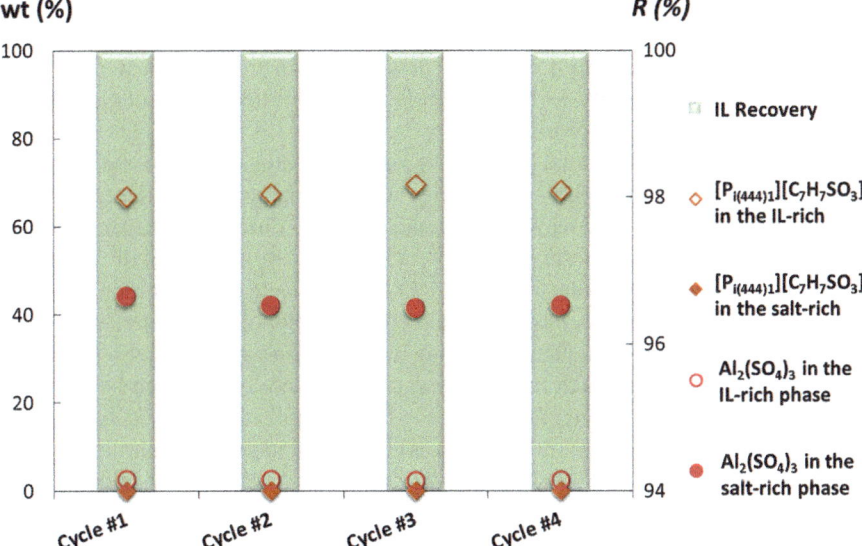

Fig. 12.4 Recovery of a phosphonium-based IL in several cycles using the inorganic salt $Al_2(SO_4)_3$. The concentrations of both salt and IL in both aqueous phases at the end of each cycle are also depicted (Adapted from Ref. [53]. Reproduced with permission from the Royal Society of Chemistry)

12.2.1.2 Other Methodologies

In this section, other methodologies to remove ILs from aqueous effluents other than using salting-out agents are described. It is not our objective to present an exhaustive list and analysis of all methodologies and processes that can be used to recover ILs from water. Instead, we are concerned in the description of the most relevant IL recycling methodologies, namely, distillation [55–61], the use of membranes or resins [62], back-extraction with organic solvents [63], the use of precipitate agents [64], the application of salting-out agents to promote the preparation of ABS [65–67], and the thermoreversibility phenomenon [68].

In 2011, Bica and coauthors [58] suggested the use of aqueous solutions of ILs (in concentrations of 20–50 wt%) to extract and purify essential fragrance oils from orange peels. They reported a comparison between the IL-based and the conventional process, regarding their capacity to extract and isolate the fragrances. ILs (1-butyl-3-methylimidazolium chloride, [C_4C_1im]Cl; 1-allyl-3-methylimidazolium chloride, [aC_1im]Cl; and 1-ethyl-3-methylimidazolium acetate, [C_2C_1im][CH_3COO]) were used as solvents to promote the dissolution of the biomass, thus allowing the extraction and separation of biopolymers and the essential oils found in the composition of the orange peels. After the dissolution of the biomass, the solution was distilled to separate the essential oils from the nonvolatile biopolymers and ILs. To this solution, water was added as anti-solvent to precipitate the biopolymers, and then water was evaporated, allowing the IL regeneration. The IL regenerated was shown to be in a "spectroscopically pure form" but "dark in color" as described by the authors [58]. The "dark" IL recovered was passed through charcoal, reaching a purity level between 90 and 95 % that would allow it to be recycled into the process [58].

Ressmann and collaborators [60] studied a scalable integrated process focused in the isolation of betulin, a pharmaceutically active compound extracted from birch bark. In this study, the IL recycling was investigated due to its relevance in the scale-up strategy. The experimental approach applied to isolate betulin from the biomass followed three main steps: First, (1) ILs were added to the crude biomass, after dissolution methanol was used to precipitate biopolymers, and then (2) a centrifugation step was conducted allowing the fast separation of the solid biomass and the supernatant, in which the betulin is concentrated (ethanol can be added to precipitate undesired biopolymers, and an extra crystallization may be applied if a higher purity level of betulin is required). This integrated process culminated with the (3) IL recovery and recycling by the distillation of the ethanol/water azeotrope.

The authors [60] tested several ILs on the extraction of betulin from the biomass, namely, 1-ethyl-3-methylimidazolium acetate ($[C_2C_1im][CH_3COO]$), 1-ethyl-3-methylimidazolium chloride ($[C_2C_1im]Cl$), 1,3-dimethylimidazolium acetate ($[C_1C_1im][CH_3COO]$), 1,3-dimethylimidazolium n-butyrate ($[C_1C_1im][C_3H_7CO_2]$), 1-ethyl-3-methylimidazolium propionate ($[C_2C_1im][C_2H_5CO_2]$), 1-ethyl-3-methylimidazolium n-butyrate ($[C_2C_1im][C_3H_7CO_2]$), and 1-ethyl-3-methylimidazolium iso-butyrate ($[C_1C_1im][CH(CH_3)_2CO_2]$. At the optimum dissolution and extraction conditions, the scale-up of the betulin isolation process was studied [60]. In the final step of purification of betulin, a filtrate stream, principally composed of water, ethanol, and IL, was obtained. This filtrate was evaporated and posteriorly dried under vacuum with stirring, at 80 °C for 24 h. As explained by the authors [60], the low content in water (20 vol.% in ethanol) was advantageous to the recovery and recycling of the IL, since through an azeotropic distillation of ethanol/water, the IL is automatically separated from the remaining solvents, with a recovery of 86–92 %. Summing up, an easier "energy-saving" recovery of the IL was achieved (the IL recovered was tested via ^{1}H NMR spectroscopy and no impurities were detected), with the possibility to reuse the IL recovered without any additional purification step [60].

Four ILs, including 1-ethyl-3-methylimidazolium hexanoate ($[C_2C_1im][C_5H_{11}CO_2]$), cholinium hexanoate ($[N_{111}C_2H_4OH][C_5H_{11}CO_2]$), cholinium octanoate ($[N_{111}C_2H_4OH][C_7H_{15}CO_2]$), and cholinium decanoate ($[N_{111}C_2H_4OH][C_9H_{19}CO_2]$), were used to extract suberin from cork, a study conducted by Ferreira and coauthors [69]. A detailed description of the methodology used to extract suberin was presented, and an accurate characterization of the extracted material was reported. The authors [69] tested the influence of the anion alkyl chain length and the basicity of the IL on the suberin separation. Taking into account the best results, found for the cholinium hexanoate IL [69], the recyclability of this IL was studied. The IL recovery followed a simple methodology, which consists in the elimination of water (the main contaminant present in the IL effluent) through evaporation under vacuum [69]. The study shows that the yield of cholinium hexanoate recovered by the methodology described was higher than 99 %. Other works also apply processes based on distillation or evaporation of water to promote the recovery and recycling of ILs used as solvents in aqueous solution [55–57, 61].

Recently, Cláudio et al. [63] proposed a process to extract caffeine from guaraná (*Paullinia cupana*, Sapindaceae) seeds using aqueous solutions of ILs (imidazolium and pyrrolidinium cations combined with the chloride, acetate, and tosylate anions) as solvents. The IL recyclability was attempted using the back-extraction of caffeine from the IL aqueous solution by applying several nonmiscible organic solvents, namely, chloroform, ethyl acetate, hexane, diethyl ether, methylene chloride, toluene, butanol, dimethylfuran, and xylene. For that purpose, the best solvents identified were chloroform and methylene chloride, being chloroform capable of a complete extraction of caffeine from the IL aqueous solution. They also show that these organic solvents can be replaced by butanol, a more environmentally benign candidate to extract caffeine from the aqueous IL solution allowing its reutilization [63].

Various are the recovery and recyclability processes for ILs described by different authors [45], here summarized in Fig. 12.5. Some of the processes (evaporation, salt precipitation) described above are capable of recovering ILs from concentrated solutions, while others can do it only with diluted aqueous effluents, with low IL contents, as the adsorption using activated carbon [43, 45] discussed hereafter.

Recent reports by Palomar and coworkers [70–74] showed various approaches for the adsorption of ILs from aqueous effluents using activated carbon. In their first work [72], the authors carried a study on the use of a commercial activated carbon as adsorbent to recover 17 imidazolium-based ILs from aqueous solutions by adsorption. The adsorption isotherms of ILs on the activated carbon were determined at different temperatures and the effects of different cations and anions analyzed. The adsorption mechanism was investigated using various activated

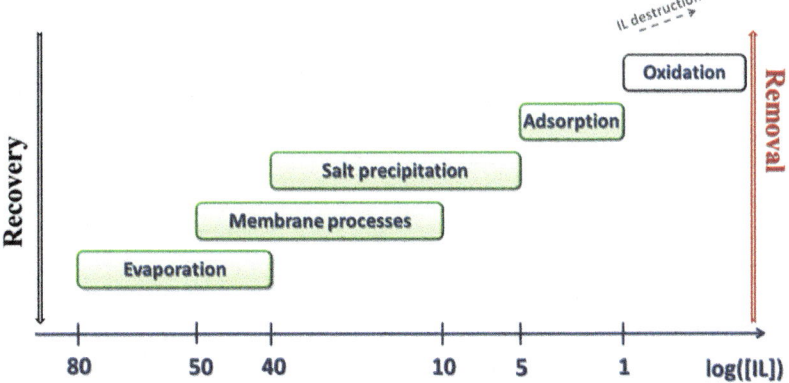

Fig. 12.5 Processes used to remove and recover ILs from aqueous effluents, taking into account the IL concentration to be treated

carbon adsorbents modified by oxidative and thermal treatments and by a computational study by COSMO-RS, which was developed to estimate molecular and thermodynamic properties of the solvent–adsorbate–adsorbent system [72]. The results obtained demonstrate that it is possible to use activated carbon to remove hydrophobic and hydrophilic ILs from water streams. They concluded that the adsorption of the hydrophilic ILs may be improved by the formation of more hydrogen-bonding interactions and by changing the number and nature of oxygen groups on the activated carbon surface [72]. Finally, the authors [72] proposed the use of acetone to regenerate the solid adsorbent. Later, they attempted at the development of heuristics for the choice of the activated carbon regarding the recovery of a given IL from aqueous media [73]. The conclusions suggested that the selection of a specific adsorbent is dependent on the physical and chemical properties of both IL and activated carbon. Microporous and/or narrow mesoporous activated carbons, i.e., with high amount of pores and of diameters lower than 8 nm, present the best adsorption capacities, with maximum values of 1 g of IL *per* gram of activated carbon. Moreover, the authors [73] concluded that the adsorption process was improved by the proper modification of the chemical surface of the absorbent, in which the recovery of hydrophilic ILs is more efficient by applying activated carbons with a higher number of polar groups on their surface, and the hydrophobic ILs are more efficiently recovered from water when low polar materials are applied (such as the thermally treated adsorbents) [73]. Finally, the regeneration of the activated carbons was also successfully achieved by applying acetone as a solvent to remove the IL adsorbed. An additional evaporation was used to remove the acetone to complete the process of the IL recovery [73].

Lemus et al. [70] described a systematic analysis of the influence of different IL alkyl chain lengths, head groups, and anions and also the presence of functional groups on their adsorption onto commercial activated carbons from water effluents. In this work, 21 ILs were studied, including imidazolium-, pyridinium-, pyrrolidinium-, piperidinium-, phosphonium-, and quaternary ammonium-based cations, those conjugated with several hydrophobic and hydrophilic anions. The range of ILs tested was here increased, and in this context, it was possible to improve the understanding of the adsorption mechanism of these IL structures. The results evidenced the significant effect of the IL cation (or family) on the adsorption phenomenon of ILs into the activated carbon. Increased uptakes were described considering the following tendency: quaternary ammonium > phosphonium > pyridinium > imidazolium > pyrrolidinium > piperidinium, promoted by van der Waals interactions between the IL and the surface of the adsorbent [70]. The increase of the alkyl chain length in the anion or cation and the inclusion of substituents can promote different and increased (in the case of the long alkyl side chains) adsorption coefficients, due to high number of interactions between the IL and the activated carbon. As the adsorption was more difficult for the hydrophilic ILs, the authors further proposed the use of a *salting-out* salt to enhance the adsorption of hydrophilic ILs onto activated carbon [71]. The effect of several concentrations of Na_2SO_4 on the adsorption profile of five different ILs of imidazolium and pyridinium families conjugated with hydrophilic and fluorinated anions was studied. The results suggest that the adsorption of the studied ILs onto activated carbon is increased 5.5 times with the addition of Na_2SO_4 [71].

In 2016, e Silva and collaborators [64] aimed at the development of a process for the recovery of ibuprofen from pharmaceutical residues. The ibuprofen solid–liquid extraction and the drug polishing were the main tasks developed. Moreover, also the excipient elimination and the aqueous solution recycling were proposed. The optimization of the solid–liquid extraction and drug isolation tasks was assessed. Regarding the solid–liquid extraction of ibuprofen, different conditions were tested, namely the type of IL and the ratio of IL *versus* citrate buffer. From this particular study, the tetrabutylammonium chloride ($[N_{4444}]Cl$) was selected for further process optimization. Moreover, through the results obtained for the IL/citrate buffer ratio, the mixture 45 wt% of $[N_{4444}]Cl + 5$ wt % of citrate buffer + 50 wt% of water was adopted as the most efficient extracting ibuprofen (extraction efficiency of around 98 %). The recovery of ibuprofen from aqueous media was evaluated by precipitation promoted by the addition of different amounts of KCl and water as precipitating agents. Different volumes of extract/volume ratios of an aqueous solution of KCl (25 wt% and proportions of 1:1, 1:2, 1:3, 1:4, and 1:5) and water were added. The recovery results suggested a significant influence of citrate buffer used in the solid–liquid extraction step, represented by higher recovery efficiencies of ibuprofen (around 97 % and 92 %) when using as precipitating agents KCl aqueous solutions (1:5) and water (1:3), respectively. However, the authors demonstrated that, when applying KCl or water in the same ratios, in the aqueous solutions with IL and 5 wt% of citrate buffer initially applied in the solid–liquid extraction, the precipitation of ibuprofen decreases for 88 % and 35 %, respectively. The authors justify that water has a great capacity to precipitate ibuprofen from the solutions free of citrate buffer due to the hydrophobicity of the drug [64]. Meanwhile, the authors explain that the ionic speciation of the drug in solution and the pH induced by the addition of KCl in the system decreased the main interactions between the drug and the components of the extractive solution, thus promoting its precipitation in aqueous solution. By applying these precipitation strategies, it was possible to obtain precipitates with high level of purity, at around 80 %, considering the IL as the main contaminant. Due to the presence of IL in the ibuprofen precipitate, extra steps were included in the integrated process defined in this work (Fig. 12.6), namely, a washing step with cold water eliminating the IL [64].

Another very recent strategy to recover and recycle some ILs is the use of thermosensitive polymers that, conjugated with aqueous solutions of some specific ILs, namely, the protic ones, can promote the two-phase formation by the exclusive effect of the temperature increase [68]. In this study, and despite the different aims of the authors, the thermoreversibility of the systems based in aqueous solutions of N,N-dimethyl-N-ethylammonium acetate, $[N_{1120}][C_1CO_2]$; N,N-diethyl-N-methylammonium methane sulfonate, $[N_{1220}][C_1SO_3]$; N,N-dimethyl-N-(N′,N′dimethylaminoethyl)ammonium acetate, $[N_{11\,2(N110)]0}][C_1CO_2]$; N,N-dimethyl-N-(N′,N′dimethylaminoethyl)ammonium chloride, $[N_{11}$

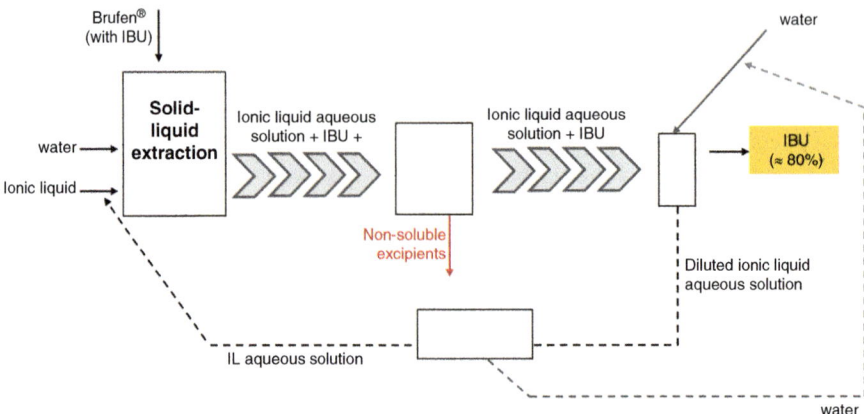

Fig. 12.6 Process developed in Ref. [64] to recover and purify ibuprofen from pharmaceutical wastes by using aqueous solutions of different ILs and the citrate buffer

[2(N110)]0]Cl; N,N-dimethyl-N-ethylammonium phenylacetate, $[N_{1120}][C_7H_7CO_2]$; and N,N-dimethyl-N-(N',N'dimethylaminoethyl)ammonium octanoate, $[N_{11[2(N110)]0}][C_7CO_2]$ and polypropylene glycol (PPG) was studied. In general, this work allows to understand the region of thermoreversibility of these systems. Briefly, if aqueous solutions of these protic ILs are used to extract biomolecules from biomass or fermentation broth, the IL can be regenerated by adding PPG and increasing the temperature to a certain value, which will allow the formation of two phases, one rich in IL and the other richer in the PPG, thus promoting the regeneration of the IL-rich solution and allowing its reuse in other steps of purification.

12.2.2 Recovery and Recycle of ILs from IL-Based ABS Extraction Processes

The application of ABS as extractive platforms for the extraction and/or purification of biomolecules has been extensively described and discussed [6]. However, the molecule polishing and the recycle of the phase-forming components are yet deficiently explored [6].

In 2012, Li and coauthors [23] developed a work focused on the use of ABS composed of cholinium-based ILs and polypropylene glycol 400 (PPG400) for the recovery of distinct proteins, namely, bovine serum albumin (BSA), trypsin, papain, and lysozyme. The phase diagrams were determined for various aqueous systems: PPG400 + tri-cholinium citrate, di-cholinium oxalate, cholinium glycolate, cholinium lactate, cholinium butyrate, cholinium formate, cholinium propionate, and cholinium acetate [23]. It was observed the protein preference for the IL-rich phase and extraction efficiencies between 86.4 % and 99.9 % were obtained. After the partition studies, the authors [23] investigated the recovery of the phase-forming agents, in particular the polymer recyclability. PPG, being a thermosensitive polymer, can be properly recovered by changing the temperature of the medium. Thus, and in order to investigate the recovery and reuse of PPG from aqueous solutions, different temperatures were tested, and the cloud point curves of PPG400 aqueous solutions were defined. With the cloud point curve of PPG at different temperatures, it was possible to determine the PPG400 lower critical solution temperature (LCST) to be approximately 46 °C that further decreases by about 20 °C in the presence of 2 % of IL (cholinium propionate) [23]. The temperature of 26.6 °C was then adopted to study the optimization of the recycling process. Aqueous solutions tested were exposed to different temperatures, above and below the critical temperature as the authors show. In the image shown in this work, the system becomes turbid at around 26.6 °C, and at 35 °C two distinct phases are already formed being the separation even more intense at 45 °C. The recycling process proposed is thus based on a temperature increase that drives the polymer from the aqueous solution of cholinium propionate + PPG400, promoting the concentration of PPG400 in the polymer-rich phase to circa 78 wt% at 35 °C and 90 wt % at 45 °C, finally allowing the recyclability of 90 % of PPG400 [23].

Another approach to recover ILs from ABS was proposed by Cláudio et al. [65] for systems based on carbonate and sulfate salts. The authors [65] used ABS composed of imidazolium-based ILs and two distinct salts, Na_2CO_3 and Na_2SO_4, being those applied on the extraction of gallic acid [65]. They developed a sequential two-step cycle of product extraction and IL recyclability. The first task consisted of the extraction of gallic acid in IL-ABS formed with sodium sulfate. Based on an extensive study on different ILs, the best results were obtained with $[C_4C_1im][CF_3SO_3]$ and $[C_4C_1im][N(CN)_2]$. The gallic acid concentrated in the IL-rich phase was then back-extracted into a saline solution using sodium carbonate [65]. The overall process is depicted in Fig. 12.7. Following the back-extraction, the IL-rich phase, without gallic acid, can be recycled into the process. The IL recovery efficiencies obtained were of about 95 %.

Still in 2014, Ferreira and collaborators have described a new methodology to recover textile dyes from aqueous media, based in the use of IL-based ABS [75]. In this work, the extraction efficiency and the partition of three common textile dyes, chloranilic acid, indigo blue, and Sudan III, were investigated regarding the use of ABS based in phosphonium and imidazolium ILs and the salts aluminum sulfate and potassium citrate. Conditions like the IL chemical structure, the nature of the salt, and the pH of the aqueous medium were investigated. The results achieved in this work revealed that with the adequate selection of the IL structure and salt, it is possible to obtain the complete extraction toward the IL-rich phase. Moreover, the dye recovery was also studied, taking into account, more than the recovery of the dyes, the recycling and reuse of the IL-rich phase, since the IL is the most expensive solvent used. These tests were done using the ABS based in the tributylmethylphosphonium methylsulfate $[P_{4441}][CH_3SO_4]$ and the potassium citrate salt. After the complete recovery for the IL layer of both Sudan III and indigo blue dyes, each phase of the ABS was carefully separated, being the IL-rich layer

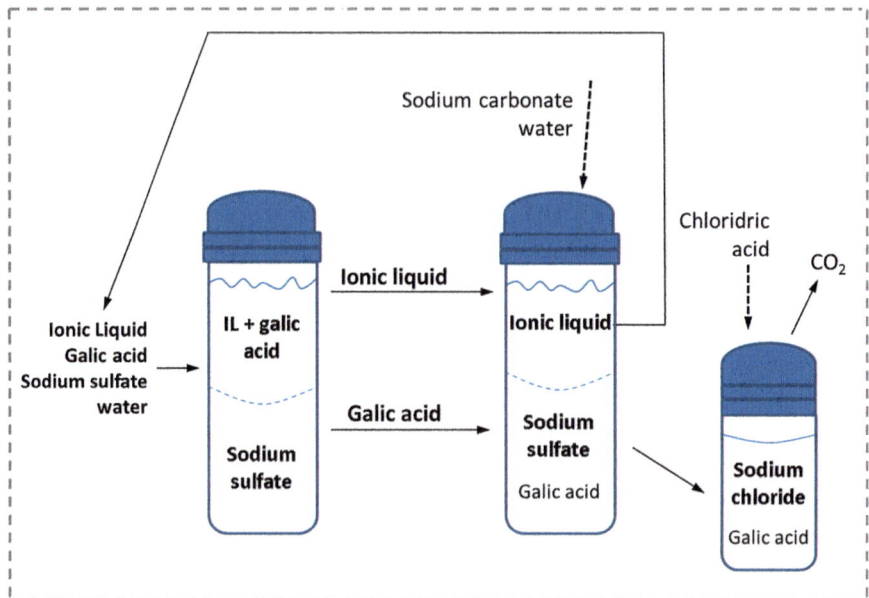

Fig. 12.7 Process diagram representing the integrated extraction and back-extraction of gallic acid and the IL recovery and reuse developed in Ref. [65]

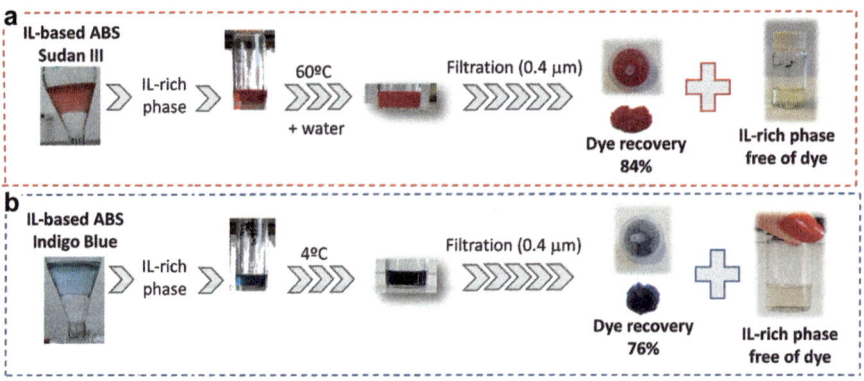

Fig. 12.8 Schematic representation of the process developed in [75] to recover the dyes and to reuse the IL-rich phase

then used to precipitate the dyes. For each dye a specific via of precipitation was applied, for (i) Sudan III a slow process of water evaporation at 60 °C was used, and for the recovery of (ii) indigo blue, a decrease in temperature until 4 °C was defined (Fig. 12.8). After the temperature strategy was applied, each solution was filtered and each dye recovered as a solid. Then, the removal of each dye was quantified in the IL-rich phase proving the IL solution regeneration (i.e., the IL phase is free of dye). As the authors mentioned, aiming at the development of a cost-effective and sustainable process, the possibility to reuse the IL-rich phase was also checked, by means of the identification of the saturation limit of the IL-rich layer and the number of cycles in which this phase can be used. Thus, in the method defined for Sudan III, after the dye filtration, water was added in the IL-rich phase to regenerate it. Since the saturation limit of the IL-rich phase was defined as being well above the saturation limit of each dye in water, the authors concluded that the IL-rich layer may be several times reused, at least up to 800 times without reaching saturation [75].

One year after in 2015, ABSs based in phosphonium (tetrabutylphosphonium bromide, $[P_{4444}]Br$; tetrabutylphosphonium chloride, $[P_{4444}]Cl$; tri(isobutyl) methylphosphonium tosylate, $[P_{i(444)1}][C_7H_7SO_3]$; tributylmethylphosphonium methylsulfate, $[P_{4441}][MeSO4]$) and ammonium (tetrabutylammonium chloride, $[N_{4444}]Cl$) families combined with the potassium citrate buffer at pH 7.0 were explored in the extraction of bovine serum albumin (BSA) [76]. The phase diagrams of the five IL-based ABSs were determined including the tie-lines and tie-line length assessment, and then the partition of BSA through the two aqueous phases was tested. The main results have shown that the majority of the systems under study allowed the complete extraction of BSA for the IL-rich phase in just one single step of extraction, being $[P_{4444}]Br$ the only exception. In this case, the BSA precipitated due to the hydrophobic nature of the IL. Other conditions were tested considering the BSA partition to the IL-rich layer, namely, the influence of the (i) IL and citrate salt content aiming at the reduction of the costs associated with the amount of phase components used to form the ABS and (ii) the protein content. These conditions were tested for the system composed of $[P_{i(444)1}][C_7H_7SO_3]$ (20–30 wt%) + citrate buffer (20–30 wt%) + water with concentrations of BSA varying from 0.5 to 10.0 $g·L^{-1}$. The main results suggested the possibility to decrease both the amounts of IL and salt and still maintain the complete extraction efficiency of the protein up to 10 $g·L^{-1}$. As a final step, the authors assessed the recovery of the protein and the IL-rich phase recycle, being this used to perform a second cycle of extraction (sequential ABSs were prepared) to infer about the possibility to perform sequential extractions of BSA using the same IL-rich phase. This allows the definition of a more sustainable process, in which not only its economic impact is decreased but also the environmental impact is lowered. In this context, three steps of sequential extractions were carried considering the system based in 30 wt% of $[P_{i(444)1}][C_7H_7SO_3]$ + 30 wt% of citrate buffer and 40 wt% of an aqueous solution containing BSA at 0.5 $g·L^{-1}$. In this work, BSA was removed for the IL-rich phase by dialysis. Then, the IL-rich phase was cleaned through the same treatment for each cycle of extraction. First, the IL-rich phase was dried under vacuum conditions and at 60 °C to remove water (the volatile component); then, both salt and IL components were reintroduced in the system to form a new ABS. In this last step, specific amounts of salt and BSA (aqueous solution at 0.5 $g·L^{-1}$) were added to prepare exactly the same systems defined in the first step of extraction. This procedure was repeated to perform the three cycles of extraction. The final results regarding the isolation of BSA and the recycle of IL and salt

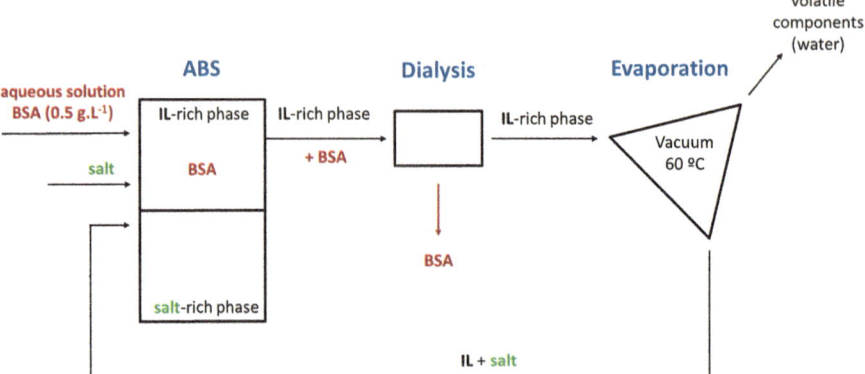

Fig. 12.9 Integrated process developed in Ref. [76] to fractionate and isolate BSA and to reuse the IL-rich phase

(described schematically in Fig. 12.9) indicated the complete recovery of BSA from the IL-rich phase. Moreover, and considering the overall results obtained for the various steps of extraction, it was confirmed the complete extraction of BSA with its structure integrity was maintained, in all cycles, consequently indicating the potential reuse of the IL, however, with small losses of 1.74 wt% $[P_{i(444)1}]$ $[C_7H_7SO_3]$ *per* cycle, in a total of 9.96 wt% of IL loss regarding the entire process [76].

In 2016, some progresses have been carried out. Almeida and collaborators [77] have shown the feasibility of several IL-based ABSs in the recovery of quinolones from the aqueous media. They have selected ABS based in imidazolium and phosphonium families conjugated with the aluminum-based salt, a salt commonly used in the water treatment plants. The one-step extraction approach was applied in the extraction of six fluoroquinolones (ciprofloxacin, enrofloxacin, moxifloxacin, norfloxacin, ofloxacin, and sarafloxacin), being the extraction efficiency data achieved up to 98 %. In addition to the extraction study, the authors also performed studies considering the recyclability and reusability of the solvents. To accomplish that, the example of the ciprofloxacin was taken and its precipitation studied, considering the alteration of the pH of the IL-rich phase, the layer in which the fluoroquinolone was concentrated. The authors used the alteration of the pH of the system to induce the precipitation of this fluoroquinolone. At pH 7.2 (the pH of the system), the acidic dissociation constants of ciprofloxacin are neutral or in its zwitterionic form. The authors explain that, because these species are nonionic, the fluoroquinolone precipitation from the IL-rich phase is possible to occur, a phenomenon justified by the decrease of its solubility in aqueous media for the non-charged species. In this sense, this fluoroquinolone was initially dissolved in an aqueous solution in which HCl and then NaOH were added to change the pH of the solution from 5 to 9. The results obtained have shown that the maximum precipitation of the ciprofloxacin occurred at pH 7.2, being this condition selected for further studies. The authors showed that the percentage of fluoroquinolone precipitated increased from pH 5.23 to 7.23 (the compound is in its neutral form), but decreased in the pH range of 7.23–9.53, due to the presence of its zwitterionic and neutral forms. A second study was done aiming at the recycling of the IL and the isolation of the fluoroquinolone. In this step, the same precipitation procedure was tested but this time in an aqueous solution of the salt used in the ABS formation, $Al_2(SO_4)_3$. Three distinct salts were investigated regarding their capacity to precipitate ciprofloxacin by changing the pH of the saline solution to an alkaline solution, KOH, K_3PO_4, and K_2CO_3. As anticipated, K_3PO_4 was the most efficient (96.97 ± 0.35 of compound precipitated) salt promoting the ciprofloxacin precipitation, not only because the fixation of the pH around 7.2 but also due to its higher salting-out capacity. Moreover, the authors identify also the formation of another crystal, $AlPO_4$, precipitating together with the ciprofloxacin, and according to the authors it corresponds to the white crystals [77]. Considering all the process details discussed, Almeida and collaborators developed a process diagram in which each step is evidenced, namely, the fractionation of the fluoroquinolones from the aqueous to the IL-rich phase and the treatment of this phase enclosing both steps of ciprofloxacin isolation and IL recycling, as described in Fig. 12.10. Finally, aiming to evaluate the viability of the process developed, consecutive cycles of fluoroquinolone isolation and IL recycling and reuse were tested, being the results indicating that the capacity of $[C_2C_1im][CF_3SO_3]$ to remove fluoroquinolones from water and concentrate them in the IL-rich phase is maintained during the four cycles with low cross contamination of the phases and low losses of IL between the four cycles of extraction.

Recently, Zawadzki and collaborators [78] have developed a process for the recovery of the antidepressant drug amitriptyline hydrochloride from their pharmaceutical residues by applying ABSs based in five ILs, these belonging to the phosphonium and ammonium families. In this study, the removal of the excipients

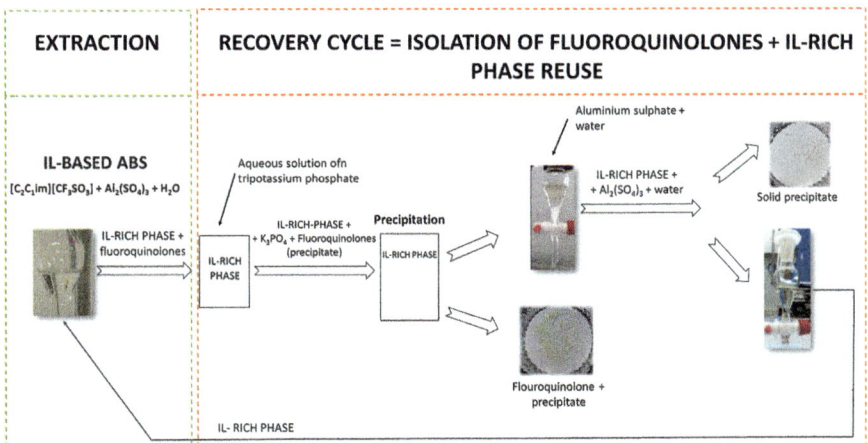

Fig. 12.10 Schematic representation of the process developed in Ref. [77] to recover fluoroquinolones and recover and reuse the IL-rich phase regarding various cycles

used in the antidepressant formulation was attained. The systems under study were based in the ILs, tetrabutylammonium bromide ($[N_{4444}]Br$), tetrabutylammonium chloride ($[N_{4444}]Cl$), tetrabutylphosphonium bromide ($[P_{4444}]Br$), tributylmethylphosphonium methylsulfate ($[P_{4441}][CH_3SO_4]$), and tri(isobutyl)methylphosphonium tosylate ($[P_{i(444)1}][C_7H_7SO_3]$), and salts, the potassium phosphate tribasic (K_3PO_4), potassium phosphate monobasic (KH_2PO_4), and potassium phosphate buffer (prepared with KH_2PO_4 and potassium phosphate dibasic – K_2HPO_4). The experimental study started with the optimization of the main conditions associated with the use of ABS as purification platforms using the commercial standard of the antidepressant. The effects of various conditions of the process of purification, namely, the IL type, the pH of the system, and the composition of the mixture applied in the extraction step, were evaluated and then used in the process optimization, aiming to maximize the extraction of the antidepressant drug. The main results obtained by the application of IL-based ABS indicated the high capacity of these ABSs to concentrate the amitriptyline toward the IL-rich phase (the most hydrophobic layer), a conclusion represented by logarithmic functions of the partition coefficients higher than 2.5 and extraction efficiencies between 93 % ± 3 % and 100 %. Then, the pH of the systems and different mixture points (meaning distinct compositions of the ABS) were tested and the best systems and conditions adopted in the development of the integrated process considering the purification of the antidepressant from the real pharmaceutical residue.

The three-step process was proposed [78] including the solid–liquid extraction of amitriptyline hydrochloride from ADT 25 pills, a purification step using the IL-based ABS with high purification performance defined in the optimization study, and finally the polishing or isolation of the drug by precipitation with anti-solvent (Fig. 12.11). Water was used as solvent in the solid–liquid extraction step, resulting in an aqueous extract rich in amitriptyline hydrochloride. This extract was centrifuged and filtrated aiming at the removal of the insoluble excipients present, the final amount of target antidepressant measured, and the most efficient IL-based ABS prepared using, as basis, the drug-rich aqueous extract. In general, the results of partition coefficients obtained in this second step were similar to those achieved in the optimization step. Then, the isolation of the target antidepressant from the IL layer was investigated by changing the pH of the system, promoting the appearance of the uncharged form of amitriptyline, which consequently decreased its solubility in water/aqueous media. For that purpose, an aqueous solution of KOH was added to the IL-rich phase considering the ABS based in the potassium phosphate buffer (pH 6.6) and only water in the K_3PO_4-based ABS (the neutral form of the drug is guaranteed in the characteristic pH of ABS based in this salt). The step of precipitation was performed at low temperature (277 ± 1 K), again to decrease even more the drug's solubility. The results evidenced that higher precipitations were achieved (95% ± 2 %) for the ABS constituted by $[N_{4444}]Br + K_2HPO_4/KH_2PO_4$ and $[N_{4444}]Cl + K_3PO_4$. As a final step, the authors proposed the two phases of recycling. The IL-rich phase should be neutralized by the addition of phosphoric acid, and the salt-rich phase should pass by an ultrafiltration to remove the high molecular weight excipients present.

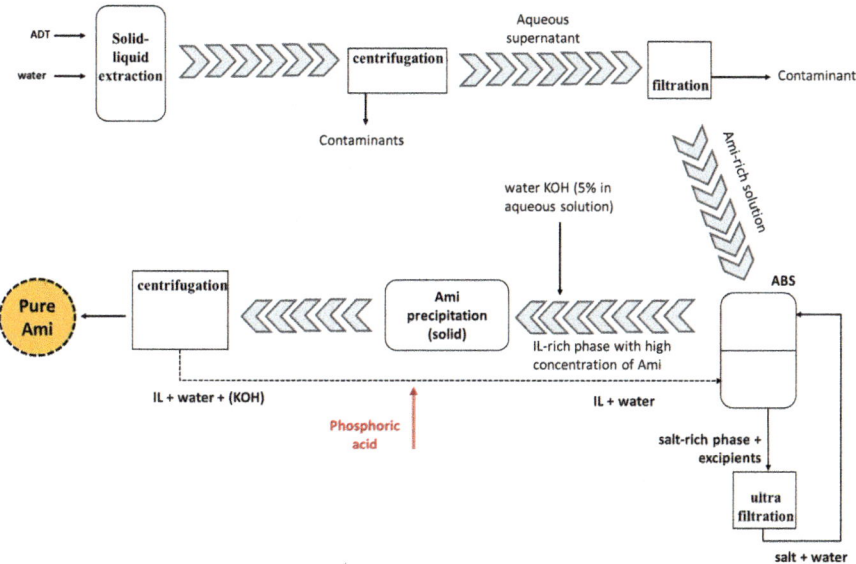

Fig. 12.11 Representation of the integrated process developed in Ref. [78] to recover amitriptyline from pharmaceutical residues

Ionic liquid three-phase partitioning (ILTPP) systems were recently developed by Alvarez-Guerra et al. [66, 67, 79], considered as a new type of extraction/separation technology for proteins. The three-phase partitioning (TPP) systems represent an emerging technique for protein separation/extraction, involving the accumulation of the target protein at the liquid–liquid interface created between two liquid phases, as depicted in Fig. 12.12.

TPP can be applied to purify and/or concentrate proteins, commonly used as a one-step purification methodology, and it stands out for being simple, inexpensive, scalable, and rapid procedure that may lead to high purification factors (>100-fold) and great final purity levels (between 70 % and 85 %) [80, 81]. Some benefits emerge from the ILTPP, namely, the combination of the advantages of IL-based ABS and the TPP technique. The authors applied this technique to extract lactoferrin, a bovine whey protein with recognized biological properties [82], purified at industrial scale by means of cation exchange chromatography techniques, suffering from high costs and relatively low yields [83]. They tested different combinations of ILs and salts to prepare the ILTPP systems as well as the conventional system based in t-butanol, for comparative terms. The system based on 1-butyl-3-methylimidazolium trifluoromethanesulfonate, $[C_4C_1im]$ $[CF_3SO_3]$, at moderate acidic conditions, leads to high protein accumulation in the interphase and high lactoferrin–BSA selectivity. The authors have also addressed the IL recovery and recycling [66], considering the use of various alternatives [67]. The first approach considered the thermodynamic characterization of the ILTPP systems of interest, based on the use of $[C_4C_1im][CF_3SO_3]$ and

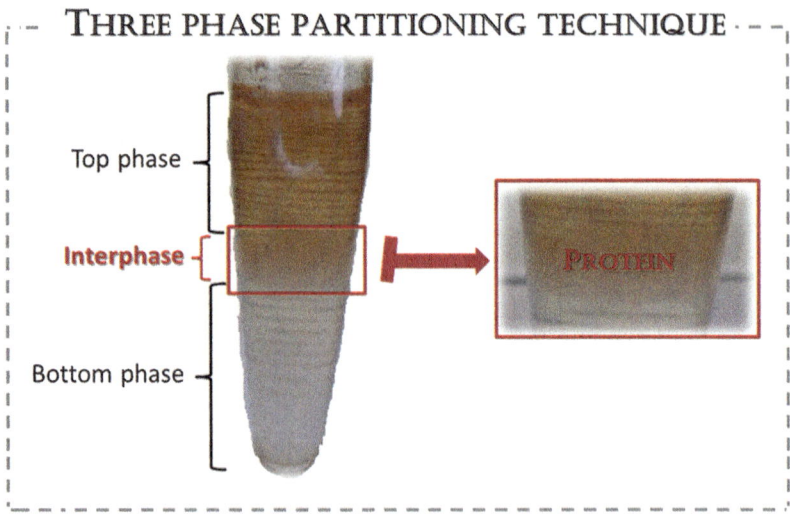

Fig. 12.12 Schematic representation of the three-phase partitioning technique applied to the extraction of proteins

NaH_2PO_4. The fraction of IL in the salt-rich phase, impossible to recycle to the ILTPP process, was determined as well as the influence of the amount of protein treated in the ILTPP systems studied. In contrast to previous studies on the recovery of the IL from IL-ABS based only on experimental data of the thermodynamic equilibrium [5, 46, 47, 53], here the ILTPP was modeled, allowing the operational variables and constraint characteristic of the process to be taken into account in the analysis and optimization of the operating conditions, able to minimize the IL losses in the overall process. The determination of the IL fraction possible to be recycled was described by the parameter R. After the thermodynamic analysis, some simplifications were taken into account, namely, the capacity to recycle the entire IL-rich phase, meaning the high IL recovery, since most of it was present in the top phase (richer in IL) – Fig. 12.13 – and that, in this case, only a fraction of the salt-rich phase is possible to be recovered and reused, being this dependent on the concentration of lactoferrin present in the initial feed and the salt added to prepare each new cycle of extraction.

In general, it was noticed that the distribution coefficient of the IL between the two phases decreases with the amount of salt introduced in the system, favoring the IL recyclability with the maximum R obtained being around 99 % [66].

For the ILTPP systems studied, the fraction of IL that is not possible to be recycled ($1-R$) was monitored in each cycle of extraction, taking into account different lactoferrin loads [66]. It was concluded that this parameter varied from almost 5 %, at low protein concentrations in the feed stream, to values around 0.8 %, when high protein mass fractions were considered [66]. It was finally demonstrated that the ILTPP process can lead to a potential reuse of more than 99 % of the IL. Later, Alvarez-Guerra et al. [67] have proposed two other alternatives to maximize the fraction of IL being recycled. Figure 12.14 depicts the block

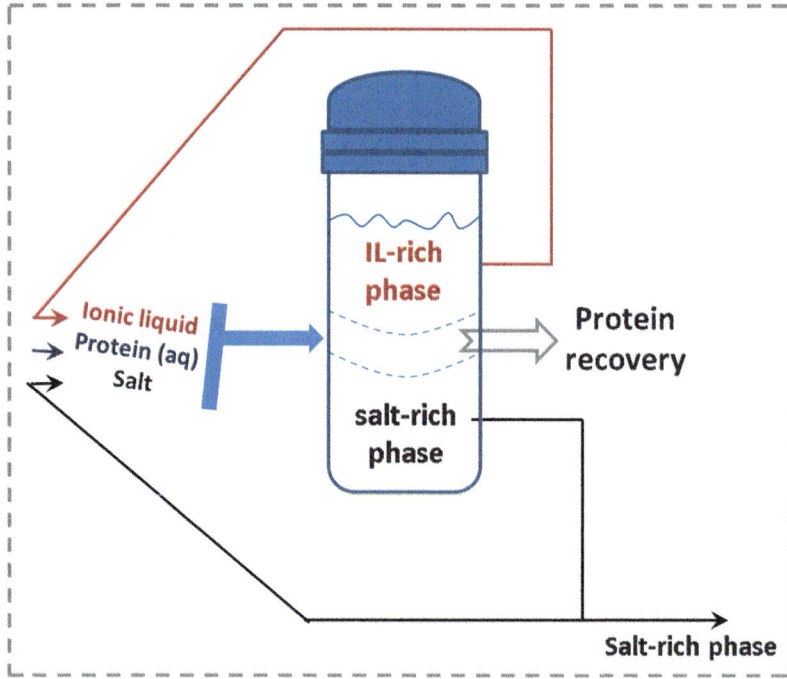

Fig. 12.13 Schematic diagram of the IL recycling approach developed in [66] considering the use of [C_4C_1im][CF_3SO_3] TPP system to recover lactoferrin

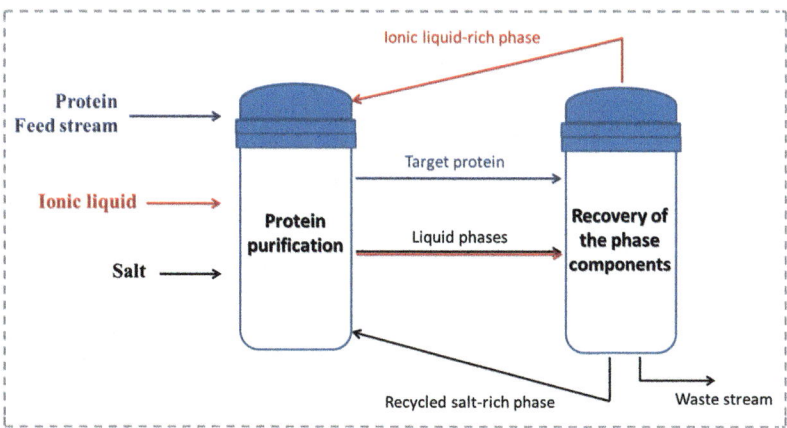

Fig. 12.14 Representation of the block diagram of the ILTPP process proposed in [67] in which an additional step of recovery to reduce the fraction of IL loss in the process is presented

diagram of the ILTPP process based on the $[C_4C_1im][CF_3SO_3] + NaH_2PO_4$ and $[C_4C_1im][CF_3SO_3] + NaH_2PO_4/Na_2HPO_4$, in which the additional recovery step to reduce the IL mass fraction loss investigated by the authors is presented [67].

Two alternatives were investigated through modeling, aiming at increasing the IL recyclability: (1) the addition of extra amounts of salt increasing its concentration (Fig. 12.15a) and (2) the concentration of the salt-rich phase by evaporation of some water, as depicted in Fig. 12.15b. Considering the first approach, two sequential separation units were considered, the first representing the lactoferrin recovery and the second the IL recycling.

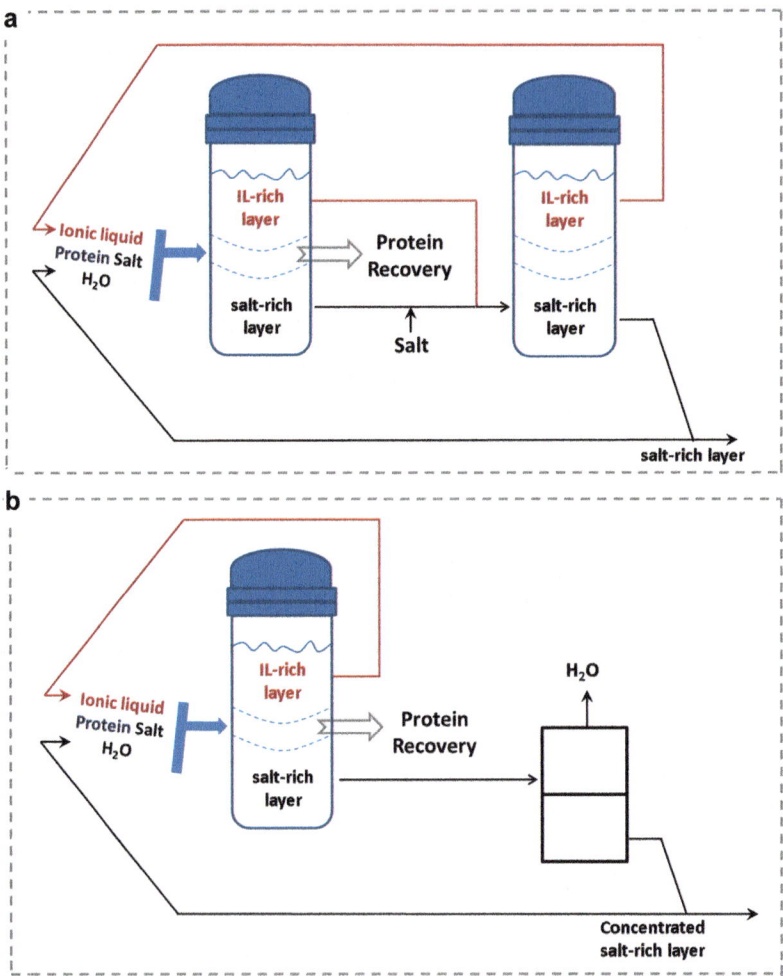

Fig. 12.15 Schematic representation of the two processual approaches developed in [67] regarding the IL recyclability: (**a**) represents the addition of extra amounts of salt and (**b**) represents the concentration of the salt-rich phase by means of vacuum evaporation of some of the water content

The introduction of the lactoferrin feed corresponds to the addition of significant amounts of water to the process, which is very relevant in the process of recyclability of the phase promoters. This water prevents the complete recycling of the salt-rich phase, due to the fact that, if the salt-rich phase is directly recirculated, the extra amounts of water will be changing the individual composition of each component. However, if the same amount of water included in the lactoferrin feed and introduced in the system is previously removed by evaporation from the salt-rich phase, IL and salt can be easily recycled, without any change in the composition of the ILTPP system [67]. It was also demonstrated that the salt concentration increase has a limited effect on the recovery of lactoferrin for high protein concentrations, while the application of evaporation of water could promote the IL's and salt's complete recycling [67].

As suggested by the authors [67], the conclusions of this work can be extended to IL-ABS, with the required development of specific strategies for the recovery of the solute, since ILTPP and IL-based ABS are based on the same thermodynamic systems and rules.

Despite the absence of experimental results describing and analyzing the strategies of IL recycling and isolation of the target biomolecules, other works [7, 84, 85] are applying IL-based ABS in the purification or extraction of various biomolecules using distinct extractive approaches and identifying theoretical possibilities for the recycling and reuse of the main components. Souza and co-workers [84], for example, have shown the application of polymer + salt ABS based in different imidazolium ILs used as adjuvants. Adjuvants are compounds used in small amounts, for example, 5 wt% as used in this work, to promote the alteration of the physicochemical properties of the original ABS, which normally are described by increases in the purification and extraction performances of various biomolecules. In this work, the authors applied these systems to promote the purification of an enzyme produced by submerged fermentation by *Bacillus* sp. ITP-001. The first optimization step was performed using the commercial lipase B from *Candida antarctica* (CaLB), in which a purification factor of 5.22 ± 0.65 is much better than the result (purification factor of 3.25 ± 0.65) obtained for the common polymeric ABS (without IL). After a careful optimization of the pretreatment task (salt precipitation and dialysis) and the ABS type and operational conditions, a purification factor of 245 for the lipase from *Bacillus* sp. ITP-001 was obtained using the system PEG 1500 + phosphate buffer + 1-hexyl-3-methylimidazolium chloride ($[C_6C_1im]Cl$) at pH 7. In this work [84], the authors explain that the use of the IL as adjuvant is promoting the manipulation of the contaminant proteins' migration to the opposite phase in which the lipase is concentrated, allowing the concentration of the lipase in the salt (phosphate buffer)-rich phase. The authors close this work describing a potential strategy to isolate the enzyme and contaminant proteins, from the salt- and polymer-rich phases, respectively, by using a dialysis process in each phase. According to their description, after the dialysis of both layers, the contaminant proteins and lipase can be collected in their most pure state and each phase reintroduced in the system to promote new cycles of purification [84].

Still in the purification of macromolecules appeared the work of Lee and coauthors [85]. Briefly, a new set of self-buffering ILs based in Good's buffer anions were synthesized, applied in the formation of different ABSs composed by these GB-ILs and potassium citrate and found to be significantly selective for the partition of the lipolytic enzyme from *Pseudomonas cepacia* toward the GB-IL-rich phase. The results allowed the development of a sustainable integrated catalytic process in which the self-buffer nature of GB-ILs and their capacity to maintain the enzyme stable are conjugated. In this sense, the authors meant that because the enzyme participating as a catalyst in the reaction is concentrated in the GB-IL-rich phase (with buffer properties), if the product formed in the reaction is hypothetically concentrated in the saline-rich phase, it will be possible to promote the reaction and the removal of the product in real time and as simultaneous steps, being the enzyme always pure and free to continue its catalytic action. Summing up, the authors attest that the process will allow the recovery and reuse of the top phase enriched in GB-ILs which will diminish the economic and environmental effects of the process [85].

Later in 2016, Santos and collaborators [7] studied the extraction and purification of capsaicin from pepper *Capsicum frutescens*, by applying various ABSs composed of acetonitrile and three cholinium-based ILs (cholinium chloride, cholinium bitartrate, and cholinium dihydrogenocitrate). The phase diagrams were prepared and the partition of capsaicin optimized considering the cholinium anion, the composition of the system, and the temperature of fractionation. After a careful optimization of the process of partition, and with a simple and more sustainable technology, capsaicin was extracted from pepper and purified through the application of acetonitrile + cholinium IL-based ABS, being defined an extraction efficiency of 90 % and a purification factor of 3.26, considering as main contaminants the phenolic compounds extracted from the fresh biomass. Again, experimental tests were not performed, but the authors suggest a methodology to isolate capsaicin from the acetonitrile-rich phase and the phenolic compounds from the cholinium-rich phase, allowing thus the reutilization of the main phase formers. In this context, an evaporation step is described for the recovery of acetonitrile and isolation of capsaicin. In what concerns the recycle of the cholinium-rich phase contaminated with the phenolic compounds, an acidic precipitation step using HCl was defined to precipitate the contaminants, followed by the neutralization (NaOH) of the cholinium-rich phase to be reincorporated in the process [7].

12.3 Conclusions

Much attention has been given to ILs and their applications as solvents in different fields [6, 86]. The crescent number of works evidencing the use of ILs as solvents [86], and in particular as phase formers of ABS, is evident [6]; however, the number of studies reporting the scale-up of these processes is indeed very limited. The limited industrial impact of this technique results from a limited knowledge about the market and industry needs by the academic groups working on this field. These techniques, based on special solvents, can only be economically applied to the extraction of molecules or compounds with high value. The use of ILs for the extraction and purification of compounds of low cost that can be obtained from synthetic pathways, or molecules with low economic/industrial value, is not recommendable [86]. Another aspect is the extraction of compounds that are only needed in a small scale and that can be purified by chromatographic techniques and for which the use and application of ABS does not make sense, since these processes are by far more efficient regarding their purification capacities; however, compounds, for which chromatographic techniques are not adequate, are improved candidates to be purified with alternative techniques such as ABS. Taking these factors into account and focusing on both economic and environmental problems/ benefits still strongly associated with ILs, it is crucial to study the ILs' effective recovery, removal, and recyclability [6, 45, 86] and to encourage the use of ILs in separation processes. Until today, and taking into account the studies reviewed in this chapter, the evaporation of water or other organic solvents used in the back-extraction for the isolation of the target compounds extracted, or their precipitation by anti-solvents, is the process most used. The results reported indicate that it is possible to regenerate and reintroduce in new cycles of extraction the IL or IL-rich phase recovered while maintaining the IL integrity, the compound integrity, and the initial efficient extractions. However, the number of works studying the recyclability of the phase components is only occasional, and we are now limited, principally when ILs are part of an ABS used to purify a specific compound, to a few examples in which the purification of one enzyme [66, 67, 84, 85], four proteins [23, 76], three antioxidants [65], two dyes [75], and four active pharmaceutical ingredients [77, 78] was attempted. The need for more studies contemplating the purification of a variety of other (bio)compounds from a diversity of relevant raw materials integrated with studies regarding the recyclability of ILs and phase-forming components in general is apparent.

12.4 Critical Analysis and Future Perspectives

The use of ILs in technologies of purification is often criticized due to their high cost, when compared with other and more traditional solvents. It is thus mandatory to perform studies in which several variables are taken into account, namely:

- The cost of the target product, consumables, equipment, human resources, and licensing
- The purity required for the target product since the price of each technology will increase with the purity level required
- The number of potential molecules to be extracted, since the higher the number of compounds recovered, the higher the economic return of the entire process

- The study of processes with industrial importance, i.e., processes planned to purify compounds with real industrial relevance
- The possibility to recover and reuse the phase-forming components to decrease the environmental impact and to increase the viability and sustainability of IL-based ABS

As mentioned, there are some processes/strategies that can be applied to reduce the overall cost of a process of purification by the use of IL-based ABS. Among these, only nondestructive processes can be used, while there are some good strategies to remove ILs in low concentration, and while others are excellent to be applied in the recovery of ILs present in high concentrations in aqueous media. Still, there is much more open space to explore and develop novel strategies for the recovery of ILs while envisaging their use in scaled-up technologies.

This review, while still not very common, shows that some authors have started looking into the recovery and recycling of ILs applied in extraction steps, which is not surprising since they are normally the most expensive components used. However, the effect of impurities (both from contaminants and from IL degradation) is not addressed neither is their accumulation on the system that may be very important for cyclic processes. The recovery of other phase-forming compounds is also seldom addressed. The idea conveyed in most works that IL-based ABS display superior extraction performance is only supported by a limited number of extraction/recyclability cycles that do not allow to satisfactorily conclude about the success of the actual extraction process in the long run. To soundly establish ABS as sustainable technologies, the study of a higher number of cycles of extraction and recyclability must be carried out as this is the only way to understand the viability and sustainability of any extraction process. Other important issues to be focused in future publications are a set of properties of ILs, namely, purity, stability, biodegradability, and toxicity, since the control and deeper understanding of these properties can be determinant in the sustainability nature of IL-based separation processes. A life cycle assessment should eventually be carried out to the phase-forming components used in IL-based ABS to understand the impact of each individual component used in the purification processes.

There are other conditions that can be considered to design more cost-efficient and sustainable extraction strategies, namely, the design of purification processes with a lower number of separation steps, which can be translated into less components to be removed and less amounts of aqueous effluents to be treated and recycled, which finally can be translated into a reduced waste content and a lower environmental footprint.

Summing up, in order to replace conventional extraction approaches by a process using ILs, it is mandatory to initially evaluate its costs, environmental impact, and scale-up viability.

References

1. Seddon KR (1997) Ionic liquids for clean technology. J Chem Technol Biotechnol 68:351–356
2. Holbrey JD, Rogers RD, Mantz RA, Trulove PC, Cocalia VA, Visser AE, Anderson JL, Anthony JL, Brennecke JF, Maginn EJ, Welton T, Mantz RA (2008) Physicochemical properties. In: Ionic liquids in synthesis. Wiley-VCH Verlag GmbH & Co. KGaA, Weinheim, pp 57–174
3. Wasserscheid P, Welton T (2003) Some liquids in synthesis. Wiley-VCH, Weinheim
4. Schubert TJS (2011) Ionic Liquids Today, Issue 2–11. www.iolitec.com
5. Gutowski KE, Broker GA, Willauer HD, Huddleston JG, Swatloski RP, Holbrey JD, Rogers RD (2003) Controlling the aqueous miscibility of ionic liquids: aqueous biphasic systems of water-miscible ionic liquids and water-structuring salts for recycle, metathesis, and separations. J Am Chem Soc 125:6632–6633
6. Freire MG, Claudio AFM, Araujo JMM, Coutinho JAP, Marrucho IM, Lopes JNC, Rebelo LPN (2012) Aqueous biphasic systems: a boost brought about by using ionic liquids. Chem Soc Rev 41:4966–4995
7. Santos PL, Santos LNS, Ventura SPM, de Souza RL, Coutinho JAP, Soares CMF, Lima ÁS Recovery of capsaicin from Capsicum frutescens by applying aqueous two-phase systems based on acetonitrile and cholinium-based ionic liquids. Chem Eng Res Des 112:103–112
8. Vicente FA, Malpiedi LP, e Silva FA, Pessoa A Jr, Coutinho JAP, Ventura SPM (2014) Design of novel aqueous micellar two-phase systems using ionic liquids as co-surfactants for the selective extraction of (bio)molecules. Sep Purif Technol 135:259–267 (accepted)
9. Freire MG, Neves CMSS, Marrucho IM, Canongia Lopes JN, Rebelo LPN, Coutinho JAP (2010) High-performance extraction of alkaloids using aqueous two-phase systems with ionic liquids. Green Chem 12:1715–1718
10. Passos H, Trindade MP, Vaz TSM, da Costa LP, Freire MG, Coutinho JAP (2013) The impact of self aggregation on the extraction of biomolecules in ionic-liquid-based aqueous two-phase systems. Sep Purif Technol 108:174–180
11. Li S, He C, Liu H, Li K, Liu F (2005) Ionic liquid-based aqueous two-phase system, a sample pretreatment procedure prior to high-performance liquid chromatography of opium alkaloids. J Chromatogr B 826:58–62
12. Pereira JFB, Vicente F, Santos-Ebinuma VC, Araújo JM, Pessoa A, Freire MG, Coutinho JAP (2013) Extraction of tetracycline from fermentation broth using aqueous two-phase systems composed of polyethylene glycol and cholinium-based salts. Process Biochem 48:716–722
13. e Silva FA, Sintra T, Ventura SPM, Coutinho JAP (2014) Recovery of paracetamol from pharmaceutical wastes. Sep Purif Technol 122:315–322
14. Han J, Wang Y, Yu C, Li C, Yan Y, Liu Y, Wang L (2011) Separation, concentration and determination of chloramphenicol in environment and food using an ionic liquid/salt aqueous two-phase flotation system coupled with high-performance liquid chromatography. Anal Chim Acta 685:138–145
15. Li C-X, Han J, Wang Y, Yan Y-S, Xu X-H, Pan J-M (2009) Extraction and mechanism investigation of trace roxithromycin in real water samples by use of ionic liquid–salt aqueous two-phase system. Anal Chim Acta 653:178–183
16. Wang Y, Xu X-h, Han J, Yan Y-s (2011) Separation/enrichment of trace tetracycline antibiotics in water by [Bmim]BF4–(NH4)2SO4 aqueous two-phase solvent sublation. Desalination 266:114–118
17. Zhang C, Huang K, Yu P, Liu H (2013) Ionic liquid based three-liquid-phase partitioning and one-step separation of Pt (IV), Pd (II) and Rh (III). Sep Purif Technol 108:166–173
18. Huaxi L, Zhuo L, Jingmei Y, Changping L, Yansheng C, Qingshan L, Xiuling Z, Urs W-B (2012) Liquid-liquid extraction process of amino acids by a new amide-based functionalized ionic liquid. Green Chem 14:1721–1727

19. Pei YC, Li ZY, Liu L, Wang JJ (2012) Partitioning behavior of amino acids in aqueous two-phase systems formed by imidazolium ionic liquid and dipotassium hydrogen phosphate. J Chromatogr A 1231:2–7
20. Ventura SPM, de Barros RLF, de Pinho Barbosa JM, Soares CMF, Lima AS, Coutinho JAP (2012) Production and purification of an extracellular lipolytic enzyme using ionic liquid-based aqueous two-phase systems. Green Chem 14:734–740
21. Deive FJ, Rodriguez A, Pereiro AB, Araujo JMM, Longo MA, Coelho MAZ, Lopes JNC, Esperanca JMSS, Rebelo LPN, Marrucho IM (2011) Ionic liquid-based aqueous biphasic system for lipase extraction. Green Chem 13:390–396
22. Bai Z, Chao Y, Zhang M, Han C, Zhu W, Chang Y, Li H, Sun Y (2013) Partitioning behavior of papain in ionic liquids-based aqueous two-phase systems. J Chem 2013:938154
23. Li ZY, Liu XX, Pei YC, Wang JJ, He MY (2012) Design of environmentally friendly ionic liquid aqueous two-phase systems for the efficient and high activity extraction of proteins. Green Chem 14:2941–2950
24. Taha M, e Silva FA, Quental MV, Ventura SPM, Freire MG, Coutinho JAP (2014) Good's buffers as a basis for developing self-buffering and biocompatible ionic liquids for biological research. Green Chem 16:3149–3159
25. de Souza RL, Campos VC, Ventura SPM, Soares CMF, Coutinho JAP, Lima ÁS (2014) Effect of ionic liquids as adjuvants on PEG-based ABS formation and the extraction of two probe dyes. Fluid Phase Equilib 375:30–36
26. Ventura SM, Santos-Ebinuma V, Pereira JB, Teixeira MS, Pessoa A, Coutinho JP (2013) Isolation of natural red colorants from fermented broth using ionic liquid-based aqueous two-phase systems. J Ind Microbiol Biotechnol 40:507–516
27. Cláudio AFM, Freire MG, Freire CSR, Silvestre AJD, Coutinho JAP (2010) Extraction of vanillin using ionic-liquid-based aqueous two-phase systems. Sep Purif Technol 75:39–47
28. Cláudio AFM, Ferreira AM, Freire CSR, Silvestre AJD, Freire MG, Coutinho JAP (2012) Optimization of the gallic acid extraction using ionic-liquid-based aqueous two-phase systems. Sep Purif Technol 97:142–149
29. Matzke M, Arning J, Ranke J, Jastorff B, Stolte S (2010) Design of inherently safer ionic liquids: toxicology and biodegradation. In: Handbook of green chemistry. Wiley-VCH Verlag GmbH & Co. KGaA, Weinheim, pp 225–290
30. Jastorff B, Molter K, Behrend P, Bottin-Weber U, Filser J, Heimers A, Ondruschka B, Ranke J, Schaefer M, Schroder H, Stark A, Stepnowski P, Stock F, Stormann R, Stolte S, Welz-Biermann U, Ziegert S, Thoming J (2005) Progress in evaluation of risk potential of ionic liquids-basis for an eco-design of sustainable products. Green Chem 7:362–372
31. Ranke J, Stolte S, Stormann R, Arning J, Jastorff B (2007) Design of sustainable chemical products – the example of ionic liquids. Chem Rev 107:2183–2206
32. Ventura SPM, Gonçalves AMM, Gonçalves F, Coutinho JAP (2010) Assessing the toxicity on [C_3mim][Tf_2N] to aquatic organisms of different trophic levels. Aquat Toxicol 96:290–297
33. Ventura SPM, Gonçalves AMM, Sintra T, Pereira JL, Gonçalves F, Coutinho JAP (2013) Designing ionic liquids: the chemical structure role in the toxicity. Ecotoxicology 22:1–12
34. Ventura SPM, Marques CS, Rosatella AA, Afonso CAM, Gonçalves F, Coutinho JAP (2012) Toxicity assessment of various ionic liquid families towards *Vibrio fischeri* marine bacteria. Ecotoxicol Environ Saf 76:162–168
35. Pereira JL, Mendes CD, Gonçalves F (2007) Short- and long-term responses of *Daphnia spp.* to propanil exposures in distinct food supply scenarios. Ecotoxicol Environ Saf 68:386–396
36. Petkovic M, Ferguson JL, Gunaratne HQN, Ferreira R, Leitao MC, Seddon KR, Rebelo LPN, Pereira CS (2010) Novel biocompatible cholinium-based ionic liquids-toxicity and biodegradability. Green Chem 12:643–649
37. Pham TPT, Cho C-W, Yun Y-S (2009) Environmental fate and toxicity of ionic liquids: a review. Water Res 44:352–372
38. Pernak J, Sobaszkiewicz K, Mirska I (2003) Anti-microbial activities of ionic liquids. Green Chem 5:52–56

39. Pretti C, Chiappe C, Baldetti I, Brunini S, Monni G, Intorre L (2009) Acute toxicity of ionic liquids for three freshwater organisms: Pseudokirchneriella subcapitata, Daphnia magna and Danio rerio. Ecotoxicol Environ Saf 72:1170–1176
40. Zhao D, Liao Y, Zhang Z (2007) Toxicity of ionic liquids. Clean Soil Air Water 35:42–48
41. Pisarova L, Steudte S, Dörr N, Pittenauer E, Allmaier G, Stepnowski P, Stolte S (2012) Ionic liquid long-term stability assessment and its contribution to toxicity and biodegradation study of untreated and altered ionic liquids. Proc Inst Mech Eng J J Eng Tribol 226:903–922
42. Ventura SPM, Gardas RL, Gonçalves F, Coutinho JAP (2011) Ecotoxicological risk profile of ionic liquids: octanol-water distribution coefficients and toxicological data. J Chem Technol Biotechnol 86:957–963
43. Mai NL, Ahn K, Koo Y-M (2014) Methods for recovery of ionic liquids—a review. Process Biochem 49:872–881
44. Siedlecka EM, Czerwicka M, Neumann J, Stepnowski P, Fernández JF, Thöming J (2011) Ionic liquids: methods of degradation and recovery. In: Kokorin A (ed) Ionic liquids: theory, properties, new approaches. ISBN 978-953-307-349-1
45. Fernandez JF, Neumann J, Thoming J (2011) Regeneration, recovery and removal of ionic liquids. Curr Org Chem 15:1992–2014
46. Deng Y, Long T, Zhang D, Chen J, Gan S (2009) Phase diagram of [Amim]Cl + salt aqueous biphasic systems and its application for [Amim]Cl recovery†. J Chem Eng Data 54:2470–2473
47. Li C, Han J, Wang Y, Yan Y, Pan J, Xu X, Zhang Z (2009) Phase behavior for the aqueous two-phase systems containing the ionic liquid 1-butyl-3-methylimidazolium tetrafluoroborate and kosmotropic salts. J Chem Eng Data 55:1087–1092
48. Wu B, Zhang Y, Wang H (2008) Phase behavior for ternary systems composed of ionic liquid + saccharides + water. J Phys Chem B 112:6426–6429
49. Wu B, Zhang YM, Wang HP (2008) Aqueous biphasic systems of hydrophilic ionic liquids + sucrose for separation. J Chem Eng Data 53:983–985
50. Hofmeister F (1888) Zur Lehre von der Wirkung der Salze, Archiv f. experiment. Pathol u Pharmakol 24:247–260
51. Wu B, Liu W, Zhang Y, Wang H (2009) Do we understand the recyclability of ionic liquids? Chem Eur J 15:1804–1810
52. Wu C, Wang J, Pei Y, Wang H, Li Z (2010) Salting-out effect of ionic liquids on poly (propylene glycol) (PPG): formation of PPG + ionic liquid aqueous two-phase systems. J Chem Eng Data 55:5004–5008
53. Neves CMSS, Freire MG, Coutinho JAP (2012) Improved recovery of ionic liquids from contaminated aqueous streams using aluminium-based salts. RSC Adv 2:10882–10890
54. American Water Works Association (AWWA) (1990) Water quality and treatment: a handbook of community water supplies, 4th edn. McGraw-Hill, New York
55. Chowdhury SA, Vijayaraghavan R, MacFarlane DR (2010) Distillable ionic liquid extraction of tannins from plant materials. Green Chem 12:1023–1028
56. Jiao J, Gai Q-Y, Fu Y-J, Zu Y-G, Luo M, Wang W, Zhao C-J (2013) Microwave-assisted ionic liquids pretreatment followed by hydro-distillation for the efficient extraction of essential oil from Dryopteris fragrans and evaluation of its antioxidant efficacy in sunflower oil storage. J Food Eng 117:477–485
57. Jiao J, Gai Q-Y, Fu Y-J, Zu Y-G, Luo M, Zhao C-J, Li C-Y (2013) Microwave-assisted ionic liquids treatment followed by hydro-distillation for the efficient isolation of essential oil from Fructus forsythiae seed. Sep Purif Technol 107:228–237
58. Bica K, Gaertner P, Rogers RD (2011) Ionic liquids and fragrances – direct isolation of orange essential oil. Green Chem 13:1997–1999
59. Rothenberger OS, Krasnoff SB, Rollins RB (1980) Conversion of (+)-Limonene to (−)-Carvone: an organic laboratory sequence of local interest. J Chem Educ 57:741
60. Ressmann AK, Strassl K, Gaertner P, Zhao B, Greiner L, Bica K (2012) New aspects for biomass processing with ionic liquids: towards the isolation of pharmaceutically active betulin. Green Chem 14:940–944

61. Yansheng C, Zhida Z, Changping L, Qingshan L, Peifang Y, Welz-Biermann U (2011) Microwave-assisted extraction of lactones from Ligusticum chuanxiong Hort. using protic ionic liquids. Green Chem 13:666–670
62. Usuki T, Yasuda N, Yoshizawa-Fujita M, Rikukawa M (2011) Extraction and isolation of shikimic acid from Ginkgo biloba leaves utilizing an ionic liquid that dissolves cellulose. Chem Commun 47:10560–10562
63. Cláudio AFM, Ferreira AM, Freire MG, Coutinho JAP (2013) Enhanced extraction of caffeine from guarana seeds using aqueous solutions of ionic liquids. Green Chem 15:2002–2010
64. e Silva F, Caban M, Stepnowski P, Coutinho JAP, Ventura SPM (2016) Recovery of ibuprofen from pharmaceutical wastes using ionic liquids. Green Chem 18:3749–3757
65. Cláudio AFM, Marques CFC, Boal-Palheiros I, Freire MG, Coutinho JAP (2014) Development of back-extraction and recyclability routes for ionic-liquid-based aqueous two-phase systems. Green Chem 16:259–268
66. Alvarez-Guerra E, Ventura SPM, Coutinho JAP, Irabien A (2014) Ionic liquid-based three phase partitioning (ILTPP) systems: ionic liquid recovery and recycling. Fluid Phase Equilib 371:67–74
67. Alvarez-Guerra E, Irabien A, Ventura SPM, Coutinho JAP (2014) , Ionic liquid recovery alternatives in ionic liquid-based three-phase partitioning (ILTPP). AIChE J 60:3577–3586
68. Passos H, Luís A, Coutinho JAP, Freire MG (2016) Thermoreversible (ionic-liquid-based) aqueous biphasic systems. Sci Rep 6:20276
69. Ferreira R, Garcia H, Sousa AF, Petkovic M, Lamosa P, Freire CSR, Silvestre AJD, Rebelo LPN, Pereira CS (2012) Suberin isolation from cork using ionic liquids: characterisation of ensuing products. New J Chem 36:2014–2024
70. Lemus J, Neves CMSS, Marques CFC, Freire MG, Coutinho JAP, Palomar J (2013) Composition and structural effects on the adsorption of ionic liquids onto activated carbon. Environ Sci Process Impacts 15:1752–1759
71. Neves CMSS, Lemus J, Freire MG, Palomar J, Coutinho JAP (2014) Enhancing the adsorption of ionic liquids onto activated carbon by the addition of inorganic salts. Chem Eng J 252:305–310
72. Palomar J, Lemus J, Gilarranz MA, Rodriguez JJ (2009) Adsorption of ionic liquids from aqueous effluents by activated carbon. Carbon 47:1846–1856
73. Lemus J, Palomar J, Heras F, Gilarranz MA, Rodriguez JJ (2012) Developing criteria for the recovery of ionic liquids from aqueous phase by adsorption with activated carbon. Sep Purif Technol 97:11–19
74. Lemus J, Palomar J, Gilarranz MA, Rodriguez JJ (2013) On the kinetics of ionic liquid adsorption onto activated carbons from aqueous solution. Ind Eng Chem Res 52:2969–2976
75. Ferreira AM, Coutinho JAP, Fernandes AM, Freire MG (2014) Complete removal of textile dyes from aqueous media using ionic-liquid-based aqueous two-phase systems. Sep Purif Technol 128:58–66
76. Pereira MM, Pedro SN, Quental MV, Lima ÁS, Coutinho JAP, Freire MG (2015) Enhanced extraction of bovine serum albumin with aqueous biphasic systems of phosphonium- and ammonium-based ionic liquids. J Biotechnol 206:17–25
77. Almeida HFD, Freire MG, Marrucho IM (2016) Improved extraction of fluoroquinolones with recyclable ionic-liquid-based aqueous biphasic systems. Green Chem 18:2717–2725
78. Zawadzki M, e Silva FA, Domanska U, Coutinho JAP, Ventura SPM (2016) Recovery of an antidepressant from pharmaceutical wastes using ionic liquid-based aqueous biphasic systems. Green Chem 18(12):3527–3536
79. Alvarez-Guerra E, Irabien A (2014) Ionic liquid-based three phase partitioning (ILTPP) systems for whey protein recovery: ionic liquid selection. J Chem Technol Biotechnol 90:939–946
80. Harde SM, Singhal RS (2012) Extraction of forskolin from Coleus forskohlii roots using three phase partitioning. Sep Purif Technol 96:20–25

81. Dennison C, Lovrien R (1997) Three phase partitioning: concentration and purification of proteins. Protein Expr Purif 11:149–161
82. Adlerova L, Bartoskova A, Faldyna M (2008) Lactoferrin: a review. Vet Med 53:457–468
83. Ndiaye N, Pouliot Y, Saucier L, Beaulieu L, Bazinet L (2010) Electroseparation of bovine lactoferrin from model and whey solutions. Sep Purif Technol 74:93–99
84. Souza RL, Ventura SPM, Soares CMF, Coutinho JAP, Lima AS (2015) Lipase purification using ionic liquids as adjuvants in aqueous two-phase systems. Green Chem 17:3026–3034
85. Lee SY, Vicente FA, e Silva FA, Sintra TE, Taha M, Khoiroh I, Coutinho JAP, Show PL, Ventura SPM (2015) Evaluating self-buffering ionic liquids for biotechnological applications. ACS Sustain Chem Eng 3:3420–3428
86. Passos H, Freire MG, Coutinho JAP (2014) Ionic liquid solutions as extractive solvents for value-added compounds from biomass. Green Chem 16:4786–4815

Printed by Printforce, the Netherlands